**metric
affine
geometry**

metric affine geometry

Ernst Snapper
Dartmouth College

Robert J. Troyer
Lake Forest College

Academic Press
New York and London

ACADEMIC PRESS, INC.
111 Fifth Avenue, New York, New York 10003

United Kingdom Edition published by
ACADEMIC PRESS, INC. (LONDON) LTD.
Berkeley Square House, London W1X 6BA

LIBRARY OF CONGRESS CATALOG CARD NUMBER: 76-154402

AMS (MOS) 1970 Subject Classification: 50A05, 50A10,
50A15; 50B10, 50B35; 50C25; 50D05, 50D10

PRINTED IN THE UNITED STATES OF AMERICA

To our parents

Contents

Preface

For a long time, the axioms of Euclid and Hilbert were the best for Euclidean geometry, but not any more. The modern source for the axiom systems of all affine and projective geometries, both metric and nonmetric, is linear algebra. Examples of metric affine geometries are the Euclidean geometries of various dimensions, the Lorentz plane, and Minkowski space. Examples of metric projective geometries are the elliptic and hyperbolic planes.

Linear algebra is the algebraic theory of vector spaces. How can this provide us with axioms for all these different geometries?

Nonmetric affine space of dimension n consists of a set X on which the additive group of an n-dimensional vector space V acts (Chapter 1). The elements of X are the points of the space and the action of a vector of V on these points is a translation of the space. A "metric" is introduced in the affine space X by choosing an inner product (metric) for V (Chapters 2 and 3). Different choices of V and inner product give rise to the different n-dimensional metric affine spaces. For example, if V is a vector space over the real numbers and the inner product is positive definite, X is n-dimensional Euclidean space.

Nonmetric projective space of dimension n has the lines (one-dimensional subspaces) of an $(n+1)$-dimensional vector space V as its points. The geometry of these lines is n-dimensional projective geometry. If V has been given an inner product, one obtains an n-dimensional metric projective space. For instance, if V is a three-dimensional vector space over the real numbers and the inner product is positive definite, one obtains the elliptic plane as the projective plane.

In this way, the axioms for a vector space give rise to axiom systems for all affine and projective, nonmetric and metric geometries. Such unity and simplicity cannot be claimed by the classical axiom systems of Euclid and Hilbert, nor by the Erlangen program of Felix Klein. Furthermore, the axioms systems which come from linear algebra are eminently suited to handle concrete problems in geometry, a virtue the axioms of Euclid and Hilbert do not possess. For these reasons, our book makes no use of the Euclid–Hilbert system or the Erlangen program and only mentions them to give background material or historical tidbits. Nevertheless, it should be realized that *we are dealing with precisely the same geometries as are treated by the classical systems.* The geometries have not changed, only the choice of axioms.

Chapter 1 develops nonmetric affine geometry, Chapter 2 studies inner products of vector spaces, and Chapter 3 treats metric affine geometry. Chapters 1 and 2 are independent and either can be used for a one-semester (term) course. It will take two semesters (terms) to cover the entire book.

Our main disappointment in writing this book has been the omission of a proper treatment of projective geometry because it would have made the book too long. Consequently, we have completely omitted projective geometry, except for background remarks for those readers who already have some knowledge of the subject. Others will probably want to omit our remarks on projective spaces.

The prerequisites for this book are a course in linear algebra, not necessarily including the study of inner products, and an elementary course in modern algebra which includes the concepts of group, normal subgroup, and quotient group. Although these prerequisites are light, we make heavy use of them. We assume that the reader can think in terms of vector spaces over an arbitrary field as scalar domain. We judge the level of our book to be advanced undergraduate or beginning graduate.

The exercises have been placed in the text where we felt they were most appropriate; they have not been held until the end of each section. A starred exercise is one to which a specific reference is made in the text or in another exercise and should not be omitted. Many exercises have hints and since text material is often covered in the exercises, many of the hints are actually short solutions. Of course, the reader should use the hints only in case of "emergency"; with these hints most of the exercises can be worked by any reader.

The numbering system for theorems, propositions, definitions, etc., is based on the principle that the number should tell where the statement can be found. Hence Proposition 378.2 means "the second proposition on page 378." Do not fear that 377 propositions precede it! Exercises are numbered consecutively in each section and Exercise 12.46a means "Part a of Exercise 46 which begins on page 12."

Most of the second chapter of this book can be found in E. Artin's book, *Geometric Algebra.* Our main effort has been to slow Artin's exposition and thereby make the tremendous geometric wealth which his masterpiece harbors available to the advanced undergraduate and beginning graduate student. We hope that anyone who has worked through our second chapter can read *Geometric Algebra* from the geometric viewpoint which Artin found so important.

We fear the level of our book is too high for many high school teachers. Nevertheless, this does not mean that discussions of questions related to teaching geometry in high school are out of place in a book such as ours. For example, what does it imply for high school geometry that the proper axioms for Euclidean geometry come from linear algebra? Several questions of this kind have been discussed and we hope the reader finds these discussions of interest. Both authors feel that the geometry curriculum for future high school teachers should be based on linear algebra. However, the level should be modified from that presented in this book.

A course along the lines of this book was taught by Snapper at the first Cooperative Summer Seminar held at Cornell University in 1964. Excellent notes ("Metric Geometry Over Affine Spaces") were written by Professor J. T. Buckley of Western Michigan University and distributed by the Mathematical Association of America. A grant from the Course Content Improvement Section of the National Science Foundation made it possible for Troyer to spend the academic year 1965–66 at Dartmouth College where Snapper taught an improved version of the course and Troyer wrote the notes "Metric Geometry of Vector Spaces." These notes were then taught by Professor Sherwood Ebey, Mercer University; Jack Graver, Syracuse University; Roger Lyndon, University of Michigan; E. R. Mullins, Grinnell College; Paul Rygg, Western Washington State College; Jon Sicks, University of Massachusetts; Seymour Schuster, University of Minnesota; and R. J. Troyer, University of North Carolina. The present book is a rewrite of these notes, based on the many fine suggestions by these teacher-friends and their students, and on the careful reviews of the notes by Professors Jean Dieudonné and Roger Lyndon.

Even with all this assistance, our book would never have seen the light of day had it not been for the unselfish help and expert typing of Mrs. Ann Hackney and Mrs. Mary Lou Troyer; the generous support from Dartmouth College, Lake Forest College, the University of North Carolina, and the National Science Foundation; and the infinite patience of Academic Press.

We are grateful to all who have assisted with the painful birth of our book.

SYMBOLS

chapter 1
affine geometry

1. INTUITIVE AFFINE GEOMETRY

Our subject is geometry and we shall begin with the study of affine geometry. To follow a more historical order, we would begin with Euclidean geometry. However, it seems useful to separate the axioms of affine geometry from those of Euclidean geometry. Roughly speaking, affine geometry is what remains after practically all ability to measure length, area, angles, etc., has been removed from Euclidean geometry. One might think that affine geometry is a poverty-striken subject. On the contrary, affine geometry is quite rich. Even after almost all ability to measure has been removed from Euclidean geometry, there still remains the concept of parallelism. Consequently, the whole theory of homothetic figures lies within affine geometry. The notions of translation and magnification (these are the dilations) are in the domain of affine geometry and, more generally, as the name suggests, affine transformations (one-to-one, onto functions which preserve parallelism) constitute an affine notion.

The question of how to obtain a rigorous approach to affine geometry arises. We choose to base our axiom system on linear algebra, that is, on the theory of vector spaces. First, let us discuss, on an intuitive level, the connection between plane affine geometry and vector spaces. What is the intuitive notion of the affine plane P over the field of real numbers? The points of P can be viewed as the points of any physical plane. If we fix a particular point $\mathbf{0}$ of P, we can construct a vector space in a natural way. The vectors are the ordered pairs $(\mathbf{0}, A)$ where $A \in P$ and we picture such a vector as an arrow from $\mathbf{0}$ to A (see Figure 2.1). We often write A for the vector $(\mathbf{0}, A)$,

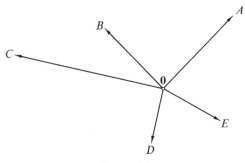

Figure 2.1

but this can only be done if it is clear from the context which point is being used as $\mathbf{0}$, that is, as the common initial point of the vectors. Vectors are added in the usual way by "completing the parallelogram" and we perform scalar multiplication by stretching or contracting the vector in the proper direction. Both operations are pictured in Figure 2.2.

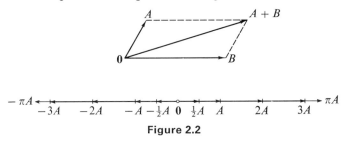

Figure 2.2

Thus we have associated a real, two-dimensional vector space with the real affine plane. The main thing to realize is that the construction of this vector space is based solely on notions of parallelism, that is, on affine concepts, and not on Euclidean concepts such as distance. However, this vector space is not unique because it depends on the choice of the point $\mathbf{0}$ in P. It is true that a different point $\mathbf{0}'$ gives rise to an isomorphic vector space, but it is equally true that the vector spaces coming from **0** and **0'** are not identical. The ordered pair (**0**, A) is not the same as the ordered pair (**0'**, A).

Exercises

*1. In the vector space relative to the point **0**, the zero element or "origin" is the vector **0**. Choose two points **0** $\neq$ **0'** in P. A point A of P determines the vector (**0**, A) in the vector space with origin **0** and the vector (**0'**, A) in the vector space with origin **0'**. In this way, we obtain a one-to-one mapping (**0**, A) → (**0'**, A) from the vector space with origin **0** onto the vector space with origin **0'**. Show by means of pictures that this mapping is *not* a linear isomorphism of vector spaces.

2. We can also associate the vector (**0'**, A') to the vector (**0**, A) where the point A' is chosen so that **00'**$A'A$ is a parallelogram (see Figure 3.1).

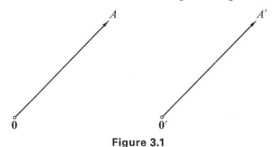

Figure 3.1

Show by pictures that the mapping (**0**, A) → (**0'**, A') is a linear isomorphism between the vector spaces in question.

There is also a *unique* vector space, not depending on the choice of a point, associated with the affine plane P. This is the vector space of the translations of P onto itself. Again, it is important to realize that this vector space is an affine concept; the reason is that translations and their compositions can be defined in terms of the notions of parallelism alone. A translation T of the plane is completely known as soon as the image $T(A)$ of one point $A \in P$ is known. Namely, T is then the one-to-one mapping of P onto itself such that, if B is a point of P, the four points A, B, $T(B)$, $T(A)$ form a parallelogram when taken in proper order as shown in Figure 4.1. If T_1 and T_2 are two translations of P, their sum $T_1 + T_2$ is defined as the composite mapping $T_1 T_2 : P \to P$, meaning first $T_2 : P \to P$ and then $T_1 : P \to P$.

It is also easy to see how a translation T is multiplied by a real number c. Suppose T sends the point A of P onto the point $T(A)$, that is, T is determined

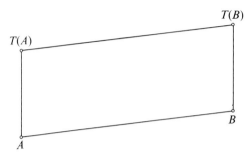

Figure 4.1

by the vector $(A, T(A))$. Then cT is determined by the vector $c(A, T(A))$. Clearly, $0T$ is the identity mapping of the plane.

Exercises

3. Show by pictures that $T_1 + T_2 = T_2 + T_1$, that is, $T_1 T_2 = T_2 T_1$. Show, furthermore, that the translations form a commutative group under addition.

4. Show by pictures that the translation cT does not depend on the choice of the point A.

5. Show by pictures that the set of translations under the above definitions of addition and scalar multiplication is a two-dimensional vector space over the real numbers.

6. Choose a point $\mathbf{0}$ in P and denote the vector space which has $\mathbf{0}$ as origin by $P(\mathbf{0})$. For every vector $A \in P(\mathbf{0})$, we denote the translation which sends the point $\mathbf{0}$ to the point A by T_A. (As pointed our earlier, the letter A is used to describe both the point A and the vector $(\mathbf{0}, A)$.) Show by pictures that the mapping $A \rightarrow T_A$ is an isomorphism of vector spaces.

7. In Exercise 6 above, an isomorphism α was described from the vector space $P(\mathbf{0})$ with origin $\mathbf{0}$ onto the vector space W of the translations. Choose another point $\mathbf{0}' \in P$ and denote the isomorphism from the vector space $P(\mathbf{0}')$ with origin $\mathbf{0}'$ onto W by β. We then have the diagram in Figure 4.2. Show by pictures that the resulting isomorphism

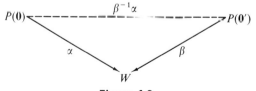

Figure 4.2

of vector spaces $\beta^{-1}\alpha: P(0) \to P(0')$ ($\beta^{-1}\alpha$ means "first α and then β^{-1}") is the isomorphism of Exercise 2 above.

The affine plane is not completely bereft of all measurement. A rudiment of measurement is left as is evidenced by the scalar multiplication of translations by real numbers. Consider the two parallel line segments AB and $A'B'$ pictured in Figure 5.1. Let T be the translation which sends A to B and T' the

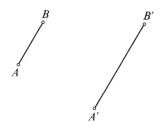

Figure 5.1

translation which sends A' to B'. In the vector space of the translations, $T' = 2T$ which is the same as to say the line segment $A'B'$ is twice as long as the line segment AB. The upshot is that, even in the affine plane, one can compare lengths of *parallel* line segments. If $A \neq B$ are points of P, it is perfectly all right to choose the line segment AB as a "unit length" for the affine plane. However, this unit length has only limited power because it can only be used *to measure the length of line segments which are parallel to AB.* In order to compare the length of line segments which are not parallel, we need a measuring rod which will work for all directions at the same time. Such a measuring rod is provided by Euclidean geometry or by one of the non-Euclidean, metric geometries.

In n-dimensional affine space $\mathscr{A}^n$, $n = 1, 2, 3, \ldots$, the situation is entirely analogous to the case of the plane ($n = 2$). For every fixed point $\mathbf{0}$ in $\mathscr{A}^n$, the ordered pairs of points $(\mathbf{0}, A)$, where A is an arbitrary point of $\mathscr{A}^n$, form an n-dimensional vector space with $\mathbf{0}$ as origin. We can visualize these vector spaces when n is 1 or 3; $\mathscr{A}^1$ is visualized as any physical line and $\mathscr{A}^3$ as the space which surrounds us. If we go over from a point $\mathbf{0}$ to another point $\mathbf{0}'$ of $\mathscr{A}^n$, the vector space changes to an isomorphic vector space (how is this isomorphism defined?). The translations of $\mathscr{A}^n$ form an n-dimensional vector space which is determined intrinsically by the affine structure of $\mathscr{A}^n$. Basically, the usual axioms of an n-dimensional vector space are those of the translations of an n-dimensional affine space. Of course, no matter how one defines translations axiomatically, they must come from the functions of the space $\mathscr{A}^n$ onto itself.

2. AXIOMS FOR AFFINE GEOMETRY

Since we shall consider vector spaces over division rings as well as fields, we first review the standard definitions of these terms. A **division ring** k is a set with two binary operations called addition and multiplication and denoted by $a + b$ and ab, respectively. The following axioms are satisfied.

1. k is a commutative group with respect to addition.
2. The nonzero elements of k form a group under multiplication.
3. For all a, b, c in k,

$$(a + b)c = ac + bc \quad \text{and} \quad c(a + b) = ca + cb.$$

A division ring for which multiplication is commutative is called a **field**. In other words, we reserve the term field for *commutative* division rings.

Note that *all fields are division rings*. The quaternions are an example of a division ring which is not a field. See MacLane and Birkhoff, *Algebra* §7, [16] for a discussion of quaternions.

We strongly suggest that readers who have only studied linear algebra over fields restrict themselves from the outset to fields.

The axiom system for *n*-**dimensional affine space** over a division ring k consists of a nonempty set X, an n-dimensional left vector space V over the division ring k, and an "action" of the additive group of V on X. The elements of X are called points and are denoted by $x, y, z, \ldots$; the elements of V are called vectors and are denoted by $A, B, C, \ldots$; the elements of k are called scalars and are denoted by $a, b, c, \ldots$. Scalars are always written on the left of the vectors. We now define what is meant by an action of the additive group of V on the set X.

Definition 6.1.   To say that **the additive group of the vector space V acts on the set X** means that, for every vector $A \in V$ and every point $x \in X$, there is defined a point $Ax \in X$ such that:

1. If $A, B \in V$ and $x \in X$, then $(A + B)x = A(Bx)$.
2. If $\mathbf{0}$ denotes the zero vector, $\mathbf{0}x = x$ for all $x \in X$.
3. For every ordered pair (x, y) of points of X, there is one and only one vector $A \in V$ such that $Ax = y$.

The unique vector A such that $Ax = y$ will be denoted by $\overrightarrow{x, y}$.

The dimension n of the vector space V is also called the **dimension of the affine space** X.

The affine space defined by X, V, k and the action of the additive group of V on X will be denoted by (X, V, k). In case k is the field of the real numbers $\mathbf{R}$, $(X, V, \mathbf{R})$ is called **real affine space**.

Throughout this book, $\mathbf{R}$ denotes the field of real numbers, $\mathbf{Q}$ the field of rational numbers, and $\mathbf{C}$ the field of complex numbers.

It is very important that appropriate pictures be drawn while studying geometry. Pictures indicate what, probably, the correct theorem is and how one might attempt to prove it. Of course, the actual proof must stand on its own logical legs and be independent of figures. In geometry, in fact, in all of mathematics, one should be aware of the danger of figures and then live dangerously. Draw a lot of pictures! In order to help the reader get started with this important work, we have drawn the pictures for Exercises 4, 6, and 10 which follow; Exercises 5, 7, and 11 certainly require pictures also.

Exercises

Prove the statements listed below; x, y, z denote points of X and A, B vectors in V.

1. If $Ax = x$ for some $x \in X$, then $A = \mathbf{0}$.
2. $\overrightarrow{x, Ax} = A$.
3. $(\overrightarrow{x, y})x = y$.

*4. $\overrightarrow{x, By} = \overrightarrow{x, y} + B$.

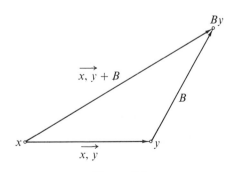

Figure 7.1

5. $\overrightarrow{Bx, y} = \overrightarrow{x, y} - B$.
6. $\overrightarrow{x, y} + \overrightarrow{y, z} = \overrightarrow{x, z}$.

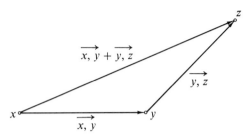

Figure 8.1

7. $\overrightarrow{Bx, By} = \overrightarrow{x, y}$.
8. $-(\overrightarrow{x, y}) = \overrightarrow{y, x}$.
9. $Ax = y$ if and only if $x = (-A)y$.
*10. $\overrightarrow{Ax, By} = -A + \overrightarrow{x, y} + B$.

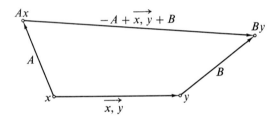

Figure 8.2

*11. $(\overrightarrow{x, y})z = (\overrightarrow{x, z})y = (\overrightarrow{x, y} + \overrightarrow{x, z})x$.

The pictures for Exercises 4, 6, and 10 are not intended to convey that the vectors of V lie in the set X. On the contrary, V and X usually have nothing in common. A lettered arrow A connecting two points x and y of X (see Figure 8.3) simply records pictorially that the vector A of V sends the point

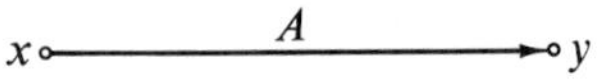

$$x \overset{A}{\longrightarrow} y$$

Figure 8.3

x to the point y.

3. A CONCRETE MODEL FOR AFFINE SPACE

We now look at a model for n-dimensional affine space. In fact, the model we are going to describe is often used as the definition of affine space. Let V

be an n-dimensional left vector space over a division ring k. For the set X, we choose the vectors of V, that is, $X = V$ where V is considered only as a set. The action of the additive group of V on the set V is defined as follows:

$$\text{If } C \in V \text{ and } A \in V, \text{ then } AC = A + C.$$

Some care must now be taken so that no confusion arises from the fact that we have defined X to be the set V.

Exercise

1. Verify that (V, V, k), as defined above, is a model for n-dimensional, affine space. In other words, prove that the three conditions of Definition 6.1 are satisfied.

The reader has become accustomed in his linear algebra courses to visualize the vector A of the vector space V as an arrow starting at $\mathbf{0}$. When V is regarded as an affine space, that is, $X = V$, the *point A* should be visualized as the end of the arrow. Always visualize and draw pictures!

Another concrete model for n-dimensional affine space will be introduced later after the notion of a coordinate system has been discussed.

4. TRANSLATIONS

We now introduce translations into our affine space (X, V, k).

Definition 9.1. Let $A \in V$. The function $T_A : X \to X$ defined by $T_A(x) = Ax$ for all $x \in X$ is called the **translation of X associated with A.**

From our intuitive notion of translations, we certainly expect each translation to be a one-to-one function from X onto X. This fact is, indeed, true; we shall eventually prove much more about the set of translations.

Proposition 9.1. For each $A \in V$, the translation T_A is a one-to-one mapping of X onto itself.

Proof. We first show that T_A is one-to-one. Select x, y in X and assume $T_A(x) = T_A(y)$. By definition, $T_A(x) = T_A(y) \Leftrightarrow Ax = Ay$. (A two-pointed double arrow ($\Leftrightarrow$) stands for " if and only if.") But $Ax = Ay \Leftrightarrow -A(Ax) = -A(Ay) \Leftrightarrow (-A + A)x = (-A + A)y \Leftrightarrow 0x = 0y \Leftrightarrow x = y$.

We now prove T_A is onto. For any $x \in X$, $x = 0x = (A + (-A))x = A(-Ax) = T_A(-Ax)$. Done.

So, with every vector $A \in V$, there is associated a one-to-one mapping of X onto itself. Let S_X denote the set of all one-to-one mappings of X onto X, that is, the set of all permutations of the set X. As we know, S_X is a group under the operation of composition of mappings and is called the permutation group of X.

Proposition 10.1.   The mapping from V to S_X which sends the vector A onto the translation T_A is a one-to-one homomorphism of the additive group of V into the permutation group S_X.

Proof. In order to show that the mapping is a group homomorphism, we must show that $T_{A+B} = T_A T_B$, equivalently, $T_{A+B}(x) = (T_A T_B)(x)$ for all $x \in X$. By definition, $T_{A+B}(x) = (A + B)x = A(Bx) = T_A(T_B(x)) = (T_A T_B)(x)$. We shall show that the mapping is one-to-one by proving that its kernel is $\{0\}$. Since the identity element of the group S_X is the identity mapping of X, we assume that T_A is the identity mapping of X. Then $T_A(x) = x$ for all $x \in X$, that is, $Ax = x$ for all $x \in X$. We conclude from Exercise 10.1 that $A = 0$. Done.

It follows from Proposition 10.1 that the mapping $A \to T_A$ is an isomorphism from the additive group of V onto the (multiplicative) subgroup of S_X which consists of the translations of X. Consequently, the translations of X form a group under the operation of composition of mappings. This group is called the **translation group** of X. We collect our results.

Proposition 10.2.   The (multiplicative) translation group of X is isomorphic to the additive group of V and hence is a commutative subgroup of S_X.

We end this section with a very useful, although geometrically evident, fact about translations.

Proposition 10.3.   A translation T is completely determined by the image of one point. If T leaves one point fixed, T is the identity mapping 1_X of X.

Proof. Suppose we are given that $T: X \to X$ is a translation and that $T(x) = y$ for two points x and y in X. This means that $T = T_A$ for some vector $A \in V$ and that $Ax = y$. Consequently, A is the unique vector $\overrightarrow{x, y}$ and T is completely known. If T leaves a point x fixed, the translation 1_X and T have the same action on x. Hence $T = 1_X$. Done.

5. AFFINE SUBSPACES

In our study of the n-dimensional affine space (X, V, k), we must say what is meant by lines, planes, and, more generally, by the affine subspaces of X. An affine subspace $S(x, U)$ of X is determined by a point $x \in X$ and a linear subspace U of V.

Definition 11.1.   The **affine subspace** $S(x, U)$ of X consists of the points Ax where A is an arbitrary vector of U. In short,

$$S(x, U) = \{Ax \mid A \in U\}.$$

The dimension of the linear subspace U of V is also called the **dimension of the affine subspace** $S(x, U)$ of X. Affine subspaces of dimension one are called **lines** while those of dimension two are called **planes**. The affine subspaces of dimension $n - 1$ are called **hyperplanes**.

Exercises

1. Prove that the affine subspaces of dimension zero are the points of X.
2. Prove that X has only one affine subspace of dimension n, namely, X itself.

Let us say a few words about the pictures we associate with the affine space X and its affine subspaces when the dimension of X is two and three. For dimension two, we picture X as a physical plane of points with no special

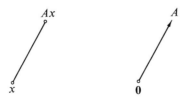

Figure 11.1

points singled out. The sets V and X have nothing in common and we picture the elements of V as arrows beginning at the origin **0**. If $x \in X$ and $A \in V$, we picture the result of the action of A on x as in Figure 11.1.

Continuing with dimension two, we suppose that U is a one-dimensional linear subspace of V and that $x \in X$. We picture U and $S(x, U)$ as in Figure 12.1.

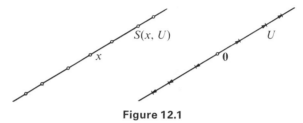

Figure 12.1

Similarly, if the dimension of X is three, we visualize X as the space which surrounds us. If U is a two-dimensional linear subspace of V and $x \in X$, we picture U and the plane $S(x, U)$ as in Figure 12.2.

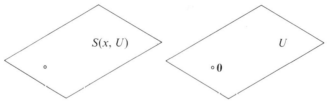

Figure 12.2

It is convenient to introduce the abbreviation "dim" to stand for dimension.

Even if $\dim X > 3$, one should force oneself to visualize. Then X is viewed as a very roomy three-space in which two planes may possibly have only a point in common ($\dim X = 4$) which cannot happen when $n = 3$. Also, V is viewed as an equally roomy bundle of arrows, all starting at the origin **0**. One-dimensional linear subspaces of V are viewed as lines through **0** and two-dimensional linear subspaces of V are viewed as planes through **0**. Lines and planes of X are again viewed as ordinary physical lines and planes. As shown in the following exercise, the affine subspace $S(x, U)$ must pass through the point x.

Exercise

*3. Prove that $x \in S(x, U)$.

Suppose we are given a subset S of X and that we are told there is a point $x \in X$ and a linear subspace U of V such that S is the affine subspace $S(x, U)$ of X. Can we recover U by just knowing the pointset S? Equivalently, does the pointset S uniquely determine U? The answer is yes, as is seen from Proposition 13.1 where the vectors of U are described intrinsically in terms of the points of S. Recall that, if $x, y \in X$, $\overrightarrow{x, y}$ denotes the unique vector A in V such that $Ax = y$.

Proposition 13.1.   Let $S = S(x, U)$ be the affine subspace determined by the point x and the linear subspace U. Then $U = \{\overrightarrow{y, z} \mid y, z \in S\}$. Consequently, U is uniquely determined by $S(x, U)$.

Proof. Let $A \in U$. Then $x \in S$ (Exercise 12.3) and, by definition of S, $y = Ax \in S$. Consequently, $A = \overrightarrow{x, y}$ which shows that $U \subset \{\overrightarrow{y, z} \mid y, z \in S\}$. Conversely, let $y, z \in S.\,\vert$ Then there exist vectors $A, B \in U$ such that $y = Ax$ and $z = Bx$. Figure 13.1 below illustrates the situation when S is a plane.

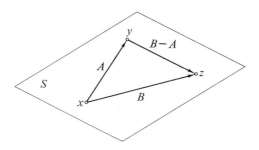

Figure 13.1

Now $y = Ax \Leftrightarrow x = (-A)y$; thus $z = Bx \Leftrightarrow z = B(-A)y = (B - A)y$. Consequently, $B - A = \overrightarrow{y, z}$ and, since $B - A \in U$, $\overrightarrow{y, z} \in U$. This proves that $U \supset \{\overrightarrow{y, z} \mid y, z \in S\}$. Done.

Of course, the pointset $S(x, U)$ could not possibly determine the point x uniquely unless $\dim U = 0$. All points of $S(x, U)$ play the same role! We now say precisely when two affine subspaces $S(x, U)$ and $S(y, W)$ are equal.

Proposition 14.1.   Let x, y be points of X and U, W linear subspaces of V. Then $S(x, U) = S(y, W)$ if and only if $U = W$ and $S(x, U) \cap S(y, W) \neq \varnothing$.

Proof. If $S(x, U) = S(y, W)$, then $U = W$ by Proposition 13.1 while both x and y belong to $S(x, U) \cap S(y, W)$. Conversely, suppose that $U = W$ and that $z \in S(x, U) \cap S(y, U)$. We shall prove that $S(x, U) = S(y, W)$ by showing that both of these spaces are equal to $S(z, U)$. It is sufficient to show that $S(x, U) = S(z, U)$, the proof that $S(y, U) = S(z, U)$ being the same. Since $z \in S(x, U)$, $z = Ax$ for $A \in U$. If $B \in U$, $Bz = B(Ax) = (B + A)x \in S(x, U)$ since $B + A \in U$. This shows that $S(z, U) \subset S(x, U)$. For the inverse conclusion, we use the fact that $x = (-A)z \in S(z, U)$. Done.

The following exercise tells us when $S(x, U) = S(y, W)$ in terms of the points x and y.

Exercise

4. Let x, y be points of X and U, W linear subspaces of V. Prove that $S(x, U) = S(y, W)$ if and only if one of the following two equivalent conditions is satisfied:
   a. $U = W$ and $x \in S(y, W)$.
   b. $U = W$ and $y \in S(x, U)$.

Definition 14.1.   The linear subspace U of V is called the **direction space** of $S(x, U)$.

The dimension of $S(x, U)$ is, therefore, the dimension of its direction space U. Clearly, the direction space of X is V.
The term "affine subspace" will now be justified by showing that an affine subspace is an affine space in its own right. For convenience, we often write S for $S(x, U)$ when there is no danger of confusion.

Proposition 14.2.   Let $S = S(x, U)$ be a d-dimensional affine subspace of X. Then the triple (S, U, k), together with the action inherited from the affine space (X, V, k), is a d-dimensional affine space.

Proof. Let $A \in U$ and $y \in S$. According to Definition 6.1, we must first show that $Ay \in S$. However, $y = Bx$ for $B \in U$ and hence $Ay = A(Bx) = (A + B)x \in S$

since $A + B \in U$. The action of U on S immediately inherits Properties 1 and 2 of Definition 6.1 from the action of V on X. There remains to be shown only that, if $y, z \in S$, there exists one and only one vector $A \in U$ such that $Ay = z$. The vector $A = \overrightarrow{y, z}$ belongs to U (Proposition 13.1) and $Ay = z$. Moreover, A is the only vector in all of V such that $Ay = z$. Done.

Many questions arise. Is the intersection of two affine subspaces an affine subspace? When are two affine subspaces parallel? How does one assign a coordinate system to an affine subspace? Questions of this nature will now be considered.

6. INTERSECTION OF AFFINE SUBSPACES

Since the intersection of two affine subspaces can be empty and an affine subspace is never empty, the intersection of two affine subspaces is not always an affine subspace. For example, let V be a two-dimensional vector space and U a one-dimensional subspace of V. Select $x, y \in X$ so that $\overrightarrow{x, y}$ does not belong to U; then $S(x, U)$ and $S(y, U)$ have an empty intersection. For, if there were a point $z \in S(x, U) \cap S(y, U)$, then $S(x, U) = S(y, U)$ by

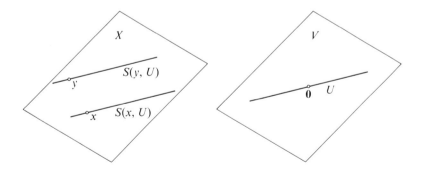

Figure 15.1

Proposition 14.1. It then follows from Proposition 13.1 that $\overrightarrow{x, y}$ belongs to U, contrary to assumption. The situation is pictured in Figure 15.1.

However, with the exception of the empty intersection, the intersection of affine subspaces is an affine subspace.

Proposition 16.1. Let $S(x, U)$ and $S(y, W)$ be affine subspaces of X. If there is a point z such that $z \in S(x, U) \cap S(y, W)$, then

$$S(x, U) \cap S(y, W) = S(z, U \cap W).$$

Proof. Since $z \in S(x, U) \cap S(y, W)$, we know from Proposition 14.1 that $S(x, U) = S(z, U)$ and $S(y, W) = S(z, W)$. Thus we must prove that $S(z, U) \cap S(z, W) = S(z, U \cap W)$. If $z' \in S(z, U) \cap S(z, W)$, the vector $\overrightarrow{z, z'}$ belongs to U and W whence $\overrightarrow{z, z'} \in U \cap W$. But $(\overrightarrow{z, z'})z = z'$, so $z' \in S(z, U \cap W)$ and, therefore, $S(z, U) \cap S(z, W) \subset S(z, U \cap W)$. Conversely, suppose that $z' \in S(z, U \cap W)$. Then there is a vector $A \in U \cap W$ such that $z' = Ax$. Consequently, z' belongs to both $S(z, U)$ and $S(z, W)$ whence $S(z, U \cap W) \subset S(z, U) \cap S(z, W)$. Done.

We see that, if $S(x, U)$ and $S(y, W)$ are affine subspaces with at least a point in common, then their intersection is again an affine subspace. In fact, we know that $U \cap W$ is the direction space of that intersection. Consequently, the dimension of $S(x, U) \cap S(y, W)$ equals the dimension of $U \cap W$.

Exercise

*1. Let $\{S_i = S(x_i, U_i) | i \in I\}$ be a collection of affine subspaces of X where I is not necessarily finite. Prove that, if $\bigcap_{i \in I} S_i \neq \varnothing$, this intersection is an affine subspace of X whose direction space is $\bigcap_{i \in I} U_i$.

(Hint: Use the proof of Proposition 16.1 as a model.)

7. COORDINATES FOR AFFINE SUBSPACES

Before we can say what we mean by a coordinate system for an affine subspace, we must recall what is meant by a coordinate system for the n-dimensional vector space V. A coordinate system for the vector space V consists of an *ordered* basis $A_1, \ldots, A_n$. Given such an ordered basis, we can associate with each vector A the unique n-tuple of coordinates $(a_1, \ldots, a_n)$ such that $A = a_1 A_1 + \ldots + a_n A_n$. This sets up a one-to-one correspondence between the vectors in V and the elements in k^n which is the Cartesian product of the division ring k with itself n times. (A one-to-one correspondence between two sets B and C means that there is a one-to-one function from B onto C.) Of course, each linear subspace of V is a vector space in its own right and hence has its own bases and coordinate systems.

Let $S(x, U)$ be a d-dimensional affine subspace of X.

Definition 17.1. An **affine coordinate system** for $S(x, U)$ consists of a point $x_0 \in S(x, U)$ and a coordinate system $A_1, \ldots, A_d$ for U. If $y \in S(x, U)$, the **affine coordinates of the point** y are the coordinates of the vector $\overrightarrow{x_0, y}$. The point x_0 is called the **origin** of the affine coordinate system.

We shall usually say coordinate system (coordinates) instead of affine coordinate system (affine coordinates).

Select $A \in V$. When $A \neq \mathbf{0}$, $\langle A \rangle = \{tA \mid t \in k\}$ is the line through $\mathbf{0}$ which contains A; if $A = \mathbf{0}$, $\langle A \rangle = \{\mathbf{0}\}$.

Let $x_0, A_1, \ldots, A_d$ be a coordinate system for $S(x, U)$ and put $\langle A_i \rangle = \{tA_i \mid t \in k\}$; that is, $\langle A_i \rangle$ is the one-dimensional subspace of U generated by A_i. Then $S(x_0, \langle A_i \rangle)$, $i = 1, 2, \ldots, d$ is the "ith coordinate axis" of the coordinate system $x_0, A_1, \ldots, A_d$ of $S(x, U)$. When S has dimension two, we have Figure 17.1.

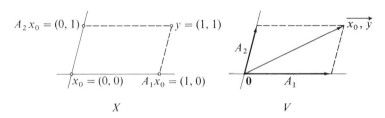

$$A_2 x_0 = (0, 1) \quad\quad\quad\quad y = (1, 1)$$
$$x_0 = (0, 0) \quad\quad A_1 x_0 = (1, 0)$$
$$X$$

$$\overrightarrow{x_0, y}$$
$$A_2$$
$$0 \quad\quad A_1$$
$$V$$

Figure 17.1

Exercises

1. Let $x_0, A_1, \ldots, A_d$ be a coordinate system for the d-dimensional affine subspace $S(x, U)$. What are the coordinates of each of the $d + 1$ points $x_0, A_1 x_0, \ldots, A_d x_0$?
2. What are the coordinates at the point $(A_1 + \cdots + A_d) x_0$?

Proposition 17.1. Assume that we are given a coordinate system $x_0, A_1, \ldots, A_d$ for a d-dimensional affine subspace $S(x, U)$. Then the mapping from $S(x, U)$ to k^d, which associates with each point $y \in S(x, U)$ its d-tuple of affine coordinates $(y_1, \ldots, y_d)$, is a one-to-one mapping from $S(x, U)$ onto k^d.

Proof. The given mapping $S(x, U) \to k^d$ is the composition of two mappings. The first is the mapping $S(x, U) \to U$ which associates with each $y \in S(x, U)$

the vector $\overrightarrow{x_0, y}$. The second is the mapping $U \to k^d$ which associates the d-tuple $(a_1, \ldots, a_d)$ with the vector A where $A = a_1 A_1 + \cdots + a_d A_d$. From linear algebra, we know that the second mapping is one-to-one and onto. Hence we need to prove only that the first mapping is one-to-one and onto. Suppose that $\overrightarrow{y, z} \in S(x, U)$ and $\overrightarrow{x_0, y} = \overrightarrow{x_0, z}$. Then $(\overrightarrow{x_0, y})x_0 = (\overrightarrow{x_0, z})x_0 \Leftrightarrow y = z$. So the mapping is one-to-one. Select $A \in U$; then $Ax_0 \in S(x, U)$ and $A = \overrightarrow{x_0, Ax_0}$. Therefore, the mapping is also onto. Done.

We use the notation $|W|$ for the cardinality of a set W. Observe that, as a consequence of the above proposition, $|S| = k^d \geq 2^d$. This inequality holds because every division ring contains at least the two elements 0 and 1. Hence affine subspaces of the same dimension have the same number of points and an affine space of dimension ≥ 2 contains at least four points.

Exercises

3. If $\dim X = 0$, prove that X consists of just one point. In this case, V has no basis (why not?) and hence X has no coordinate system.
4. If $\dim X \geq 1$, prove that X contains at least two points. For what n and k is $|X| = 2$?
*5. Let $\dim X \geq 1$ and let $x \neq y$ be two points of X. Prove that there is one and only one line in X which passes through x and y and that this line is

$$S(x, \langle \overrightarrow{x, y} \rangle) = S(y, \langle \overrightarrow{x, y} \rangle).$$

6. Let $\dim X \geq 1$. Then $\dim V \geq 1$ (why?) and we can find a nonzero vector $A \in V$. If $x \in X$, prove that $x \neq Ax$ and that $S(x, \langle A \rangle)$ is the unique line which passes through x and Ax.
7. Let $\dim X \geq 2$. Prove that X contains at least three noncollinear points, that is, three points which do not lie on a line. (Hint: Since $\dim V \geq 2$, there are two linearly independent vectors $A, B \in V$. Choose a point $x \in X$ and show that the three points x, Ax, and Bx are not collinear.)

The following proposition describes the action of U on $S(x, U)$ in terms of a coordinate system. It states that the action of U on $S(x, U)$ is given by "adding coordinates." Precisely:

Proposition 18.1. Let $x_0, A_1, \ldots, A_d$ be a coordinate system for the d-dimensional affine subspace $S(x, U)$. Suppose the vector $B \in U$ has coordinates $(b_1, \ldots, b_d)$ and the point $z \in S(x, U)$ has (affine) coordinates $(z_1, \ldots, z_d)$. Then the affine coordinates of the point Bz are

$$(z_1 + b_1, \ldots, z_d + b_d).$$

Proof. We put $A = \overrightarrow{x_0, z}$ and hence the coordinates of the vector A are $(z_1, \ldots, z_d)$. Furthermore, $z = Ax_0 \Rightarrow Bz = B(Ax_0) = (B + A)x_0$. It follows that the coordinates of the point Bz are those of the vector $B + A$; we know from linear algebra that $B + A$ has coordinates $(z_1 + b_1, \ldots, z_d + b_d)$. Done.

Exercise

*8. Let $x_0, A_1, \ldots, A_n$ be a coordinate system for X. Let $x = (x_1, \ldots, x_n)$ and $y = (y_1, \ldots, y_n)$ be points of X. (Here $x = (x_1, \ldots, x_n)$ is an abbreviation for "the point x has coordinates $(x_1, \ldots, x_n)$.") Prove that the vector $\overrightarrow{x, y}$ has the coordinates $(y_1 - x_1, \ldots, y_n - x_n)$.

The last proposition makes it possible to introduce a second concrete model for affine space. Let $A_1, \ldots, A_n$ be a coordinate system for V. We know from linear algebra that the resulting one-to-one mapping from V onto k^n enables us to regard k^n as a model for V. Choose a point $x_0 \in X$ and consider the coordinate system $x_0, A_1, \ldots, A_n$ of X. Each point $x \in X$ now has a unique n-tuple of coordinates $(x_1, \ldots, x_n)$ and the resulting one-to-one mapping from X onto k^n enables us to regard k^n also as a model for X. The action of $V = k^n$ on $X = k^n$ is dictated by Proposition 18.1; namely, if $B = (b_1, \ldots, b_n) \in V = k^n$ and $x = (x_1, \ldots, x_n) \in X = k^n$, then

$$Bx = (x_1 + b_1, \ldots, x_n + b_n).$$

We ask the reader to complete this investigation by doing the following exercise.

Exercise

9. Let $X = V = k^n$. If $B = (b_1, \ldots, b_n) \in V$ and $x = (x_1, \ldots, x_n) \in X$, define the action of B on x by $Bx = (x_1 + b_1, \ldots, x_n + b_n)$. Prove that, under this action of V on X, the triple (k^n, k^n, k) is a model for n-dimensional affine space over k. This model is often denoted simply by k^n.

8. ANALYTIC GEOMETRY

We shall now show, by examples, that the usual analytic geometry of n dimensions can be carried out in n-dimensional affine space over the division ring k. Since we have not yet introduced measurement into our spaces, we can only treat that part of analytic geometry which does not involve measurement. This includes equations of lines and hyperplanes and we shall concentrate on them.

Special care must be taken because k is a division ring and, therefore, multiplication is not necessarily commutative. Nevertheless, the attitude one should take is that all of analytic geometry can be done for division rings, but the computation is less familiar and a little more involved. This will become apparent as we proceed.

Example 1

Let x_0, A_0, ..., A_n be a coordinate system for X and select a line l in X. Then $l = S(x, U)$ where $x \in l$ and U is the one-dimensional direction space of l. How can we find an equation for this line with respect to the given coordinate system?

Select a nonzero vector B in U; then $U = \{tB \mid t \in k\}$ and $S(x, U) = \{(tB)x \mid t \in k\}$. If B has coordinates $(b_1, \ldots, b_n)$ and x has coordinates $(x_1, \ldots, x_n)$, then $(tB)x$ has coordinates $(x_1 + tb_1, \ldots, x_n + tb_n)$ by Proposition 18.1. Thus, given the coordinates of one point x on l and the coordinates of one nonzero vector B in U, the **parametric equations of the line** $S(x, U)$ are:

$$z_1 = x_1 + tb_1$$
$$\vdots$$
$$z_n = x_n + tb_n.$$

For different choices of x and B, the system of equations will change, but the form remains the same. The coordinates $(b_1, \ldots, b_n)$ of the vector B are called the direction numbers for the line l since they determine the direction of l in n-space. Clearly, not all direction numbers are zero since $B \neq \mathbf{0}$; moreover, they are unique, except for a nonzero proportionality constant.

If the dimension of X is two, the parametric equations of the line l are

$$z_1 = x_1 + tb_1$$
$$z_2 = x_2 + tb_2.$$

Since $B \neq 0$, either $b_1 \neq 0$ or $b_2 \neq 0$. If $b_1 \neq 0$, we can solve the first equation for t, giving $t = (z_1 - x_1)b_1^{-1}$. If we substitute this value for t in the second equation, we obtain

$$z_2 - x_2 = (z_1 - x_1)b_1^{-1}b_2.$$

This linear equation for the line l may seem strange because we are used to working with scalars in a field. In case k is a field, we obtain the linear equation

$$z_2 - x_2 = (b_2/b_1)(z_1 - x_1)$$

which one recognizes as the equation of the line which passes through the point $x = (x_1, x_2)$ with slope b_2/b_1.

Now suppose that $b_1 = 0$. Then $b_2 \neq 0$ and we can solve the second equation for t, giving $t = (z_2 - x_2)b_2^{-1}$. If we substitute this value for t in the second equation, we obtain

$$z_1 - x_1 = (z_2 - x_2)b_2^{-1}b_1.$$

In this case, however, $b_1 = 0$; consequently, the above equation reduces to $z_1 - x_1 = 0$. We recognize this as the linear equation of a line which passes through the point $x = (x_1, x_2)$ and is "parallel to the second coordinate axis."

Example 2

We show in this example that the hyperplanes of n-dimensional affine space are given by the usual linear equations in n variables. Again, special care has to be taken since k may not be commutative. Everything is based on the following well-known result from linear algebra.

A mapping $f: V \to k$ from our n-dimensional vector space V into k is called a **functional** if f is a linear mapping where k is considered as a one-dimensional left vector space over itself. In other words, f is a functional if

$$f(A + B) = f(A) + f(B) \quad \text{and} \quad f(tA) = tf(A)$$

for all $A, B \in V$ and $t \in k$.

Proposition 21.1. An $(n - 1)$-dimensional linear subspace U of V is the kernel of a nonzero functional $f: V \to k$ and, conversely, the kernel of every nonzero functional f is an $(n - 1)$-dimensional linear subspace of V. Moreover, f is determined by U up to a nonzero right factor of k.

Proof. Let f be a nonzero functional on V. The image of f is denoted by $\operatorname{im} f$ and the kernel of f by $\ker f$. (This notation is used throughout this book for all linear transformations.) Since $f \neq 0$, there is a vector $A \in V$ such that $f(A) = a \neq 0$. Then the image of f is k and, since $\dim(\ker f) + \dim(\operatorname{im} f) = n$ and $\dim k = 1$, $\ker f$ is an $(n-1)$-dimensional linear subspace of V.

Conversely, let U be an $(n-1)$-dimensional linear subspace of V. Select a vector $A \notin U$; then $V = U \oplus \langle A \rangle$ where $\oplus$ stands for the direct sum of vector spaces. If $C \in V$, then $C = B + tA$ where $B \in U$ and $t \in k$. It is clear that $f(C) = t$ defines a functional of V with kernel U.

There only remains to show that, if f and g are two functionals with the same $(n-1)$-dimensional linear subspace U as kernel, then $g = fc$ for some nonzero $c \in k$; this means that $g(C) = f(C)c$ for all $C \in V$ (see Exercise 26.15). Hereto, we again put $V = U \oplus \langle A \rangle$. Then, if $C = B + tA$ as above, $f(C) = tf(A)$ and $g(C) = tg(C)$. Since $f, g \neq 0$ (their kernel is U and not V), $f(A) \neq 0$ and $f(B) \neq 0$; let $f(A) = a$, $g(A) = b$ and $a^{-1}b = c$. Then

$$g(C) = tb = (ta)(a^{-1}b) = f(C)c.$$

Thus $g = fc$. Done.

So far we have made no use of a coordinate system for V. We now choose a coordinate system $A_1, \ldots, A_n$ for V and recall that a functional $f \colon V \to k$ is completely determined by its values at $A_1, \ldots, A_n$. If $f(A_i) = \alpha_i$ for $i = 1, \ldots, n$ and $B \in V$ has coordinates $(b_1, \ldots, b_n)$, then

$$f(B) = f(b_1 A_1 + \cdots + b_n A_n)$$
$$= b_1 f(A) + \cdots + b_n f(A_n) = b_1 \alpha_1 + \ldots + b_n \alpha_n.$$

In particular,

$$\ker f = \{B \mid B \in V, \ b_1\alpha_1 + \cdots + b_n\alpha_n = 0, \quad B \text{ has coordinates } (b_1, \ldots, b_n)\}.$$

We now make use of functionals to discuss the equation of a hyperplane in our n-dimensional affine space (X, V, k). Let $x_0, A_1, \ldots, A_n$ be a coordinate system for X. To say that a hyperplane S is determined by a linear equation $z_1\alpha_1 + \cdots + z_n\alpha_n = a$ where $\alpha_1, \ldots, \alpha_n, a \in k$, means that a point y belongs to S if and only if its coordinates $(y_1, \ldots, y_n)$ satisfy the equation

$$z_1\alpha_1 + \cdots + z_n\alpha_n = a.$$

Proposition 22.1. A hyperplane S of X is determined by a linear equation $z_1\alpha_1 + \cdots + z_n\alpha_n = a$ where not all α_i are zero. Conversely, every linear equation $z_1\alpha_1 + \cdots + z_n\alpha_n = a$ where not all α_1 are zero determines a hyperplane of X.

Proof. Let $S(x, U)$ be a hyperplane of X and assume $x = (x_1, \ldots, x_n)$. We choose a functional $f: V \to k$ which has U as kernel and put $f(A_i) = \alpha_i$ for $i = 1, \ldots, n$ and $x_1 \alpha_1 + \cdots + x_n \alpha_n = a$. The scalar a can be zero, but at least one $\alpha_i \neq 0$ since f has kernel U and not V. We show that $S(x, U)$ is given by the equation $z_1 \alpha_1 + \cdots + z_n \alpha_n = a$.

Suppose that a point y of X belongs to $S(x, U)$; that is, $y = Bx$ for some $B \in U$. If the coordinates of B are $(b_1, \ldots, b_n)$, those of y are

$$(x_1 + b_1, \ldots, x_n + b_n).$$

Since $B \in U$, $b_1 \alpha_1 + \cdots + b_n \alpha_n = 0$; it follows that

$$(x_1 + b_1)\alpha_1 + \cdots + (x_n + b_n)\alpha_n = a$$

which says that the coordinates of y satisfy the equation in question. Conversely, suppose $y = (y_1, \ldots, y_n)$ is a point in X such that

$$y_1 \alpha_1 + \cdots + y_n \alpha_n = a.$$

We put $y_i = x_i + b_i$ for $i = 1, \ldots, n$ and consider the vector $B \in V$ with coordinates $(b_1, \ldots, b_n)$. Then $y = Bx$ (Proposition 18.1) and

$$b_1 \alpha_1 + \cdots + b_n \alpha_n = 0$$

whence $B \in U$. This shows that $y \in S(x, U)$ and, therefore, that

$$z_1 \alpha_1 + \cdots + z_n \alpha_n = a$$

is an equation of $S(x, U)$.

Finally, we consider an arbitrary linear equation $z_1 \alpha_1 + \cdots + z_n \alpha_n = a$ where not all α_i are zero and show that it is the equation of a hyperplane of X. Define the functional $f: V \to k$ by $f(A_i) = \alpha_i$ and denote the kernel of f by U. Since not all α_i are zero, $f \neq 0$ and, consequently, U is a hyperplane of V. Now we choose a point $x = (x_1, \ldots, x_n)$ in X such that $x_1 \alpha_1 + \cdots + x_n \alpha_n = a$. (Why does such a point exist?) $S(x, U)$ is a hyperplane of X and the first part of this proof shows immediately that $S(x, U)$ is given by the equation $z_1 \alpha_1 + \cdots + z_n \alpha_n = a$. Done.

Exercise

1. Let $x_0, A_1, \ldots, A_n$ be a coordinate system for X. Suppose that the hyperplanes $S(x, U)$ and $S(y, W)$ of X have, respectively, the equations $z_1 \alpha_1 + \cdots + z_n \alpha_n = a$ and $z_1 \beta_1 + \cdots + z_n \beta_n = b$. Prove that $S(x, U) = S(y, W)$ if and only if there is a nonzero scalar $\gamma \in k$ such that $\beta_i = \alpha_i \gamma$ for $i = 1, \ldots, n$ and $b = a\gamma$. In short, the equation of a hyperplane is unique except for a nonzero right factor in k.

In case k is a field, the coefficients $\alpha_1, \ldots, \alpha_n$ in the equation

$$z_1 \alpha_1 + \cdots + z_n \alpha_n = a$$

may be written on either side of the variables $z_1, \ldots, z_n$. However, if k is a noncommutative division ring, the coefficients $\alpha_1, \ldots, \alpha_n$ must be placed on the right of the variables.

Exercises

2. (This is a special exercise for those readers interested in noncommutative division rings.) Let k be noncommutative. Consider the equation $\alpha_1 z_1 + \cdots + \alpha_n z_n = a$ where not all α_i are zero. Relative to a coordinate system $x_0, A_1, \ldots, A_n$ of X, the solutions of this equation determine a subset F of X. By analyzing the proof of Proposition 22.1, explain why F may fail to be a hyperplane of X.

In each of the following exercises, it is assumed that a coordinate system has been chosen for X. If x is a point of X, $x = (x_1, \ldots, x_n)$ is an abbreviation for "the coordinates of x are $(x_1, \ldots, x_n)$."

*3. Assume dim $X = 2$. Find a set of parametric equations for the line through the points x and y in each of the following cases. (Hint: Use Exercises 18.5 and 19.8.)
   a. $x = (0, 0)$ and $y = (3, 0)$.
   d. $x = (7, -1)$ and $y = (2, 1)$.
   b. $x = (0, 5)$ and $y = (0, 3)$.
   e. $x = (3, 1)$ and $y = (3, 2)$.
   c. $x = (1, 2)$ and $y = (4, 0)$.

4. When dim $X = 2$, the lines are the hyperplanes of X. For each of the lines of Exercise 3, find a linear equation of the form $z_1 \alpha_1 + z_2 \alpha_2 = a$. (Hint: Eliminate the parameter in the parametric equations.)

5. Find a set of parametric equations for the line through the ponts x and y in each of the following cases.
   a. $x = (2, -1, 7)$ and $y = (6, 4, -3)$ $(n = 3)$.
   b. $x = (8, 6, -1)$ and $y = (-1, 0, 4)$ $(n = 3)$.
   c. $x = (6, -1, 0, 4, -2)$ and $y = (2, -1, -1, 3, 0)$ $(n = 5)$.

6. Parametric equations for affine subspaces $S(x, U)$ of dimension $d > 1$ can be given just as easily as for lines. Let the vectors $B_1, \ldots, B_d$ be a coordinate system for U where the coordinates of B_i and $(b_{1i}, \ldots, b_{ni})$ for $i = 1, \ldots, d$. If $x = (x_1, \ldots, x_n)$, prove that the following set of equations with d parameters $t_1, \ldots, t_d$ is a set of parametric equations for $S(x, U)$.

$$z_1 = x_1 + t_1 b_{11} + t_2 b_{12} + \cdots + t_d b_{1d}$$
$$\vdots$$
$$z_n = x_n + t_1 b_{n1} + t_2 b_{n2} + \cdots + t_d b_{nd}.$$

*7. Find a set of parametric equations for the plane through points x, y, and z in each of the following cases.
   a. $x = (0, 1, 1)$, $y = (1, -1, 1)$, $z = (3, -2, 4)$ $(n = 3)$.
   b. $x = (1, 2, 1)$, $y = (3, 0, 2)$, $z = (4, -1, 0)$ $(n = 3)$.
   c. $x = (1, 0, 6, 1)$, $y = (2, -1, 3, 7)$, $z = (0, 0, 2, 1)$ $(n = 4)$.

8. When dim $X = 3$, the planes are the hyperplanes of X. Find a linear equation of the form

$$z_1 \alpha_1 + z_2 \alpha_2 + z_3 \alpha_3 = a$$

for the planes in Exercises 7a and 7b above. (Hint: Eliminate the parameters in the parametric equations.)

9. Find the intersection of the following pairs of affine subspaces of three-space. In general, " n-space " is short for n-dimensional affine space (X, V, k). (Hint: Use parametric equations.) The intersection may be empty or may have dimension ≥ 1. In the latter case, give a set of parametric equations for the intersection.
   a. The line through the points $x = (3, 0, 1)$, $y = (4, 2, 1)$ and the line through the points $u = (2, 0, 0)$, $v = (2, 4, 1)$.
   b. The line through the points $x = (1, 3, 1)$, $y = (2, 3, 2)$ and the plane through the points $u = (1, 1, 0)$, $v = (4, 1, 1)$, $w = (1, 2, 1)$.
   c. The plane through the points $x = (1, 1, 0)$, $y = (4, 1, -1)$, $z = (1, 2, 1)$ and the line through the points $u = (4, -1, 2)$, $v = (4, 3, -4)$.
   d. The plane through the points $x = (1, -1, 0)$, $y = (3, -1, 1)$, $z = (0, 1, 1)$ and the plane through the points $u = (-1, 2, 3)$, $v = (-1, 3, 2)$, $w = (1, 3, 3)$.

10. In three-space, let l be the line passing through the points $x = (2, 0, 1)$, $y = (0, 1, 1)$ and m the line passing through the points $u = (1, 0, 0)$, $v = (6, 1, 1)$. Show that l and m have no points in common and have different direction spaces. (They are "skew lines.")

11. Let π be the plane in three-space which passes through the points with coordinates $(1, 1, 0)$, $(0, 0, 2)$, $(0, 2, -1)$ and let l be the line which passes through the points with coordinates $(0, -1, 2)$, $(3, 1, 0)$. Show that l and π have precisely one point in common and find the coordinates of this point. Also show that the direction space of l is not contained in the direction space of π.

12. This exercise analyzes the reason it is so easy to write the parametric equations of a plane through three noncollinear points x, y, and z in n-space. (Points are said to be collinear if they lie on a line.)
   a. Prove that the vectors $\overrightarrow{x, y}$ and $\overrightarrow{x, z}$ of V are linearly independent.
   b. Prove that the points x, y, and z belong to one and only one plane of n-space.

   c. If $x = (x_1, \ldots, x_n)$, $y = (y_1, \ldots, y_n)$ and $z = (z_1, \ldots, z_n)$, find a set of parametric equations for the plane which passes through the points x, y and z.

13. This exercise extends Exercise 12 above from planes to arbitrary affine subspaces of n-space. Let $x_1, \ldots, x_d$ be $d \geq 2$ points in X which do not lie in an affine space of dimension $\leq d - 2$.

   a. Prove that $d \leq n + 1$.

   b. Prove that the $d - 1$ vectors $\overrightarrow{x_1, x_2}, \ldots, \overrightarrow{x_1, x_d}$ of V are linearly independent.

   c. Prove that the points $x_1, \ldots, x_d$ belong to one and only one $(d - 1)$-dimensional affine subspace S of X.

   d. If the coordinates of x_i are $(b_{1i}, \ldots, b_{ni})$ for $i = 1, \ldots, d$, find a set of parametric equations for S.

14. Determine whether the points $x = (1, 1, -1)$, $y = (2, 1, 1)$, $z = (3, -1, 2)$, and $u = (0, 3, -2)$ of three-space are coplanar. (Points are called coplanar if they line in a plane.)

*15. We now investigate functionals on our n-dimensional, left vector space V. Recall that a functional $f \colon V \to k$ is a linear mapping where k is considered as a left vector space over itself.

   a. Let $f \colon V \to W$ be a linear mapping from V into some left vector space W over k. Choose $c \in k$ and define a new function $cf \colon V \to W$ by $(cf)(A) = c[f(A)]$ for $A \in V$. Prove that, *if k is a field, cf is again a linear mapping from V into W.* Point out where the commutativity of k is used.

   b. Let $f \colon V \to k$ be a nonzero functional and define cf as in Part a. Suppose there is an element $b \in k$ such that $bc \neq cb$. Prove that cf is not a functional.

   c. Let $f \colon V \to k$ be a functional. Choose $c \in k$ and define $fc \colon V \to k$ by $(fc)(A) = [f(A)]c$ for $A \in V$. Prove that fc is a functional whether or not k is a field.

   d. Prove that the functionals on V form a *right* vector space over k.

9. PARALLELISM

In the plane, one calls lines l and l' parallel if $l \cap l' = \varnothing$ or $l = l'$. This definition fails in higher dimensions. For instance, one would not want to call two skew lines in three-space parallel even though they have no point in common. (**n-space** stands for "n-dimensional affine space (X, V, k).") Two affine subspaces of the same dimension should be called parallel if they have the same direction. Precisely:

Definition 27.1. Let S and S' be affine subspaces of X of the same dimension. Then S and S' are called **parallel** if they have the same direction space. Notation: $S \parallel S'$.

Definition 27.1 considers parallelism only for subspaces of the same dimension. A more general definition of parallelism for affine subspaces of different dimensions is given at the end of this section.

For each d between 0 and n, parallelism is clearly an equivalence relation for the set of affine subspaces of dimension d.

Exercise

1. The 0-dimensional subspaces of X are the points of X. Prove that every pair of points are parallel.

The usual pictures suggest that two lines in the plane are parallel if and only if there is a translation which moves one line onto the other. A few pictures quickly convince us that the same should be true for two lines or two planes in three-space. Indeed, we shall prove in Proposition 28.1 that parallelism can be defined in terms of translations.

If f is a function from X into X and S is an affine subspace of X, then $f(S)$ denotes the image of S under f in the usual sense of "sets and functions"; that is, $f(S) = \{f(x) \mid x \in S\}$. We shall be interested in the special case when f is a translation of X.

Proposition 27.1. Let $S(x, U)$ be an affine subspace of X and $A \in V$. Then $T_A(S(x, U)) = S(Ax, U)$.

Proof. We recall that, for any two vectors $A, B \in V$ and any point $x \in X$,

$$A(Bx) = (A + B)x = (B + A)x = B(Ax).$$

Since $S(x, U) = \{Bx \mid B \in U\}$, it follows that

$$T_A(S(x, U)) = \{A(Bx) \mid B \in U\} = \{B(Ax) \mid B \in U\} = S(Ax, U).$$

Done.

We see (Proposition 27.1) that $T_A(S(x, U))$ is an affine subspace of the same dimension as $S(x, U)$ and parallel to it. The situation is pictured in Figure 28.1 for the case when S has dimension two. The converse is also true.

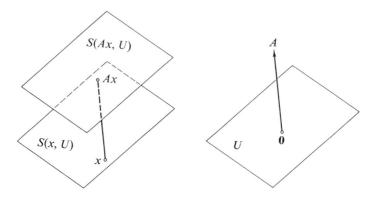

Figure 28.1

Proposition 28.1. Two affine subspaces S and S' of X of the same dimension are parallel if and only if there is a translation T such that $T(S) = S'$.

Proof. We have just seen that $T(S) \parallel S$. Conversely, suppose that dim $S =$ dim S' and $S \parallel S'$. Then, by Definition 27.1, $S = S(x, U)$ and $S' = S(y, U)$ where $x, y \in X$ and U is a subspace of V. We put $A = \overrightarrow{x, y}$ and conclude, from Proposition 27.1, that $T_A(S) = S'$. Done.

We now show that the customary definition of parallel lines in the plane (reviewed at the beginning of this section) may always be used for hyperplanes. Recall that lines in the plane are hyperplanes.

Proposition 28.2. If S and S' are parallel affine subspaces of X, then either $S = S'$ or $S \cap S' = \emptyset$. Moreover, if S and S' are hyperplanes, then $S \parallel S'$ if and only if $S = S'$ or $S \cap S' = \emptyset$.

Proof. If $S \parallel S'$, then $S = S(x, U)$ and $S' = S(y, U)$ for appropriately chosen x, y and U. If S and S' have a point in common, we conclude from Proposition 14.1 that $S = S'$.

Suppose now that $S = S(x, U)$ and $S' = S(y, W)$ are hyperplanes and that either $S = S'$ or $S \cap S' = \emptyset$. If $S = S'$, they have the same direction space and hence are parallel. There only remains to be shown that, if $S \cap S' = \emptyset$, then $S \parallel S'$, or, equivalently, if $S \nparallel S'$, then $S \cap S' \neq \emptyset$.

Hence we assume that $U \neq W$ and prove that $S \cap S' \neq \emptyset$. Since U and W are distinct $(n - 1)$-dimensional linear subspaces of V, $U + W = V$. Consequently, the vector $A = \overrightarrow{x, y}$ can be written as $A = B + C$ where $B \in U$ and $C \in W$. Then

$$y = Ax = (B + C)x \Rightarrow (-C)y = Bx$$

and this point clearly belongs to both S and S'. (A one-pointed arrow $\Rightarrow$ stands for "implies.") Done.

The following proposition states that Euclid's famous fifth parallel axiom holds for arbitrary n ($= $ dimension of X) and all affine subspaces of X, no matter what their dimension may be. Moreover, the proof is trivial!!

Proposition 29.1 (the fifth parallel axiom).   Let S be a d-dimensional affine subspace and x a point of X. Then there exists one and only one d-dimensional affine subspace S' of X which passes through x and is parallel to S.

Proof. Let U be the direction space of S. Since $S \parallel S'$, U must also be the direction space of S'. Consequently, $S' = S(x, U)$. Done!!

Observe that the fifth parallel axiom belongs to affine geometry rather than Euclidean geometry. Euclidean geometry is affine geometry $(X, V, \mathbf{R})$ where $\mathbf{R}$ is the field of real numbers and where a very special measuring rod has been constructed (see page 126). This measuring rod changes nothing in the affine structure of the underlying space X which is why the fifth parallel axiom (as well as other affine propositions) remains valid in Euclidean geometry. The same is true for all metric affine spaces as we shall see in the next two chapters. *The measuring rod which is superimposed on the affine space (X, V, k) leaves the affine structure of X untouched; hence, in all these spaces, Euclidean or not, the fifth parallel axiom holds.* Examples of non-Euclidean metric affine spaces with $k = \mathbf{R}$ are the Lorentz plane and Minkowski space (see pages 126 and 127); in these highly non-Euclidean geometries the fifth parallel axiom holds!
The only way that one can destroy the fifth parallel axiom is by going over to a space which is not affine, such as projective space. The classical elliptical plane is obtained by constructing a metric for the projective plane while hyperbolic geometry is obtained by constructing a metric for part of the projective plane. In both cases, Euclid's fifth parallel axiom is destroyed by the transition from affine geometry to projective geometry. The slogan that the fifth parallel axiom holds solely in Euclidean geometry is correct

only if one restricts oneself to Euclidean geometry, the elliptic plane, and hyperbolic geometry. This slogan should be forgotten. The fifth parallel axiom holds in all affine spaces and hence in all metric affine spaces.

Exercises

2. Assume that dim $X = 2$ and that a coordinate system has been chosen for X. Consider the points $x = (0, -1)$, $y = (3, 1)$, $u = (3, 4)$, and $w = (0, 2)$ in X.
   a. Prove that the line through the points x and y is parallel to the line through the points u and w.
   b. Prove that the line through the points x and w is parallel to the line through the points y and u.
   c. What kind of a quadrilateral do the points x, y, u, and w form? Plot these points under the assumption that $k = \mathbf{R}$.
*3. Assume S and S' are d-dimensional affine subspaces of n-space where $d < n$. Prove that $S \parallel S'$ if and only if the following two conditions are satisfied:
   a. S and S' are contained in a $(d + 1)$-dimensional subspace of X.
   b. Either $S = S'$ or $S \cap S' = \varnothing$.
4. Let $S - S(x, U)$ and $S' = S(y, W)$ be affine subspaces of the n-space (X, V, k).
   a. Prove that, if $S \cap S' \neq \varnothing$, then

   $$\dim(S \cap S') = \dim S + \dim S' - \dim(U + W).$$

   b. Prove that, if $U + W = V$, then $S \cap S' \neq \varnothing$ and

   $$\dim(S \cap S') = \dim S + \dim S' - n.$$

   (Hint: Analyze the proof of Proposition 28.2.)
   c. Prove that, if $V = U \oplus W$, $S \cap S'$ consists of one point.
5. Let $0 \leq d \leq n$. It was pointed out that parallelism is an equivalence relation for the set H_d of d-dimensional affine subspaces of n-space. This equivalence relation partitions H_d into disjoint subsets, each of which consists of a family of parallel, d-dimensional subspaces. Prove that all these families have the same cardinality $|k|^{n-d}$ where $|k|$ denotes the cardinality of k. (Hint: The cardinality of an n-dimensional vector space, and hence of n-space, is $|k|^n$. Use Exercise 4 above to set up a correspondence between the elements of a family of parallel d-dimensional subspaces and an appropriate $(n - d)$-dimensional subspace which is one-to-one and onto.)
6. Assume a coordinate system has been chosen for n-space. Let the

hyperplane π_1 have the equation $y_1\alpha_1 + \cdots + y_n\alpha_n = a$ and the hyperplane π_2 have the equation $y_1\beta_1 + \cdots + y_n\beta_n = b$. Prove that $\pi_1 \parallel \pi_2$ if and only if there is a $\gamma \neq 0$ in k such that $\beta_i = \gamma\alpha_i$ for $i = 1, \ldots, n$.

7. Assume that a coordinate system has been chosen for the three-dimensional space X. If the plane π has the equation

$$(y_1)3 - y_2 + (y_3)2 + 6 = 0,$$

find an equation for the plane which passes through the point $x = (1, -2, 7)$ and is parallel to π.

We now generalize Definition 27.1 so that we can also speak about parallel subspaces of n-spaces which do not have the same dimension. ("Subspace" stands for "affine subspace.") For example, in Figure 31.1, every line in the plane π_2 is parallel to the plane π_1.

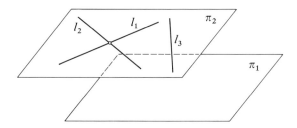

Figure 31.1

Definition 31.1. Let S and S' be affine subspaces of n-space with direction spaces U and U', respectively. Then $S \parallel S'$ if $U \subset U'$ or $U' \subset U$.

Observe that, if $\dim S = \dim S'$, Definition 31.1 becomes Definition 27.1. In the above picture, $l_1 \parallel \pi_1$ and $\pi_1 \parallel l_2$, but $l_1 \nparallel l_2$. We conclude that parallelism is no longer a transitive relation as it was when we restricted ourselves to parallel subspaces of the same dimension. Of course, parallelism is still a reflexive and symmetric relation for the subspaces of n-space. Some of the following exercises deal with the changes which occur in the propositions on parallelism now that we have extended this concept to subspaces of different dimension.

Exercises

8. Assume $\dim X = 3$ and that a coordinate system has been chosen for X. Find a set of parametric equations for the plane π which passes through the points $x = (1, -1, 2)$ and $y = (3, 1, 1)$ and is parallel to

the line through the points $u = (-1, 2, 3)$ and $w = (4, -1, 1)$. Also find an equation of the form $y_1 \alpha_1 + y_2 \alpha_2 + y_3 \alpha_3 = a$ for π.

9. Let π be a hyperplane and l a line in (X, V, k). Prove that $l \parallel \pi$ or l and π have precisely one point in common. Note that this is a generalization of the following familiar result: If l is a line and π is a plane in three-space, then either $l \parallel \pi$ or l and π have precisely one point in common.

10. Let S and S' be affine subspaces of n-space where dim $S \leq$ dim S'. Prove that:
   a. $S \parallel S'$ if and only if there exists a translation T of X such that $T(S) \subset S'$.
   b. $S \parallel S'$ if and only if there is an affine subspace S^* such that $S \subset S^*$, $S^* \parallel S'$ and dim $S^* =$ dim S'.
   c. If $S \parallel S'$, either $S \subset S'$ or $S \cap S' = \varnothing$. Show by an example that the converse is false.
   d. If S' is a hyperplane and dim $S \geq 1$, then $S \parallel S'$ if and only if $S \subset S'$ or $S \cap S' = \varnothing$.

11. Let $S = S(x, U)$ and $S' = S(y, W)$ be affine subspaces of X. Prove that $S \subset S'$ if and only if these spaces have a point in common and $U \subset W$. (This result is convenient for solving Exercises 12 and 13.)

12. Suppose that S, S', and S'' are affine subspaces of X where $S \parallel S'$ and dim $S \leq$ dim S'. If $S'' \subset S$, prove that $S'' \parallel S'$. Show by means of a picture that the condition dim $S \leq$ dim S' cannot be omitted.

(The figure below refers to Exercise 14.)

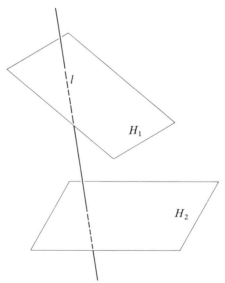

Figure 32.1

13. Let S be a d-dimensional subspace of X and x a point of X. By the fifth parallel axiom, there is one and only one d-dimensional subspace S' of X which passes through x and is parallel to S. Prove that a subspace S'' of X which passes through x is parallel to S if and only if either $S'' \subset S'$ or $S' \subset S''$.

14. Let H_1 and H_2 be hyperplanes in X. Choose a line l in X which is not parallel to either H_1 or H_2 (see Figure 32.1); that is, choose l such that the direction space of l is not contained in either the direction space of H_1 or H_2. Define what one would mean by "the projection P of H_1 onto H_2 parallel to the direction of l."

    a. Prove that P is one-to-one and onto.

    b. Let $S(x, W)$ be a d-dimensional affine subspace of H_1. Show that its projection $P(S(x, W))$ is a d-dimensional affine subspace of H_2.

10. AFFINE SUBSPACES SPANNED BY POINTS

We know, from Exercise 18.5, that two distinct points of X determine a unique line. Also, three points not lying on a line should determine a unique plane. In a certain sense, we think of three points as being "dependent" if they lie on a line and four points as being "dependent" if they lie in a plane. The general notion of independence of points will now be investigated.

Definition 33.1. $d \geq 1$ points of X are called **independent** if they do not lie in an affine subspace of X of dimension at most $d - 2$. Otherwise, these points are called **dependent**.

For $d = 1$, we see that each individual point of X is independent since X has no subspaces of dimension ≤ -1. Two distinct points are always independent; three points are independent if they do not lie on a line; $n + 1$ points are independent if they do not lie in a hyperplane; $n + 2$ or more points are always dependent. (n always stands for the dimension of X.)

Exercise

1. Let $1 \leq d \leq n + 2$. Prove that d points of X are independent if and only if they do not lie in a $(d - 2)$-dimensional subspace of X.

Can we always find d independent points where $1 \leq d \leq n + 1$? Yes, we can; the proof depends on the fact that we can always find r linearly independent vectors in V for $1 \leq r \leq n$.

Proposition 34.1. There exist d independent points in X whenever $1 \leq d \leq n + 1$.

Proof. Select $d - 1$ linearly independent vectors $A_1, \ldots, A_{d-1}$ which can be done since $d \leq n + 1$ implies that $d - 1 \leq n$. If x is a point in X, we claim that $x, A_1x, \ldots, A_{d-1}x$ are d independent points. Figure 34.1 shows the

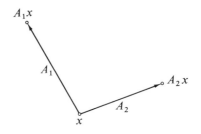

Figure 34.1

case for $d = 3$. Suppose not; that is, suppose the d points lie in $S(x, U)$ where $\dim U = d - 2$. Then the $d - 1$ vectors $A_1 = \overrightarrow{x, A_1x}, \ldots, A_{d-1} = \overrightarrow{x, A_{d-1}x}$ belong to U. But it is impossible for the $(d - 2)$-dimensional linear subspace U to contain the $d - 1$ linearly independent vectors $A_1, \ldots, A_{d-1}$. Done.

Proposition 34.2. Let $1 \leq d \leq n + 1$. If $x_1, \ldots, x_d$ are independent points, they lie in one and only one $(d - 1)$-dimensional affine subspace of X.

Proof. Let U be the linear subspace of V generated by the $d - 1$ vectors $\overrightarrow{x_1, x_2}, \ldots, \overrightarrow{x_1, x_d}$. Note that $\dim U \leq d - 1$. Then the points $x_1, \ldots, x_d$ all belong to the affine subspace $S(x_1, U)$. Since $x_1, \ldots, x_d$ are independent, they do not lie in a $(d - 2)$-dimensional affine subspace; hence $\dim S(x_1, U) = \dim U \geq d - 1$. Consequently, $\dim U = d - 1$ and $S(x_1, U)$ is a $(d - 1)$-dimensional affine subspace containing $x_1, \ldots, x_d$. Now suppose that $S(y, W)$ is any $(d - 1)$-dimensional affine subspace which contains $x_1, \ldots, x_d$. Then

$$x_1, \ldots, x_d \in S(x, U) \cap S(y, W) = S(x_1, U \cap W).$$

If $U \neq W$, $\dim(U \cap W) \leq d - 2$, contrary to the hypothesis that the points $x_1, \ldots, x_d$ are independent. Hence $U = W$ and, since $x_1 \in S(y, W)$, $S(y, W) = S(x_1, U)$. Done.

As a special case of the above proposition, we can conclude that two points determine a unique line, three noncollinear points determine a unique plane and n independent points lie on a unique hyperplane.

Let Y be a nonempty subset of X. There is at least one affine subspace of X which contains Y, namely X itself. The intersection S of all affine subspaces of X which contain Y is again an affine subspace of X (Exercise 16.1). S is called the space spanned by Y. If Y consists of the points $x_1, \ldots, x_d$, not necessarily independent, the space spanned by these points is denoted by $x_1 \vee \cdots \vee x_d$. Similarly, if Y is the union of the affine subspaces $S_1, \ldots, S_d$ of X, the space spanned by Y is denoted by $S_1 \vee \cdots \vee S_d$. Three illustrations are given below:

Figure 35.1

In Figure 35.1, $l = x_1 \vee x_2 = x_1 \vee x_3 = x_2 \vee x_3 = x_1 \vee x_2 \vee x_3 = x_1 \vee l$, etc. Any two of the three points x_1, x_2, x_3 are independent. The three points x_1, x_2, x_3 are dependent.

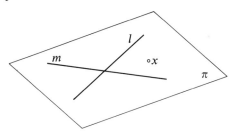

Figure 35.2

In Figure 35.2, the plane $\pi = l \vee x = m \vee x = l \vee m = \pi \vee l = l \vee m \vee x$, etc.

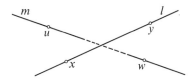

Figure 35.3

Finally, in Figure 35.3, l and m represent skew lines in the three-space X. Then $X = l \vee m = x \vee y \vee u \vee w = x \vee y \vee m$, etc.

Exercises

2. Let $x_1, \ldots, x_d$ be independent points of X. By Proposition 34.2, they lie in a unique $(d-1)$-dimensional affine subspace S of X. Prove that $S = x_1 \vee \cdots \vee x_d$.

*3. Let $x_1, \ldots, x_d$ be points in X and select some $x_i \in \{x_1, \ldots, x_d\}$. Prove that the points $x_1, \ldots, x_d$ are independent if and only if the $d-1$ vectors $\overrightarrow{x_i, x_1}, \ldots, \overrightarrow{x_i, x_{i-1}}, \overrightarrow{x_i, x_{i+1}}, \ldots, \overrightarrow{x_i, x_d}$ are linearly independent vectors of V.

4. Let $x_1, \ldots, x_d$ be independent points in X and select some $x \in X$. Prove that $x_1, \ldots, x_d, x$ are independent if and only if $x \notin x_1 \vee \cdots \vee x_d$.

5. Let $x_1, \ldots, x_d$ be independent points of X. Prove that every nonempty subset of $\{x_1, \ldots, x_d\}$ consists of independent points.

6. Let $x_1, \ldots, x_d$ be independent points of X. If $d \leq d' \leq n+1$, prove that there is a set of d' independent points of X which contains $x_1, \ldots, x_d$.

7. Assume dim $X = 4$ and that a coordinate system has been chosen for X. Consider the points $x = (1, 0, 1, 0)$, $y = (-1, -1, 0, 0)$, $u = (2, 0, 0, 3)$, and $w = (-5, 3, 1, 1)$ of X.
   a. Show that these four points are independent.
   b. Find a linear equation $y_1 \alpha_1 + y_2 \alpha_2 + y_3 \alpha_3 + y_4 \alpha_4 = a$ for the hyperplane $x \vee y \vee u \vee w$.
   c. Find a linear equation for the hyperplane which passes through the point $e = (1, 1, 1, 1)$ and is parallel to $x \vee y \vee u \vee w$.

8. Let $S = S(x, U)$ and $S' = S(y, W)$ be affine subspaces of X. Prove that:
   a. If $S \cap S' \neq \varnothing$, the direction space of $S \vee S'$ is $U + W$.
   b. If $S \cap S' \neq \varnothing$, $\dim(S \vee S') + \dim(S \cap S') = \dim S + \dim S'$.
   c. If $S \cap S' = \varnothing$, $\dim(S \vee S') = 1 + \dim(U + W)$.
   d. If a point $x \notin S$, $\dim(S \vee x) = 1 + \dim S$.
   e. If $x_1, \ldots, x_d$ are points of X, $\dim(x_1 \vee \cdots \vee x_d) \leq d - 1$.

11. THE GROUP OF DILATIONS

It is usually worthwhile to investigate the automorphism of a mathematical structure, that is, those one-to-one mappings of the structure onto itself which preserve its structural properties. What are the automorphisms of affine n-space (X, V, k)?

Affine geometry is the geometry of parallelism and it enables us to compare the lengths of parallel line segments. Consequently, its automorphisms should, at least, preserve parallelism. More precisely, an automorphism of X

should, at least, be a one-to-one mapping f of X onto itself with the property that, if S and S' are parallel subspaces of X, then $f(S)$ and $f(S')$ are parallel subspaces of X. These mappings are the "semiaffine transformations" of X and will be studied in Section 15. Actually, a true automorphism of X should do more than preserve parallelism; it should also preserve the relative lengths of parallel line segments. These true automorphisms are appropriately called "affine transformations" and are also the subject of Section 15.

At present, we study one-to-one mappings f of X onto itself which do more than preserve parallelism and relative length of line segments. They preserve direction; that is, for every affine subspace S of X, $S \parallel f(S)$. If X is a line ($n = 1$), its subspaces are the points and X itself; hence every one-to-one mapping of a line onto itself preserves direction. This is the reason lines have to be treated separately (see page 58). When $n \geq 2$, it turns out that the condition $S \parallel f(S)$ has to be postulated only for the lines of X.

Definition 37.1.   Let $n \geq 2$. A **dilation** of X is a one-to-one mapping of X onto itself which maps every line of X onto a parallel line.

In most books, the term dilatation is used instead of dilation. Since darkness dilates the pupils of our eyes and does not dilatate them, we see no reason for the extra "ta."

Until we have discussed dilations of lines, it is understood that $n \geq 2$ whenever we mention dilations.

The identity mapping of X is clearly a dilation. However, there are also interesting dilations.

Proposition 37.1.   Every translation is a dilation.

Proof. Every translation is a one-to-one mapping of X onto X (Proposition 9.1). If $S(x,U)$ is an affine subspace of X and T is a translation, then $T(S(x,U)) = S(Tx,U)$ whence $S \parallel T(S)$. So, in particular, if S is a line, $S \parallel T(S)$. Done.

The question arises whether there are dilations which are not translations. We expect an affirmative answer from our experience with plane geometry. Let X be the real affine plane, c a point in X and r a nonzero real number. Consider the mapping from X to X which sends each point x onto the point $x' = Ac$ where A is the vector $r(\overrightarrow{c,x})$. The mapping is pictured for $r = 3$ in Figure 38.1. Clearly, the lines l and l' are parallel since the triangles cyx and $cy'x'$ are similar. Such mappings are called magnifications and we now define them for all $n \geq 1$.

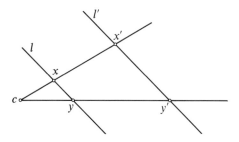

Figure 38.1

Definition 38.1.   Let $n \geq 1$. Select $c \in X$ and $r \in k$, $r \neq 0$. The **magnification of** X with **center** c and **magnification ratio** r is the function $M(c, r): X \to X$ defined by

$$M(c, r)(x) = (r(c, \vec{x}))c.$$

The formula which defines the magnification may seem a little cumbersome, but the actual mapping is very simple as Figure 38.2 shows. For a

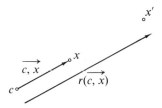

Figure 38.2

given $x \in X$, we take the vector $\vec{c, x}$, "stretch" it by the scalar r, that is, take the scalar multiple $r(\vec{c, x})$, and let the vector $r(\vec{c, x})$ act on the point c. The resulting point is the image of x. In practice, one uses mostly the fact that $M(c, r)(x) = y$ is the same as $\vec{c, y} = r(\vec{c, x})$.

We now show that magnifications are dilations and that the magnifications with fixed center c form a group isomorphic to k^* where k^* denotes the multiplicative group of the nonzero elements of k. The nicest arrangement is obtained by starting with the second statement.

Proposition 38.1.   Let c be a point in X. The magnifications with c as center form a group which is isomorphic to k^*.

Proof. We consider the mapping $\phi: k^* \to \{M(c, r) | r \in k^*\}$ defined by $\phi(r) = M(c, r)$. It is obvious that ϕ is onto.

1. Now let us prove that $\phi(rs) = \phi(r)\phi(s)$ for $r, s \in k^*$. Let $x \in X$ and put $M(c, rs)(x) = z$, $M(c, s)(x) = y$, $M(c, r)(y) = u$. We must prove that $z = u$.

$$M(c, rs)(x) = z \Leftrightarrow \overrightarrow{c, z} = rs(\overrightarrow{c, x}),$$

$$M(c, s)(x) = y \Leftrightarrow \overrightarrow{c, y} = s(\overrightarrow{c, x}),$$

$$M(c, r)(y) = u \Leftrightarrow \overrightarrow{c, u} = r(\overrightarrow{c, y}) = rs(\overrightarrow{c, x}).$$

We conclude that $\overrightarrow{c, z} = \overrightarrow{c, u}$; by letting this vector act on c, it follows that $z = u$.

2. We prove here that $M(c, r)$ is the identity mapping 1_X of X if and only if $r = 1$. $M(c, r) = 1_X$ means that $M(c, r)(x) = x$ or, equivalently, $\overrightarrow{c, x} = r(\overrightarrow{c, x})$ for all $x \in X$. We choose $x \neq c$ (Why can this be done?) in which case the vector $\overrightarrow{c, x} \neq \mathbf{0}$ and conclude that the equality $\overrightarrow{c, x} = r(\overrightarrow{c, x})$ is satisfied if and only if $r = 1$.

The fact that ϕ is onto, together with Part 1, shows that $\{M(c, r) | r \in k^*\}$ is a group. Using Part 2 also, it follows that ϕ is an isomorphism. Done.

Exercises

1. Since $\{M(c, r) | r \in k^*\}$ is a group, $M(c, r)$ must have an inverse. Show that this inverse is $M(c, r^{-1})$.

2. Prove that every magnification leaves its center fixed; that is, $M(c, r)(c) = c$.

*3. If $r \neq 1$, show that c is the only fixed point of the magnification $M(c, r)$.

4. Show that 1_X is the only function from X onto X which is both a translation and a magnification.

5. Let $M(c, r)$ be a magnification and A a vector. Prove that $M(c, r)(Ac) = (rA)c$. Note that Exercise 2 above is the special case of this exercise when $A = \mathbf{0}$.

*6. Let $M(c, r)$ be a magnification, A a vector, and x a point of X. Prove that $M(c, r)(Ax) = (rA)M(c, r)(x)$. (Hint: $x = (\overrightarrow{c, x})c$ and hence $Ax = (A + \overrightarrow{c, x})c$. Now use the previous exercise.)

*7. Interpret the result of Exercise 6 above as follows. Prove that the magnification $M(c, r)$ and the translation T_A do not commute, in general, but that, instead, $M(c, r)T_A = T_{rA} M(c, r)$.

Before showing that magnifications are dilations, we first prove the following result.

Proposition 40.1.   Let $M(c, r)$ be a magnification and $S = S(x, U)$ an affine subspace of X. Then $M(c, r)(S)$ is an affine subspace of X which is parallel to S. Precisely,

$$M(c, r)(S(x, U)) = S(y, U)$$

where $y = M(c, r)(x)$.

Proof. By the definition of a function acting on a set,

$$M(c, r)(S) = \{M(c, r)(Ax) \,|\, A \in U\}.$$

According to Exercise 39.6, this set is equal to $\{(rA)y \,|\, A \in U\}$ where $y = M(c, r)(x)$. When A runs through U, rA also runs through U since $r \neq 0$ and hence $\{(rA)y \,|\, A \in U\} = \{Ay \,|\, A \in U\} = S(y, U)$. Done.

Proposition 40.2.   Every magnification is a dilation.

Proof. Let $M(c, r)$ be a magnification. $M(c, r)$ is one-to-one and onto because it has the inverse $M(c, r^{-1})$ and the only functions which have inverses are one-to-one and onto.

As a special case of Proposition 40.1, if l is a line, then $M(c, r)(l)$ is a line parallel to l. Done.

Exercise

8. If l is a line which passes through the point c, prove that $M(c, r)(l) = l$. Show also that, if $r \neq 1$, $M(c, r)$ leaves no points of l fixed except c.

It is not surprising that translations and magnifications are dilations. Neither is the next proposition surprising, even though it is very important. Every dilation is a one-to-one mapping of X onto itself. In other words, the set Di of the dilations of X is a subset of the group S_X of permutations of X.

Proposition 40.3.   The set Di of dilations of X is a subgroup of S_X.

Proof. Since 1_X is a dilation, the set Di is nonempty. Assume D_1, $D_2 \in Di$ and let us prove that $D_1 D_2 \in Di$. We choose a line l of X and have to show that $D_1 D_2(l)$ is a line parallel to l. Since D_2 is a dilation, $D_2(l)$ is a line parallel to l. Since D_1 is a dilation, $D_1(D_2(l))$ is a line parallel to $D_2(l)$. By the transitivity of parallelism of lines, $D_1 D_2(l) \parallel l$.

Suppose $D \in Di$ and let us prove that $D^{-1} \in Di$. Here D^{-1} denotes the inverse of D in S_X, that is, the set-theoretic inverse of the dilation D. We choose a line l and must prove that $D^{-1}(l)$ is a line parallel to l. The argument is based on the fifth parallel axiom. We choose a point x on l and conclude from the fifth parallel axiom that there is a unique line l' which is parallel to l and passes through $D^{-1}(x)$; see Figure 41.1. Since D is a dilation, $D(l')$ is a

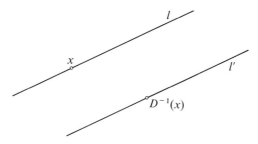

Figure 41.1

line which is parallel to l' and passes through the point $D(D^{-1}(x)) = x$. By the transitivity of parallelism of lines $D(l') \parallel l$ and hence, by the fifth parallel axiom, $D(l') = l$. It follows set-theoretically from $D(l') = l$ that $l' = D^{-1}(l)$. Thus D^{-1} is a dilation.

The three conditions just verified say precisely that Di is a subgroup of S_X. Done.

Now comes a surprise! Consider a magnification $M(c, r)$ and a translation T_A of X. The product $M(c, r)T_A$ is again a dilation by Proposition 40.3. No one expects that the product of a magnification with a translation is always either a magnification or a translation. Hence everyone expects that there are plenty of dilations which are neither magnifications nor translations. Well, everyone is wrong. There are no dilations other than magnifications and translations. In fact, we shall show that dilations can be classified as follows:

Classification of Dilations
1. A dilation has more than one fixed point if and only if it is the identity mapping 1_X of X.
2. A dilation has precisely one fixed point c if and only if it is a magnification $M(c, r)$ with $r \neq 1$.

3. A dilation has no fixed points if and only if it is a translation different from 1_X.

Every dilation falls in precisely one of the above three classes and hence we conclude that every dilation is either a translation or a magnification. In particular, the product of a magnification and a translation is always either a magnification or a translation.

We begin by proving Statement 1 of our classification.

Proposition 42.1.   A dilation D is completely determined by the images of two points. If D leaves two points fixed, $D = 1_X$.

Proof. Let D be a dilation and assume that the images $D(x) = x'$ and $D(y) = y'$ of two points x, y are known. We must show that the image of any $z \in X$ is known. We recall that the unique line through the points x and y is denoted by $x \vee y$.

Case 1. $z \notin x \vee y$. Then $z \neq x$ and $z \neq y$ and we consider the two lines $x \vee z$ and $y \vee z$; see Figure 42.1. We first observe that $z \in (x \vee z) \cap (y \vee z)$. If

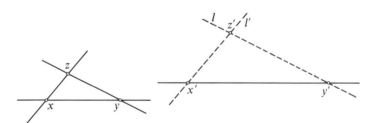

Figure 42.1

these lines had a point in common besides z, they would be equal (Exercise 18.5); then we could conclude that $x \vee z = y \vee z = x \vee y$; but this is impossible because $z \notin x \vee y$. Therefore, $\{z\} = (x \vee z) \cap (y \vee z)$.

The fact that D is one-to-one and onto implies, for set-theoretic reasons alone, that

$$D((x \vee z) \cap (y \vee z)) = D(x \vee z) \cap D(y \vee z);$$

equivalently,

$$\{D(z)\} = D(x \vee z) \cap D(y \vee z).$$

In other words, the lines $D(x \vee z)$ and $D(y \vee z)$ have precisely one point in common, namely, the point $D(z)$ for which we are looking. The line $D(x \vee z)$ is completely known since it is the unique line which passes through the point

x' and is parallel to $x \vee z$. Similarly, the line $D(y \vee z)$ is the unique line which passes through the point y' and is parallel to $y \vee z$. Since $D(z)$ is the unique point of intersection of $D(x \vee z)$ and $D(y \vee z)$, the point $D(z)$ is completely determined by x' and y'.

Case 2. $z \in x \vee y$. If z is x or y, we are given $D(z)$, so we may assume $z \neq x$ and $z \neq y$. Since dim $X \geq 2$ (we still speak about dilations only when $n \geq 2$), X contains a point $p \notin x \vee y$. The point $z \notin x \vee p$. This should not be concluded from Figure 43.1, but should be proved; we leave the proof to the

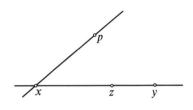

Figure 43.1

reader. As shown in Case 1, the point $D(p)$ is known. Hence, using the line $x \vee p$ instead of the line $x \vee y$, we conclude from Case 1 that $D(z)$ is known.

There only remains to be shown that $D = 1_X$ if D leaves two distinct points x, y fixed. In that case, the dilations 1_X and D have the same effect on the two points x, y. Hence $D = 1_X$. Done.

As we shall soon see, Proposition 42.1 is very powerful.

Exercises

9. Let x, y, z be points of X. Use the reasoning of the proof of Proposition 42.1 to show that the following statements are equivalent:
   a. The points x, y, z are not collinear.
   b. $x \neq y$, $x \neq z$, $y \neq z$ and $\{x\} = (x \vee y) \cap (x \vee z)$,
      $\{y\} = (x \vee y) \cap (y \vee z)$, $\{z\} = (x \vee z) \cap (y \vee z)$.
*10. Let c, x be two points of X and put $l = c \vee x$.
   a. If $y \in l$, prove that $\overrightarrow{c, y} = r(\overrightarrow{c, x})$ where $r \in k$.
   b. Prove that the mapping $y \to r$ defined by the equation $\overrightarrow{c, y} = r(\overrightarrow{c, x})$ in Part a is a one-to-one mapping from l onto k.
   c. Prove that $r = 0$ if and only if $y = c$ and that $r = 1$ if and only if $y = x$.

11. Assume c, x, and y are collinear points of X where $c \neq x$ and $c \neq y$. Prove that there is one and only one magnification $M(c, r)$ of X such that $M(c, r)(x) = y$.

We now turn to Statement 2 of the classification of dilations.

Proposition 44.1.   A dilation has precisely one fixed point c if and only if it is a magnification $M(c, r)$ with center c and magnification ratio $r \neq 1$.

Proof. A magnification $M(c, r)$ with $r \neq 1$ has c as its only fixed point (Exercise 39.3).

Assume D is a dilation which has point c as its only fixed point and let us prove that $D = M(c, r)$ for some $r \neq 1$. Select a point $x \neq c$ and put $l = c \vee x$ and $y = D(x)$. Since $D(l) = l$, $y \in l$ while $x \neq c$ implies $y \neq D(c)$; see Figure

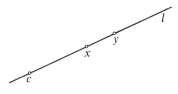

Figure 44.1

44.1. We conclude, from Exercise 43.10, that $\overrightarrow{c, y} = r(\overrightarrow{c, x})$ where $r \in k$ and $r \neq 0$. The magnifications $M(c, r)$ and D are dilations which have the same effect on the two points c and x. Hence $D = M(c, r)$.

Finally, $r \neq 1$ since $M(c, 1) = 1_X$ and D has only one fixed point. Done.

Statement 3 of the classification of dilations is now considered. First, some exercises are given whose results are used in proving the main proposition.

Exercises

*12. Let D be a dilation of X. Suppose x is a point of X and $x \neq D(x)$. If $l = x \vee D(x)$, prove that $D(l) = l$.

*13. Let D be a dilation of X. Assume the two lines l and m of X intersect in precisely one point c. Assume further that l contains a point x such that $D(x) \neq x$ and $D(x) \in l$ while m contains a point y such that $D(y) \neq y$ and $D(y) \in m$. Prove that c is a fixed point of D.

*14. Let D be a dilation of X. If x and y are points of X, prove that the points $x, y, D(x)$ and $D(y)$ lie in a plane. (Hint: Use Exercise 30.3.)

*15. Let x, x', y and y' be four distinct points of X and put $l = x \vee x'$, $m = y \vee y'$, $s = x \vee y$ and $t = x' \vee y'$. Assume that $l \parallel m$, $l \cap m = \varnothing$ and $s \parallel t$. In other words, the four points x, x', y', y form a parallelogram; see Figure 45.1. Prove that, if $A = \overrightarrow{x, x'}$, then $T_A(y) = y'$. (Hint: T_A is a dilation and, therefore, $T_A(y) \in t$. The fact that $\overrightarrow{x, x'} = \overrightarrow{y, T_A(y)}$ im-

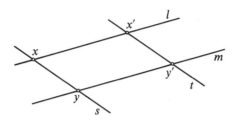

Figure 45.1

plies that the line through y and $T_A(y)$ is m. Danger: Use the picture only to find the correct proof, but in no way use it as a substitute for the proof. The statement is false without the assumption $l \cap m = \varnothing$; did this fact occur in the proof?)

Proposition 45.1. A dilation has no fixed points if and only if it is a translation different from 1_X.

Proof. According to Proposition 37.1, every translation is a dilation. Thus each translation different from 1_X is a dilation with no fixed points.

Hence we assume that D is a dilation which has no fixed points. We choose a point $x \in X$ and put $D(x) = x'$; then $x \neq x'$ and we denote $x \vee x'$ by l. Next, we choose a point $y \notin l$ and put $D(y) = y'$; then $y \neq y'$ and we denote $y \vee y'$ by m. See the diagram in Exercise 45.15.

We first show that $l \parallel m$ and $l \cap m = \varnothing$. The four points x, x', y, y' lie in a plane (Exercise 45.14) and, therefore, if l and m are not parallel, they have precisely one point in common. This point would be a fixed point of D (Exercise 44.13), contrary to our assumption. This proves $l \parallel m$. Since $y \notin l$, $l \neq m$. Consequently, $l \cap m = \varnothing$.

Since $l \cap m = \varnothing$, $x \neq x'$ and $y \neq y'$, we conclude that x, x', y, y' are four distinct points. Put $s = x \vee y$ and $t = x' \vee y'$. Then $x' = D(x)$ and $y' = D(y)$ implies that $D(s) = t$ and hence $s \parallel t$.

We now put $A = \overrightarrow{x, x'}$ and conclude, from Exercise 45.15, that $T_A(y) = y'$. By definition of A, $T_A(x) = x'$. Consequently, the dilations D and T_A have the same effect on the two points x and y; we conclude that $D = T_A$. Finally, $A \neq \mathbf{0}$ and, therefore, $T_A \neq 1_X$. Done.

We have justified the three parts of the classification of dilations. As mentioned earlier, the following result is immediate.

Proposition 46.1.   A dilation is either a translation or a magnification.

Exercises

*16. In this problem, we develop the theory of traces of a dilation D.

> **Definition.** Let D be a dilation of n-dimensional affine space (X, V, k), $n \geq 2$. A line l is called a **trace** of D if $D(l) = l$. (Note: In general, l is not left pointwise fixed.)

> a. Prove that the intersection of two nonparallel traces of D is a fixed point of D.
> b. Show that $D = 1_X$ if and only if every line in X is a trace of D.
> c. Assume c is the fixed point of a dilation $D \neq 1_X$. Prove that each line through c is a trace of D. Show also that no other line could be a trace of D.
> d. If D is a translation T_A, $A \neq \mathbf{0}$, prove that l is a trace of D if and only if l has $U = \langle A \rangle$ as direction space.

17. Let k be the field $\mathbf{Z}_2$, that is, the integers modulo two. (More generally, $\mathbf{Z}_p$ denotes the integers modulo a prime integer p.) Prove that each dilation of (X, V, k) is a translation. Thus the group of the dilations coincides with the group of the translations. Conversely, if the group of the dilations coincides with the group of the translations, show that $k = \mathbf{Z}_2$.

12. THE RATIO OF A DILATION

It has been mentioned several times that the lengths of parallel line segments can be compared in affine geometry. We now say precisely what this means.

A **line segment** is simply a pair of distinct points $\{x, y\}$ of X; in particular, it is not an ordered pair of points, that is, $\{x, y\} = \{y, x\}$. In real affine space one views the line segment $\{x, y\}$ as the set of points consisting of x, y and all points between x and y. However, the concept of "between" makes sense only if the division ring k is ordered, as is the case for the field of real numbers (see Exercise 49.6). If k is not ordered, the line segment $\{x, y\}$ should be viewed as consisting of the two points x and y and nothing else.

The line segment $\{x, y\}$ lies on the unique line $x \vee y$. If the line segments $\{x, y\}$ and $\{p, q\}$ lie on parallel lines, they are called **parallel line segments**.

An **oriented line segment** is an *ordered* pair of distinct points (x, y) of X. Of course, the oriented line segments (x, y) and (y, x) are different, even though the nonoriented line segments $\{x, y\}$ and $\{y, x\}$ are the same. The oriented line segments (x, y) and (p, q) are called parallel if the corresponding line segments $\{x, y\}$ and $\{p, q\}$ are parallel.

Observe the notation: $\{x, y\}$ stands for the nonoriented line segment while (x, y) stands for the oriented line segment. Practically all our line segments will be oriented.

Consider the parallel line segments (x, y) and (p, q). If $l = x \vee y$ and $m = p \vee q$, then, by definition, $l \parallel m$. The common direction space of l and m is $\langle \overrightarrow{x, y} \rangle$ and, since the vector $\overrightarrow{p, q}$ belongs to this direction space, $\overrightarrow{p, q} = r(\overrightarrow{x, y})$ for some $r \in k$. In Figure 47.1, $r = 2$. The scalar r is the ratio of the lengths

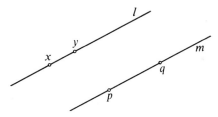

Figure 47.1

of the line segments (x, y) and (p, q); equivalently, r is the length of (p, q) when measured with (x, y) as unit length. Observe that, if k is ordered, r may very well be negative; the geometric reason for this is that we are considering here the lengths of *oriented* line segments (see Exercise 48.4). If k is not ordered, "negative" has no meaning. Also observe that, if l is not parallel to m, there is no way to compare the lengths of (x, y) and (p, q).

Instead of saying "the ratio of the lengths of the line segments," we shall usually say simply "the ratio of the line segments."

Definition 47.1. Let (x, y) and (p, q) be parallel, oriented line segments. The **ratio of** (p, q) **to** (x, y) is the scalar r with the property that $\overrightarrow{p, q} = r(\overrightarrow{x, y})$.

The scalar r of Definition 47.1 will be denoted by $(p, q)/(x, y)$.

Exercises

*1. Assume a coordinate system has been chosen for X. Let

$$x = (x_1, \ldots, x_n), \; y = (y_1, \ldots y_n), \; p = (p_1, \ldots p_n)$$

and $q = (q_1, \ldots q_n)$ be points of X where $x \neq y$ and $p \neq q$. Prove that the line segments (x, y) and (p, q) are parallel if and only if there is a scalar r such that $q_i - p_i = r(y_i - x_i)$ for $i = 1, \ldots, n$; also show that, in this case, $r = (p, q)/(x, y)$.

2. Let $k = \mathbf{R}$ and let (x, y), (p, q) be parallel line segments. Prove that, if $r = (p, q)/(x, y)$, then

$$|r| = \frac{\text{the ordinary Euclidean length of } (p, q)}{\text{the ordinary Euclidean length of } (x, y)}.$$

(Hint: Use the previous exercise and the usual distance formula for Cartesian coordinate systems.)

*3. Let L be the set of line segments (x, y) of X. Consider the binary relation $\sim$ for L defined by $(x, y) \sim (p, q)$ if the line segments (x, y) and (p, q) are parallel.

   a. Prove that $\sim$ is an equivalence relation for L.

   b. Prove that there is a one-to-one correspondence between the set of equivalence classes of the resulting partitioning of L and the set of one-dimensional subspaces of V.

*4. Assume the division ring k is ordered. (Whenever this assumption is made, readers who are not familiar with ordered division rings or ordered fields should use the field of real numbers or the field of rational numbers for k. A reference for ordered division rings is E. Artin, *Geometric Algebra*, pp. 40–47, [2].) Let (x, y) and (p, q) be parallel line segments of X and put $r = (p, q)/(x, y)$. If $r > 0$, (x, y) and (p, q) are said to have the *same orientation*; if $r < 0$, they have *opposite orientation*. Assume (x, y) and (p, q) have the same orientation.

   a. Prove that (x, y) and (q, p) have opposite orientation.

   b. Say whether the line segments (y, x) and (p, q) have the same or opposite orientation. Do the same for (y, x) and (q, p).

5. Assume that k is ordered. Again, let L be the set of line segments (x, y) of X. Consider the binary relation $\approx$ for L defined by $(x, y) \approx (p, q)$ if the line segments (x, y) and (p, q) are parallel and have the same orientation.

   a. Prove that $\approx$ is an equivalence relation.

   b. Prove that there is a one-to-one correspondence between the set of equivalence classes of the resulting partitioning of L and the set of "rays" of V with vertex $\mathbf{0}$ where rays are defined as follows. If $A \neq \mathbf{0}$, the set $\{tA \mid t \in k, t > 0\}$ is the ray determined by A.

c. Show that each equivalence class of the relation $\sim$ defined in Exercise 48.3 is the union of two equivalence classes of the relation $\approx$.

*6. Assume k is ordered. Let x, y be distinct points of X and put $l = x \vee y$. If $z \in l$, then $\overrightarrow{x, z} = r(\overrightarrow{x, y})$. The point z **is said to lie between** x **and** y if $0 \leq r \leq 1$ (Figure 49.1).

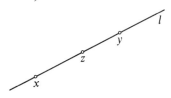

Figure 49.1

a. Prove that the points x and y are both between x and y.

b. If z lies between x and y, prove that z also lies between y and x.

c. If z lies between x and y and $z \neq x$ and $z \neq y$, prove that the line segments (x, z) and (y, z) have opposite orientation while (x, z) and (x, y) have the same orientation.

d. If $k = \mathbf{R}$, prove that the above concept of betweenness agrees with the usual concept of "a point lies between two points." (Hint: Use Exercise 48.1 along with the usual concept of "a point lies between two points"; namely, z lies between x and y if the distance from x to z plus the distance from z to y equals the distance from x to y.)

We now return to the study of dilations. It is immediate that a dilation D sends a line segment (x, y) onto the parallel line segment $(D(x), D(y))$. In fact, dilations may be defined as one-to-one mappings of X onto itself which send every line segment onto a parallel segment. Consequently, if $x \neq y$, the ratio

$$r = \frac{((D(x), D(y))}{(x, y)}$$

is well defined. We shall show that this ratio depends only on the dilation D and not on the points x and y. The term dilation comes from the fact that D "dilates" the affine space X in the sense that

1. D sends every line segment onto a parallel line segment.
2. If $x \neq y$, the ratio $(D(x), D(y))/(x, y)$ does not depend on x and y.

It is most interesting that Statement 1 implies Statement 2, as we now show.

Proposition 50.1.   Let D be a dilation. Then there is a unique $r \in k^*$ such that, for all $x, y \in X$, $\overrightarrow{D(x),\, D(y)} = r(\overrightarrow{x,\, y})$. If D is a translation, $r = 1$ and, if D is a magnification, r is the magnification ratio.

Proof. The proof is based on Identities 6, 7, and 10 on pages 7 and 8. They are
(6) $\overrightarrow{x,\, y} + \overrightarrow{y,\, z} = \overrightarrow{x,\, z}$, (7) $\overrightarrow{Bx,\, By} = \overrightarrow{x,\, y}$, and (10) $\overrightarrow{Ax,\, By} = -A + \overrightarrow{x,\, y} + B$.
Since a dilation is either a translation or a magnification, we have only two cases to consider.

   Case 1. D is a translation determined by the vector A. Let $x, y \in X$. Then $\overrightarrow{D(x),\, D(y)} = \overrightarrow{Ax,\, Ay} = \overrightarrow{x,\, y}$ by Identity 7.

   Case 2. D is the magnification $M(c, r)$. Then

$$\overrightarrow{D(x),\, D(y)} = \overrightarrow{(r(\overrightarrow{c,\, x}))c,\, (r(\overrightarrow{c,\, y}))c} = -r(\overrightarrow{c,\, x}) + c,\, c + r(\overrightarrow{c,\, y})$$

by Identity 10. Now $\overrightarrow{c,\, c} = \mathbf{0}$ and $-r(\overrightarrow{c,\, x}) = r(\overrightarrow{x,\, c})$, so

$$\overrightarrow{D(x),\, D(y)} = r(\overrightarrow{x,\, c}) + r(\overrightarrow{c,\, y}) = r(\overrightarrow{x,\, c} + \overrightarrow{c,\, y})$$

which, by Identity 6, is $r(\overrightarrow{x,\, y})$. Done.

   It is clear that different dilations can have the same ratio. (We often say ratio instead of dilation ratio.) For instance, all translations have ratio 1. However, the following is true.

Proposition 50.2.   A dilation is completely determined by its dilation ratio and the image of one point.

Proof. Let D be a dilation with ratio r. Assume that $x \in X$ and that the image $D(x)$ is known. If $y \neq x$ belongs to X, we must show that $D(y)$ is known. We

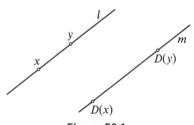

Figure 50.1

put $l = x \vee y$ and let m denote the line through $D(x)$ which is parallel to l. Since D is a dilation, $D(y) \in m$. Both l and m have the direction space $\langle \overrightarrow{x, y} \rangle$ and hence $D(y) = s(\overrightarrow{x, y})D(x)$ for some $s \in k$. But this is equivalent to $\overrightarrow{D(x), D(y)} = s(\overrightarrow{x, y})$; consequently, by definition of dilation ratio, $s = r$. We conclude that $D(y) = r(\overrightarrow{x, y})D(x)$ and hence $D(y)$ is known. Done.

Exercises

*7. Assume that D is a dilation with ratio r and that $\mu: k \to k$ is the inner automorphism of k defined by $\mu(a) = rar^{-1}$ for $a \in k$. Let x, y, p, q be points of X such that the line segments (x, y) and (p, q) are parallel. Prove that
   a. $(D(p), D(q))/(D(x), D(y)) = \mu((p, q)/(x, y))$.
   b. $(D(p), D(q))/(D(x), D(y)) = (p, q)/(x, y)$ if and only if r lies in the center of the division ring k. (The **center of** k is the subfield $\{s \,|\, s \in k, as = sa \text{ for all } a \in k\}$.) In this case, we say that the **dilation D preserves the ratio of parallel line segments**.
8. If k is a field, prove that every dilation preserves the ratio of parallel line segments.
*9. Assume k is ordered. Prove that dilations preserve betweenness. Precisely, if the point z is between the points x and y, prove that $D(z)$ is between the points $D(x)$ and $D(y)$. (Hint: Use the previous exercise. Either assume k has a field or show that $0 \le a \le 1$ implies $0 \le rar^{-1} \le 1$ for all $r, a \in k, r \ne 0$.)
10. Prove that a dilation of X may be defined as a one-to-one mapping of X onto X which maps every line segment onto a parallel line segment. For this exercise, it does not matter whether line segment means oriented or nonoriented line segment. Warning: The important part of the exercise is showing that every line is mapped onto a line.

Corresponding to an affine space, we have the group of the dilations and the group of the translations. Moreover, to each dilation we can associate an element of the multiplicative group k^*, namely, the dilation ratio of the given dilation. We are now ready to see how these groups are interrelated.

Proposition 51.1. The mapping $f: Di \to k^*$ which associates with each dilation its dilation ratio is a group homomorphism from Di onto k^*.

Proof. First we show that f is a group homomorphism. Let $D_1, D_2 \in Di$ have dilation ratios r_1 and r_2, respectively. If $x, y \in X$, then $\overrightarrow{D_2(x), D_2(y)} = r_2(\overrightarrow{x, y})$ since r_2 is the dilation ratio of D_2. Also

$$\overrightarrow{D_1(D_2(x)), D_1(D_2(y))} = r_1(\overrightarrow{D_2(x), D_2(y)});$$

hence $\overrightarrow{D_1 D_2(x), D_1 D_2(y)} = r_1 r_2(\overrightarrow{x, y})$. Thus $r_1 r_2$ is the dilation ratio of $D_1 D_2$; that is, f is a homomorphism. For any $r \in k^*$, the magnification $M(c, r)$ is a dilation with dilation ratio r (Proposition 50.2). So f is onto. Done.

A homomorphism from a group *onto* a group is often called an **epimorphism**.

Proposition 52.1. The group Tr of translations is the kernel of the epimorphism $f : Di \to k^*$. Consequently, Tr is a normal subgroup of Di and $Di/Tr \simeq k^*$.

Proof. In Proposition 50.1, we proved that the dilation ratio of a translation is 1; hence Tr is contained in the kernel of f. Conversely, if D is a dilation which is not a translation, then $D = M(c, r)$ for some $c \in X$ and $r \neq 1$. Therefore, the dilation ratio of D is $r \neq 1$ and D does not belong to the kernel of f. The isomorphism $Di/Tr \simeq k^*$ now follows from general principles and group theory. Done.

We already know that Tr is a commutative group. Now we can say more; namely, Tr is a normal, commutative subgroup of Di with k^* as quotient group.

A Side Remark about Frobenius Groups and Wedderburn's Theorem. The sequence

$$1 \to Tr \overset{i}{\to} Di \overset{f}{\to} k^* \to 1$$

of groups and group homomorphisms is very interesting. Here i denotes the inclusion function. It is an "exact sequence" because

1. i is one-to-one.
2. The image of i is the kernel of f.
3. f is an epimorphism.

It is not the exactness which makes this sequence so interesting, but the fact that Di is a "Frobenius group." This means that Di is a group of permutations of X which has the following two properties.

1. If x and y are points of X, there is a $D \in Di$ with the property that $D(x) = y$. (*Proof.* The translation $T_{\overrightarrow{x,y}}$ sends x to y.)
2. If an element $D \in Di$ leaves two points of X fixed, $D = 1_X$.

Furthermore, Tr is the so-called "Frobenius kernel" of Di; that is, Tr consists of the identity permutation 1_X along with precisely those elements of the permutation group Di which have no fixed points. The quotient of a Frobenius group by its Frobenius kernel is called the "Frobenius complement"; in our case, the Frobenius complement is k^* ($k^* \simeq Di/Tr$). The interesting thing to observe is that the *multiplicative group k^* of a division ring can always be regarded as a Frobenius complement.* This observation about division rings comes from geometry. This is the kind of application of geometry to algebra Artin had in mind when he coined the term geometric algebra.

Frobenius studied finite Frobenius groups (the term is not his, of course) in great detail. (See, for instance, W. Burnside, *Theory of Groups of Finite Order*, Chapter X, [7].) Let us assume that our division ring k is finite; then X, Di, and Tr are finite (see Exercise 53.11). Wedderburn's theorem asserts that k is then commutative and hence that k^* is a cyclic group. It is very interesting that a simple proof of Wedderburn's theorem can be given, based on the observation that k^* is a Frobenius complement and the use of some light theory of Frobenius groups. (See S. Ebey and K. Sitaram, "Frobenius Groups and Wedderburn's Theorem," *American Mathematical Monthly*, Vol. 76, No. 5, May 1969, pp. 526–528, [11].)

Exercise

*11. Let k be finite. Then k has p^m elements for some prime integer p and some integer $m \geq 1$. Prove that
   a. The set X, the vector space V and the group Tr all have p^{mn} elements ($n = \dim V$).
   b. The group Di has $p^m(p^n - 1)$ elements.

This ends our remarks about Frobenius groups. We again assume the division ring k is arbitrary, that is, not necessarily finite, and continue by proving some propositions which give further insight into the group Di of the dilations.

Proposition 53.1.   Let D be a dilation with ratio r and let $c \in X$. Then $D = TM(c, r)$ for a properly chosen translation T.

Proof. Put $A = \overrightarrow{c, D(c)}$ and $D' = T_A M(c, r)$. Then D' is a dilation with ratio r and $D'(c) = T_A(c) = D(c)$. Hence D and D' have the same ratio and act the same on c. We conclude, from Proposition 50.2, that $D = D'$. Done.

Exercises

12. Let D be a dilation and $c \in X$. Suppose that $D = TM(c, r)$ for some translation T and $r \in k^*$. Prove that T and r are uniquely determined by D and c.
13. Assume D is a dilation with ratio r and let $c \in X$. Put $D = T_A M(c, r)$ for a properly chosen vector A. One can also express D as a product $M(c, r)T_B$ for a properly chosen B. (Why?) Prove that $B = r^{-1}A$. (Hint: Use Exercise 39.7.)

By combining Proposition 53.1 with the above two exercises, we see that

$$D = T_A M(c, r) = M(c, r)T_B$$

where

$$A = \overrightarrow{c, D(c)} \quad \text{and} \quad B = r^{-1}(\overrightarrow{c, D(c)}).$$

Consequently, if k contains at least three elements, the group Di is not commutative. In Exercise 46.17, it was shown that $Tr = Di$ if and only if $k = \mathbf{Z}_2$. We now conclude that: *The group Di of dilations of X is commutative if and only if $k = \mathbf{Z}_2$.*

Every translation T comes from a vector $A \in V$. Moreover, $T = 1_X$ if and only if $A = \mathbf{0}$. If $A \neq \mathbf{0}$, the one-dimensional subspace $\langle A \rangle$ of V spanned by A is called the **direction of** T. We do not speak of the direction of the identity translation.

We have already seen that Tr is a normal subgroup of Di. Hence, if $T \in Tr$ and $D \in Di$, DTD^{-1} is again a translation.

Proposition 54.1.   Let T be a translation different from 1_X. If D is a dilation, the translations T and DTD^{-1} have the same direction.

Proof. If D is a translation, $DTD^{-1} = T$ since the group Tr is commutative. In this case, the proposition is trivially true. Now assume D is not a translation; then D is a magnification $M(c, r)$ where $c \in X$ and $r \in k^*$. We put $T = T_A$ and conclude that

$$DTD^{-1} = M(c, r)T_A M(c, r^{-1}) = T_{rA} M(c, r)M(c, r^{-1})$$

by Exercise 39.7. Since $M(c, r)M(c, r^{-1}) = 1_X$, $DTD^{-1} = T_{rA}$. Hence the direction of DTD^{-1} is $\langle rA \rangle = \langle A \rangle$. Done.

Exercises

14. Let $c \in X$, $r \in k^*$ and $A \in V$. Prove that

$$(M(c, r)T_A)^{-1} = M(c, r^{-1})T_{-rA}.$$

(Hint: Since Di is a group, $(D_1 D_2)^{-1} = D_2^{-1}D_1^{-1}$ for any two dilations D_1 and D_2.)

15. Let $c \in X$, $r \in k^*$ and $A \in V$.
   a. Prove that $T_A M(c, r)$ is a translation if and only if $r = 1$, in which case, $T_A M(c, r) = T_A$.
   b. Prove the same thing for $M(c, r)T_A$.

16. Let $c \in X$, $r \in k^*$, $r \neq 1$ and $A \in V$.
   a. Prove that $T_A M(c, r)$ is a magnification and find the center of magnification. (Hint: Show that $M(c, r)T_A$ has a fixed point.)
   b. Prove that $M(c, r)T_A$ is a magnification and find the center of magnification.

13. DILATIONS IN TERMS OF COORDINATES

Let D be a dilation and let $x_0, A_1, \ldots, A_n$ be a coordinate system of the n-dimensional affine space X. We ask how D can be described in this coordinate system. For this, we put $D = T_A M(x_0, r)$ where r is the ratio of D and $A = (a_1, \ldots, a_n)$ is the appropriate vector of V. The answer is now given.

Proposition 55.1. If $x = (x_1, \ldots, x_n)$ is a point of X, the coordinates $(y_1, \ldots, y_n)$ of the point $y = D(x)$ are

$$y_1 = rx_1 + a_1$$
$$\vdots$$
$$y_n = rx_n + a_n.$$

Proof. Now $y = (T_A M(x_0, r))(x) = T_A(M(x_0, r)(x))$. By definition, $M(x_0, r)(x) = (r\overrightarrow{x_0, x})c$. According to the definition of coordinates, the coordinates of $\overrightarrow{x_0, x}$ are $(x_1, \ldots, x_n)$. Hence the coordinates of $r\overrightarrow{x_0, x}$ are $(rx_1, \ldots, rx_n)$. Again, by definition, the coordinates of $(r\overrightarrow{x_0, x})x_0$ are also $(rx_1, \ldots, rx_n)$. Finally, the coordinates of the point $T_A(M(x_0, r)(x))$ are $(rx_1 + a_1, \ldots, rx_n + a_n)$. Done.

We see that the equations $y_i = rx_i + a_i$ for the dilation D express the new coordinates y_i as linear polynomials $rx_i + a_i$ in terms of the old coordinates $x_1, \ldots, x_n$ where the nonzero coefficient r occurs on the left of the variables $x_1, \ldots, x_n$. However, these are "left" linear polynomials of a very special type because the most general "left" linear polynomial in the variables $x_1, \ldots, x_n$ has the form $d_1 x_1 + \cdots + d_n x_n + b$ where $d_1, \ldots, d_n, b \in k$.

Exercises

1. Let $x_0, A_1, \ldots, A_n$ be a coordinate system for X. According to Proposition 55.1, every dilation is given by a set of equations $y_i = rx_i + a_i$, $i = 1, \ldots, n$ where $r, a_1, \ldots, a_n \in k$ and $r \neq 0$. Prove the converse; that is, consider a system of n equations $y_i = rx_i + a_i$ where $r \neq 0$ and prove that they are the equations of a dilation D. Furthermore,
   a. Prove that r is the dilation ratio of D.
   b. If we put $D = T_A M(x_0, r)$, prove that A has coordinates $(a_1, \ldots, a_n)$.
   c. Prove directly from the equations that D is a translation if and only if $r = 1$. (Hint: Show that $r = 1$ if and only if D has no fixed points.)
   d. If $r \neq 1$, prove that D is the magnification $M(c, r)$ where the point $c = (c_1, \ldots, c_n)$ has the coordinates $c_i = (1 - r)^{-1} a_i$ for $i = 1, \ldots, n$. (Hint: c is the unique point left fixed by D.)
   e. Prove directly from the equations that the point $D(x_0)$ has coordinates $(a_1, \ldots, a_n)$.
*2. Let $A, B \in V$, $c \in X$, and $r, s \in k^*$. Prove that

$$T_A M(c, r) T_B M(c, s) = T_{A + rB} M(c, rs).$$

(Hint: Do not use coordinates; instead, apply Exercise 39.7.)

3. Let $x_0, A_1, \ldots, A_n$ be a coordinate system for X. Consider the dilation $TM(c, r)$ where the vector A has coordinates $(a_1, \ldots, a_n)$ and the point c has coordinates $(c_1, \ldots, c_n)$. Prove that the equations of D are

$$y_1 = rx_1 + (1 - r)c_1 + a_1$$
$$\vdots$$
$$y_n = rx_n + (1 - r)c_n + a_n.$$

This means that, if $x = (x_1, \ldots, x_n)$ is a point of X, the coordinates $(y_1, \ldots, y_n)$ of the point $y = D(x)$ are given by the above system of equations.

4. Assume that a coordinate system x_0, A_1, A_2, A_3, A_4 has been chosen for the four-space X. Let $x = (1, 0, 1, 1)$, $y = (2, 1, 0, 0)$, $a = (3, 2, -1, -1)$, $p = (1, -1, 0, 1)$, $q = (8, -3, 0, 2)$, $u = (-2, -3, 1, 1)$ and $w = (-1, -9, 3, 2)$.

   a. Prove that x, y, and z are collinear.

   b. Find the equations of the dilation D_1 which leaves the point x fixed and moves the point y to the point z.

   c. Prove that the line $l = p \vee u$ is parallel to the line $m = q \vee w$.

   d. Find the equations of the dilation D_2 with the property that $D_2(p) = q$ and $D_2(u) = w$.

   e. Find the equations of the dilation $D_1 D_2$. (Hint: From the equations for D_1 and D_2, one can write $D_1 = T_A M(x_0, r)$ and $D_2 = T_B M(x_0, r)$. Now apply Exercise 56.2).

5. Let $M(c, r)$ and $M(d, s)$ be magnifications of X.

   a. Prove that $M(c, r)M(d, s)$ is a translation if and only if $rs = 1$.

   b. Prove that $M(c, r)M(d, s) = 1_X$ if and only if either $c = d$ and $rs = 1$ or $c \neq d$ and $r = s = 1$.

6. Assume a coordinate system has been chosen for X. Let $c = (c_1, \ldots, c_n)$, $d = (d_1, \ldots, d_n)$ be points of X and $r, s \in k^*$.

   a. By the previous exercise, we know that $M(c, r)M(d, r^{-1}) = T_A$ for some vector A. Express the coordinates $(a_1, \ldots, a_n)$ of the vector A in terms of the scalars c_i, d_i, and r.

   b. Assume that $rs \neq 1$ and hence $M(c, r)M(d, s)$ is a magnification $M(e, t)$. Express the scalar t and the coordinates $(e_1, \ldots, e_n)$ of e in terms of the scalars c_i, d_i, r, and s.

7. Assume that a coordinate system has been chosen for X. Let $c = (c_1, \ldots, c_n) \in X$, $r \in k^*$, $r \neq 1$ and let $(a_1, \ldots, a_n)$ be the coordinates of the vector $A \in V$. By Exercise 55.15, $T_A M(c, r)$ is not a translation and hence must be a magnification $M(d, s)$. Express the scalar s and the coordinates $(d_1, \ldots, d_n)$ of d in terms of the scalars a_i, c_i, and r.

*8. Let x and y be points of X and let $r \in k^*$, $r \neq 1$. Prove that there is a unique point $c \in X$ such that $M(c, r)(x) = y$.

9. Assume that a coordinate system for X has been chosen. Let $x = (x_1, \ldots, x_n)$, $y = (y_1, \ldots, y_n)$ be points of X and $r \in k^*$, $r \neq 1$. By the previous exercise, there is a unique point $c = (c_1, \ldots, c_n)$ in X such that $M(c, r)(x) = y$. Express the coordinates $(c_1, \ldots, c_n)$ of c in terms of the scalars x_i, y_i, and r.

10. Let $x \in X$ and $r \in k^*$, $r \neq 1$. For each $y \in X$, there is a unique point

$c \in X$ such that $M(c, r)(x) = y$ (Exercise 8 above). Prove that the mapping $M: X \to X$ defined by $M(y) = c$ is the magnification $M(x, (1 - r)^{-1})$.

We assumed that $n \geq 2$ when we defined dilations (Definition 37.1). If we had used the same definition for $n = 1$, every permutation of X would have been a dilation since X contains no lines other than itself. This we do not want. Our definitions of translation and magnification are valid for all $n \geq 1$; as in the case for higher dimensions, we also want only translations and magnifications to be dilations when $n = 1$.

Definition 58.1. Let $n \geq 1$. A one-to-one mapping D of X onto X is called a **dilation** if there is an $r \in k^*$ such that, for every ordered pair of points (x, y) of X,

$$\overrightarrow{D(x), D(y)} = r \overrightarrow{(x, y)}.$$

The scalar r is called the **dilation ratio** of D.

When $n = 2$, the above definition of dilation is equivalent to Definition 37.1 of dilation. If $n \geq 1$, both translations and magnifications are dilations according to Definition 58.1. In the following exercise, we show that a dilation of a line is either a translation or a magnification.

Exercises

11. Let $n = 1$ and let D be a dilation with ratio r. If $r = 1$, prove that D is a translation; if $r \neq 1$, prove that D is a magnification.
*12. The concept of homothetic figures will be introduced in this exercise. A configuration or a figure in X is simply a subset of X. In geometry, one is interested in special kinds of figures such as lines, planes, triangles, quadrilaterals, etc. Two figures in X are called **homothetic** if there is a dilation which maps one figure onto the other. In the diagram below, the triangles xyz and $x'y'z'$ are homothetic because they are related by a magnification with center p (Figure 58.1).

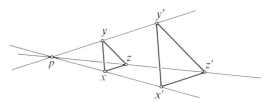

Figure 58.1

Intuitively, two triangles are homothetic if they are similar (in the high school sense) and, furthermore, are "similarly placed."

a. Let $n = 2$ and assume a coordinate system has been chosen for X. Let $x = (1, 2)$, $y = (3, -1)$, $z = (4, 1)$, $a = (3, 7)$, $b = (9, 5)$, and $c = (7, 1)$. Show that the triangles xyz and abc are homothetic by finding a dilation which maps one triangle onto the other. What is the ratio of the dilation? If it is a magnification, what are the coordinates of its center? Draw a picture analogous to the one above, assuming $k = \mathbf{R}$ and that the coordinate system is Cartesian.

b. Assume a coordinate system has been chosen for the three-space X and let $x = (1, 1, 2)$, $y = (-1, 0, -1)$, $z = (3, 4, -2)$, $w = (2, 1, 0)$, $a = (3, 6, 2)$, $b = (-1, 4, -4)$, $c = (7, 12, -6)$, and $d = (5, 6, -2)$ be points in X. Show that the "quadrilaterals" $xyzw$ and $abcd$ are homothetic.

14. THE TANGENT SPACE $X(c)$

It cannot be stressed enough that the affine space X is not a vector space. Its points cannot be added and there is no way to multiply points by scalars. No point in X is preferred; they all play the same role. In particular, there is no point in X which makes a better origin for a vector space than any other point.

The situation changes radically if we choose a point c in X and keep it fixed. It is now possible to make X into a left vector space over k by using the one-to-one mapping f from X onto V defined by $f(x) = \overrightarrow{c, x}$ for each $x \in X$. (See the proof of Proposition 17.1 when $S = S(c, V)$.) All we do is carry the vector space structure of V over to X by means of the mapping f. We observe that the inverse mapping $g: V \to X$ of f is defined by $g(A) = Ac$. Consequently, the sum of points of X and the scalar multiple of a point by a scalar are defined as below. Observe that *all definitions* are relative to the point c.

Definition 59.1.   Addition of Points of X. If $x, y \in X$,

$$x + y = g(f(x) + f(y)) = (\overrightarrow{c, x} + \overrightarrow{c, y})c.$$

Scalar Multiplication. If $x \in X$ and $r \in k$,

$$rx = g(rf(x)) = (r(\overrightarrow{c, x}))c.$$

The resulting left vector space over k is denoted by $X(c)$. Observe that the points of X are the vectors of $X(c)$. By drawing a few pictures, the above definitions will seem very natural. Drawing pictures will be helpful in solving the following exercises, but, as always, the actual proof must not depend on any picture.

Exercises

In the following exercises, it is assumed that a point $c \in X$ has been chosen. Addition of points of X and multiplication of points by scalars are carried out in the vector space $X(c)$.

1. If $x, y \in X$, prove that $x + y = (c, \overrightarrow{x})y = (c, \overrightarrow{y})x$.
2. If $x \in X$ and $r \in k^*$, prove that $M(c, r)(x) = rx$ where $M(c, r)$ is the magnification with center c and ratio r. Also prove that $0x = c$ for all $x \in X$.
3. Prove that c is the origin of the vector space $X(c)$, that is, $c + x = x + c = x$ for all $x \in X$. Also prove that $rc = c$ for all $r \in k$.
4. If $x \in X$, prove that $-x = (x, \overrightarrow{c})c$.
5. Prove that $f: X(c) \to V$ and $g: V \to X(c)$ are linear isomorphisms which are inverses of one another.

We see from Exercise 60.5 that the vector spaces $X(c)$ and V are isomorphic, equivalently, that dim $X(c) = n$. It follows that, if b is another point of X, the vector spaces $X(c)$ and $X(b)$ are isomorphic. In spite of this fact and the fact that the vector spaces $X(c)$ and $X(b)$ consist of the same vectors, namely, the points of X, these vector spaces should not be identified if $b \neq c$. The reason is that *the addition of points of X and the multiplication of points by scalars is totally different in $X(c)$ and in $X(b)$.* For instance, c is the origin of $X(c)$ while b is the origin of $X(b)$. In the accompanying pictures, we assume that X is the real affine plane. The purpose of these pictures is to bring out the difference in addition and scalar multiplication in the vector spaces.

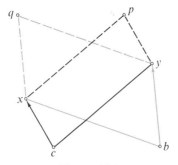

Figure 60.1

In Figure 60.1, $p = x + y$ in $X(c)$ and $q = x + y$ in $X(b)$.

Figure 61.1

In Figure 61.1, the point p is $2x$ in $X(c)$ while $q = 2x$ in $X(b)$.

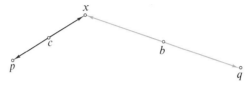

Figure 61.2

Figure 61.2 shows that $p = -x$ in $X(c)$ and $q = -x$ in $X(b)$.

Even if $k \neq \mathbf{R}$ or $n \neq 2$, the sum $x + y = p$ of the points x and y in $X(c)$ should be visualized as the fourth point of the parallelogram $cxpy$. Since a point $x \in X$ is also a vector of $X(c)$, an arrow starting at c and ending at the point x should be drawn whenever this is helpful. The terms "point of X," "vector of $X(c)$" and "point of $X(c)$" are synonymous. Similarly, the scalar multiple rx should always be visualized as the result of stretching or contracting the arrow from c to x by the "amount" r in the proper direction. We should accept the help such pictures can give us in spite of the dangers lurking in them.

The natural name for the vector space $X(c)$ is the **tangent space of X at c**. The reason is that, if $k = \mathbf{R}$, $X(c)$ is precisely the tangent space of X at c in the sense of differential geometry.

Exercises

In each of the following exercises, a point $c \in X$ has been chosen; addition of points of X and multiplication of points by scalars are carried out in the vector space $X(c)$.

*6. If $x, y \in X$, prove that $(\overrightarrow{x, y})c = y - x$. Draw an appropriate diagram which illustrates this rule.

7. If x, y, z are points of X, prove that $\overrightarrow{x + z, y + z,} = \overrightarrow{x, y}$. Draw a picture which illustrates this rule.

8. If $A \in V$ and $r \in k$, prove that $(rA)c = r(Ac)$. Draw a picture which illustrates this rule. Show by a picture that, in general, $(rA)x \neq r(Ax)$ if $x \neq c$.

9. If $x \in X$ and $r \in k$, prove that $\overrightarrow{c, rx} = r(\overrightarrow{c, x})$. Draw a picture which illustrates this rule. Also show by a picture that, in general, $\overrightarrow{y, rx} \neq r(\overrightarrow{y, x})$ if the point y is different from c.

10. If $x, y \in X$, prove that $(r(\overrightarrow{x, y}))c = ry - rx$. Draw a picture which illustrates this rule.

11. If $x, y \in X$ and $r \in k$, prove that $\overrightarrow{rx, ry} = r(\overrightarrow{x, y})$. Draw a picture which illustrates this rule.

12. Let $x, y, p, q \in X$ and $r \in k$.
    a. Prove that $\overrightarrow{p, q} = r(\overrightarrow{x, y})$ if and only if $q - p = r(y - x)$.
    b. If $x \neq y$ and $p \neq q$, show that $q - p = r(y - x)$ is equivalent to saying that the line segments (p, q) and (x, y) are parallel and that r is the ratio $(p, q)/(x, y)$.

13. Let D be a dilation with ratio r. Prove that r is the scalar with the property that

$$D(y) - D(x) = r(y - x)$$

for all points $x, y \in X$. This holds for all $n \geq 1$.

14. Let $p \in X$ and put $A = \overrightarrow{c, p}$. Prove that $T_A(x) = x + p$ for all $x \in X$. In other words, the mapping $x \to x + p$ is the translation T_A.

15. Let $A \in V$ and $r \in k^*$ and put $p = Ac$. Prove that

$$(T_A M(c, r))(x) = rx + p$$

for all $x \in X$.

16. Let D be a dilation with ratio r and put $D(c) = p$. Prove that $D(x) = rx + p$ for all $x \in X$.

17. Assume a coordinate system has been chosen for the three-space X with c as origin. Let $u = (1, 1, 2)$, $w = (3, 1, 10)$, $y = (0, 3, 0)$, and $z = (1, 5, 6)$ be points of X and assume that D is a dilation such that $D(u) = w$ and $D(y) = z$. (Then D is completely determined since a dilation is determined by the image of two points.) Find $r \in k^*$ and the coordinates (p_1, p_2, p_3) of the point p such that $D(x) = rx + p$ for all $x \in X$.

18. Let u, w be points and D_1, D_2 dilations of X where $D_1(x) = 3x + u$ and $D_2(x) = 2x + w$ for all $x \in X$.
    a. Prove that D_1 and D_2 are magnifications by finding scalars r_1, r_2 and points c_1, c_2 such that $D_1 = M(c_1, r_1)$ and $D_2 = M(c_2, r_2)$. Express the center c_1 in terms of u and the center c_2 in terms of w.

  b. Prove that $D_1 D_2$ is a magnification $M(b, r)$. Find the scalar r and
     the point p (in terms of u and w) such that $D_1 D_2(x) = rx + p$ for
     all $x \in X$. Also express the center b in terms of u and w.
  c. Do Part b for $D_2 D_1$. Do the dilations D_1 and D_2 commute?
19. Let D be the dilation $D(x) = rx + p$.
  a. Prove that D is a translation if and only if $r = 1$, in which case,
     $D = T_A$ where $A = \overrightarrow{c, p}$.
  b. If $r \neq 1$, prove that D is the magnification $M(b, r)$ where $b =$
     $(1 - r)^{-1}p$.
20. Consider the dilation D defined by $D(x) = rx + p$.
  a. Prove that D is never a linear transformation of the tangent space
     $X(c)$ into itself if $p \neq c$.
  b. If $p = c$, show that D is the magnification $M(c, r)$. Furthermore,
     prove that $M(c, r)$ is a linear transformation of $X(c)$ into itself if
     and only if r lies in the center of the division ring k. (We recall that
     the center of k is the subfield $\{s \mid s \in k,\ st = ts \text{ for all } t \in k\}$.)
     Hence if k is a field, the magnification $M(c, r)$ is always a linear
     automorphism of $X(c)$.

Let us return to a further investigation of the isomorphic vector spaces
$X(c)$ and $X(b)$. They are isomorphic because both of these vector spaces are
isomorphic with V under isomorphisms $f: X(c) \to V$ and $g: X(b) \to V$. Hence
the natural isomorphism from $X(c)$ onto $X(b)$ is $g^{-1}f$; we ask for the action of
$g^{-1}f$.

Proposition 63.1.   The isomorphism $g^{-1}f: X(c) \to X(b)$ is the translation
T_A where $A = \overrightarrow{c, b}$. It follows that $\overrightarrow{c, x} = \overrightarrow{b, T_A(x)}$ for all $x \in X$.

Proof. For each $x \in X$,

$$g^{-1}f(x) = g^{-1}(\overrightarrow{c, x}) = (\overrightarrow{c, x})b.$$

This already shows that $\overrightarrow{c, x} = \overrightarrow{b, g^{-1}f(x)}$. By Exercise 8.11, $(\overrightarrow{c, x})b = (\overrightarrow{c, b})x$
and hence $g^{-1}f(x) = T_A(x)$. Done.

We conclude from the above proposition that the line segments (c, x) and
$(b, T_A(x))$ are parallel and that their ratio $(b, T_A(x))/(c, x)$ is 1. Thus we picture
T_A as in Figure 64.1.

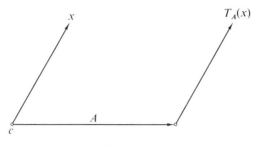

Figure 64.1

We obtain more completely the outlook of differential geometry on the affine space (X, V, k). There is a tangent space $X(c)$ at each point $c \in X$. If c and b are points of X, there is a linear isomorphism $T_A: X(c) \to X(b)$ where $A = \overrightarrow{c, b}$. In case $k = \mathbf{R}$, T_A is the "parallel displacement" of differential geometry.

Exercise

21. Let A be a vector of V and c, b points of X. Prove that the translation $T_A: X(c) \to X(b)$ is a linear transformation if and only if $A = \overrightarrow{c, b}$.

A Side Remark on High School Teaching. We feel that the best way to introduce linear algebra in the eleventh or twelfth grade in high school is by choosing a point c in the real affine plane X and identifying each point x with the arrow or vector with c as initial point and x as endpoint. Vector addition can then be defined as "completing the parallelogram" and scalar multiplication as "stretching or contracting the vector by the right amount in the right direction." The usual axioms of a two-dimensional vector space over the real numbers then appear as light theorems of elementary high school geometry. This procedure is sometimes carried out in the framework of analytic geometry, the origin of a coordinate system being used as the point c. *This is not advisable* since it misleads the student into thinking that a coordinate system is necessary to make a vector space out of the plane. Nothing, of course, could be further from the truth; all that is required to make the plane into a vector space is to fix one point c. In the resulting vector space $X(c)$, one can choose a coordinate system if one wishes, but it is very important to make certain that the concept of the tangent space $X(c)$ is not dependent upon a coordinate system.

It should, furthermore, be impressed upon the high school student that different points c and b in the plane give rise to different vector spaces which are interrelated by the translation T_A where $A = \overrightarrow{c, b}$. This can be presented in a very elementary way. In short, it is our opinion that the most efficient way to introduce linear algebra on the high school level comes from taking the attitude of differential geometry toward the plane.

15. AFFINE AND SEMIAFFINE TRANSFORMATIONS

Affine geometry is the geometry of the triple (X, V, k) and may well be called the geometry of parallelism. In affine space, we know what is meant by parallel affine spaces and by the ratio of parallel line segments. An automorphism of a mathematical structure is always a one-to-one mapping of that structure onto itself which preserves its structural properties. Hence an automorphism of the affine space X should be a one-to-one mapping of X onto itself which preserves both parallelism and the ratio of parallel line segments. Such mappings are called affine transformations and we shall define them precisely and study them in some detail.

We begin by looking at one-to-one mappings of X onto itself which preserve parallelism, but do not necessarily preserve the ratio of parallel line segments. If $n \geq 2$, such mappings are called semiaffine transformations. If $n = 1$, all permutations of the line X preserve parallelism, so semiaffine transformations have to be defined differently (see Definition 88.1).

Definition 65.1.   Let $n \geq 2$. A **semiaffine transformation** σ of X is a one-to-one mapping of X onto itself satisfying:

1. If S is a d-dimensional affine subspace of X, $\sigma(S)$ is also a d-dimensional affine subspace of X.
2. If S and S' are parallel affine subspaces of X, then $\sigma(S) \parallel \sigma(S')$.

The two conditions of the above definition express precisely what is meant by saying that σ preserves parallelism. Observe that Statement 2 does not require that S and S' have the same dimension. If we had not postulated that $n \geq 2$, it is clear that all permutations of a line would have become semiaffine transformations.

Since dilations are either translations or magnifications, all dilations are semiaffine transformations. We saw, in Exercise 51.7, that a dilation D preserves the ratio of line segments if and only if its ratio r belongs to the center of k. Therefore, if k is not commutative, there are dilations which do not preserve the ratio of line segments. Even if k is a field (in which case k equals the center of k), we shall see that, in general, there are semiaffine transformations which do not preserve the ratio of parallel line segments. Such transformations are, therefore, semiaffine transformations which are not dilations.

It is surprising that Condition 2 of Definition 65.1 is actually unnecessary.

Proposition 66.1.   Assume $n \geq 2$. Let σ be a one-to-one mapping of X onto itself satisfying: If S is a d-dimensional affine subspace of X, $\sigma(S)$ is also a d-dimensional affine subspace of X. Then σ is a semiaffine transformation.

Proof. Let S and S' be parallel affine subspaces of X. We must show that $\sigma(S) \parallel \sigma(S')$. This is trivial when $S = S'$; hence we assume $S \neq S'$.

Case 1. dim $S =$ dim S'. We denote this dimension by d and conclude that $d < n$ because $S \neq S'$. Then $S \cap S' = \varnothing$ and S and S' are contained in a $(d + 1)$-dimensional subspace S'' of X (Exercise 30.3). By hypothesis, $\sigma(S)$ and $\sigma(S')$ are d-dimensional affine subspaces of X and $\sigma(S'')$ is a $(d + 1)$-dimensional affine subspace of X. Furthermore, since σ is one-to-one, $\sigma(S) \cap \sigma(S') = \varnothing$ and both $\sigma(S)$ and $\sigma(S')$ are subspaces of $\sigma(S'')$. Therefore, $\sigma(S) \parallel \sigma(S')$ (Exercise 30.3).

Case 2. dim $S \neq$ dim S'. We assume that dim $S <$ dim S'. Then S lies in an affine subspace S'' where $S'' \parallel S'$ and dim $S'' =$ dim S. By Case 1, $\sigma(S'') \parallel \sigma(S')$ and, since $\sigma(S) \subset \sigma(S')$, we also conclude that $\sigma(S) \parallel \sigma(S')$. Done.

If k contains at least three elements, it is interesting to find that the condition of Proposition 66.1 has to be postulated only for the case where S is a line. This will be shown in the next proposition whose proof is based on the following exercises.

Exercises

*1. Assume that X is a plane and that $k = \mathbf{Z}_2$, that is, k has two elements.
   a. Prove that X has four points and six lines.
   b. Prove that there are precisely two points on each line.
   c. Prove that precisely three lines pass through each point of X.
   d. If l and m are distinct parallel lines, prove that every point lies on either l or m.
   e. If l and m are distinct lines of X which are not parallel, prove that there is a point x which does not lie on either l or m.
   f. Let l, m, and x satisfy the conditions of Part e. Prove that it is impossible to find distinct points y, x in X such that $y \in l$, $x \in m$, and x, y, z are collinear.

g. Prove that every permutation of X is a semiaffine transformation of X. Hence X has 24 semiaffine transformations.

h. If a coordinate system is chosen for X, show that the diagram in Figure 67.1 accurately depicts the four points and six lines of X. Observe that the two diagonals of the square have no point in common and hence are parallel. How dangerous and wonderful pictures can be!

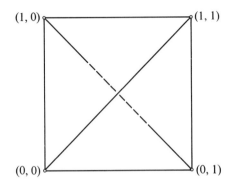

Figure 67.1

2. Let X be a plane and assume k has at least three elements. Let l and m be distinct lines of X.

a. Prove that there is a point in X which does not lie on either l or m.

b. Let $x \in X$. Prove that there are distinct points $y, z \in X$ such that $y \in l$, $z \in m$, and the points x, y, and z are collinear. (Hint: If $x \in l$ or $x \in m$, the assertion is correct even if k has only two elements. If $x \notin l$ and $x \notin m$, use the fact that l has at least three points. Conclude that l contains a point y such that $y \notin m$ and the line $x \vee y$ is not parallel to m. See Figure 67.2)

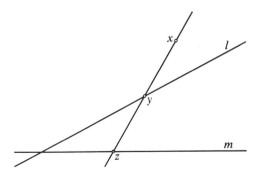

Figure 67.2

c. Observe that Part a and Part b are both false if k has just two elements and hence the above picture is invalid. Point out where the fact that k has at least three elements is used in each of the proofs.

3. In this exercise and the next one, we extend the results of Exercise 2 above to higher dimensions. In this exercise, we assume that $n \geq 2$, but there are no restrictions on k.

   a. Prove that there is a line l and a hyperplane S_0 in X which have precisely one point in common. (This is also correct when $n = 1$.)

   b. If l and S_0 satisfy the conditions of Part a, prove that there is a point $x \in Y$ which does not lie on either l or S_0.

   c. Assume l, S_0, and x satisfy the conditions of Part b. If π is the unique plane $l \vee x$, prove that $\pi \cap S_0$ is a line $m \neq l$ and that $x \notin m$. Figure 68.1 is drawn for the case $n = 3$.

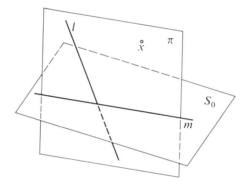

Figure 68.1

4. Assume $n \geq 1$ and that k has at least three elements. Let l be a line and S_0 a hyperplane which have precisely one point in common; finally, let $x \in X$.

   a. Prove that there are distinct points y, z in X such that $y \in l$, $z \in S_0$, and the points x, y and z are collinear.

   b. Prove that Part a is false if k has only two elements.

5. Let l be a line and S_0 a d-dimensional subspace of X where l and S_0 have precisely one point in common. (There are no restrictions on k.)

   a. Prove that $0 \leq d < n$.

   b. Prove that there is a unique $(d + 1)$-dimensional subspace S of X which contains both l and S_0.

   c. If $d \geq 1$, prove that S contains a point x which does not lie on either l or S_0. (Hint: S_0 is a hyperplane of S.)

6. Assume that k has at least three elements and that l, S_0, and S are as given in Part b of the previous exercise. Let x be a point of X. Prove

that there are distinct points y, z in S such that $y \in l$, $z \in S_0$, and the points x, y, and z are collinear. Prove that this is false if k has only two elements.

The following exercises are different in nature, but will lead up to the result in Exercise 69.9 which will be used in Proposition 69.1.

*7. Let σ be a semiaffine transformation of X. Prove that the points $x_1, \ldots, x_d$ of X are independent if and only if the points $\sigma(x_1), \ldots, \sigma(x_n)$ are independent. (Hint: If $x_1, \ldots, x_d$ are dependent, let S be an affine subspace of dimension at most $d - 2$ which contains these points and consider $\sigma(S)$. If $x_1, \ldots, x_d$ are independent, use mathematical induction on d and the fact that $x_1, \ldots, x_{d-1}$ are independent and $x_d \notin x_1 \vee \cdots \vee x_{d-1}$.)

*8. Assume σ is a semiaffine transformation of X and let $x_1, \ldots, x_d$ be points of S (not necessarily independent). Prove that

$$\sigma(x_1 \vee \cdots \vee x_d) = \sigma(x_1) \vee \cdots \vee \sigma(x_d).$$

*9. A semiaffine transformation σ of X is a one-to-one function of X onto itself and hence has an inverse σ^{-1}. Prove that σ^{-1} is again a semiaffine transformation. (Hint: One need only show that, if S is an affine subspace of x, $\sigma^{-1}(S)$ is also an affine subspace of X. Let S' be the subspace spanned by the "set" $\sigma^{-1}(S)$ and prove that $\sigma^{-1}(S) = S'$.)

Proposition 69.1. Let $n \geq 2$ and assume that k has at least three elements. Suppose that σ is a one-to-one mapping of X onto X satisfying: If l is a line of X, $\sigma(l)$ is also a line of X. Then σ is a semiaffine transformation.

Proof (by induction). We must prove that the image of each d-dimensional affine subspace of X is again a d-dimensional affine subspace of X. By hypothesis, the statement is true for $d = 1$. We suppose the assertion has been proved for all subspaces of dimension less than $d + 1$, $1 < d + 1 \leq n$; let S be a $(d + 1)$-dimensional affine subspace of X.

We select a line l in S and a d-dimensional subspace S_0 of S which have precisely one point in common. By the induction hypothesis, $l' = \sigma(l)$ is a line and $S_0' = \sigma(S_0)$ is a d-dimensional subspace of X. Since σ is one-to-one, l' and S_0' have precisely one point in common and hence they are contained in a unique $(d + 1)$-dimensional subspace S' of S. Figure 70.1 is drawn for $d = 2$.

We complete the proof by showing that $\sigma(S) = S'$.

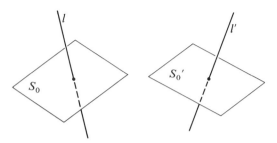

Figure 70.1

1. $\sigma(S) \subset S'$. Let $x \in S$. Since k has at least three elements, there are distinct points $y, z \in S$ such that $y \in l$, $z \in S_0$ and the points x, y, and z lie

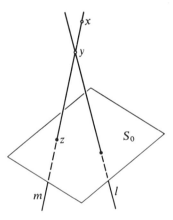

Figure 70.2

on a line m. Then $m' = \sigma(m)$ is a line of X and, since $\sigma(y)$ and $\sigma(z)$ are points of S', $m' \subset S'$. Thus $\sigma(x) \in S'$.

2. $S' \subset \sigma(S)$. By Exercise 69.9, σ^{-1} is a semiaffine transformation. The argument used on Part 1 applied to σ^{-1} allows us to conclude that $\sigma^{-1}(S') \subset S$ which is the same as $S' \subset \sigma(S)$. Done.

The condition that k must contain at least three elements cannot be omitted from the hypothesis of the above proposition. If k is the field with two elements, the lines of X are the two-element subsets of X and hence every permutation of X will send lines onto lines. If, furthermore, $n \geq 3$, it is easy to give examples of permutations of X which do not send every plane into a plane. Hence not all permutations are semiaffine transformations.

Exercise

10. Assume that $k = \mathbf{Z}_2$ and $n = 3$.
    a. Prove that X has eight points and that every plane contains four points.
    b. Assume the four points x, y, p, q form a plane of X and let u be one of the remaining points of X. Consider the permutation of X which interchanges x and u while it leaves the other six points fixed. Prove that this permutation is not a semiaffine transformation.

Using Proposition 66.1 and Exercise 69.9 it is almost trivial that the semiaffine transformations of X form a group and we leave it to the reader to supply the details of the proof.

Proposition 71.1. Let $n = 2$. The semiaffine transformations form a group Sa which contains the group Di of dilations.

We now turn to the true automorphisms of affine space, the affine transformations of X. These are the semiaffine transformations which preserve the ratio of parallel line segments. The precise definition makes use of the following exercise.

Exercise

11. Let $n \geq 2$ and let σ be a semiaffine transformation of X. Suppose that $x \neq y$ and $p \neq q$ are points of X such that the line segments (x, y) and (p, q) are parallel. Prove that the line segments $(\sigma(x), \sigma(y))$ and $(\sigma(p), \sigma(q))$ are also parallel. Hence one knows both the ratios

$$\frac{(p, q)}{(x, y)} \quad \text{and} \quad \frac{(\sigma(p), \sigma(q))}{(\sigma(x), \sigma(y))}.$$

Definition 71.1. If $n \geq 2$, an **affine transformation** is a semiaffine transformation σ which satisfies the condition: If $x \neq y$ and $p \neq q$ are points of X such that the line segments (x, y) and (p, q) are parallel, then

$$\frac{(p, q)}{(x, y)} = \frac{(\sigma(p), \sigma(q))}{(\sigma(x), \sigma(y))}.$$

If $n = 1$, an **affine transformation** is a permutation of X which satisfies: If $x \neq y$ and $p \neq q$ are points of X, then

$$\frac{(p, q)}{(x, y)} = \frac{(\sigma(p), (\sigma(q))}{(\sigma(x), \sigma(y))} \, .$$

The condition $(p, q)/(x, y) = (\sigma(p), \sigma(q))/(\sigma(x), \sigma(y))$ of the above definition says precisely that the affine transformations preserve the ratio of parallel line segments. Once we have said what semiaffine transformations exist for the case that $n = 1$, we shall see that affine transformations are also semiaffine; it will also be true that all dilations of a line are semiaffine transformations.

We conclude that all translations are affine transformations and that a magnification $M(c, r)$ is an affine transformation if and only if its ratio lies in the center of k (Exercise 51.7). Therefore, if k is a field, all dilations are affine transformations.

Exercise

12. Let $k = \mathbf{Z}_2$. Prove that all semiaffine transformations are affine. Hence when $n = 2$, conclude from Exercise 66.1 that X has 24 affine transformations.

It is practically trivial that the affine transformations of X form a group and we ask the reader to write his own proof of this. Since affine transformations are also semiaffine when $n \geq 2$, we have the following result.

Proposition 72.1. Let $n \geq 2$. The affine transformations form a group Af which is a subgroup of the group Sa of semiaffine transformations.

The above proposition will also be true for $n = 1$ when we know what semiaffine transformations exist for a line X. We already know that the affine transformations of X form a group even when $n = 1$. For $n \geq 2$, the inclusion diagram for the groups Sa, Af, and Di is shown in Figure 72.1.

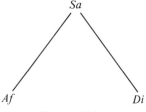

Figure 72.1

Actually, more can be said about the groups *Af* and *Sa* when $n \geq 2$. We shall learn in Exercise 93.4 that *Af* is not commutative; consequently, *Sa* is also noncommutative.

Exercise

13. Prove that the group $Af \cap Di$ consists of the translations together with those magnifications whose ratio lies in the center of k.

In the ensuing sections, we shall give a description of all affine and semiaffine transformations. The question whether there are semiaffine transformations which are not affine has nothing to do with the commutativity or noncommutativity of k. It depends on the automorphisms of k.

16. FROM SEMILINEAR TO SEMIAFFINE

We shall see that the affine transformations of X correspond to the nonsingular, linear transformations of V. Similarly, the semiaffine transformations of X correspond to the nonsingular, semilinear transformations of V. We begin by discussing semilinear mappings. (The words "transformation," "mapping," "map," and "function" are all synonymous.)

Definition 73.1. Let (V, k) and (V', k') be vector spaces over the division rings k and k', respectively. Suppose that $\mu: k \rightarrow k'$ is an isomorphism from k onto k'. A map $\lambda: V \rightarrow V'$ is called **semilinear with respect to** μ if

1. $\lambda(A + B) = \lambda A + \lambda B$ for all $A, B \in V$.
2. $\lambda(tA) = \mu(t)\lambda(A)$ for all $A \in V$ and $t \in k$.

If V and V' are both vector spaces over the same division ring k, we observe that a linear transformation from V to V' is a semilinear transformation where $\mu: k \rightarrow k$ is the identity function. We shall be interested in the case where $V = V'$, $k = k'$, and μ is an automorphism.

Example 1

Let V be the two-dimensional vector space over the field **C** of complex numbers where $V = \{(z, w) \mid z, x \in \mathbf{C}\}$. Let $\mu: \mathbf{C} \rightarrow \mathbf{C}$ be complex conjugation, that is, $\mu(z) = \bar{z}$. Then $\lambda: V \to V$, defined by $\lambda(z, w) = (\bar{z}, \bar{w})$, is a semilinear transformation with respect to μ.

Example 2

The field of the rational numbers and the field of the real numbers both have no automorphisms except for the identity function (see Artin, [2], pp. 46, 47). Assume that k is any division ring which has no automorphisms except the identity function. If (V, k) and (V', k) are vector spaces over k, the semilinear mappings from V to V' are precisely the linear mappings from V to V'.

Many of the properties of linear transformations are enjoyed by semilinear transformations. Some are included in the exercises below.

Exercises

Let (V, k) and (V', k') be vector spaces and $\mu: k \to k'$ be an isomorphism from k onto k'.

1. Let $A_1, \ldots, A_n$ be a basis for V and $B_1, \ldots, B_n$ be arbitrary vectors in V'. If $Y = y_1 A_1 + \cdots + y_n A_n$, prove that the mapping $\lambda: V \to V'$, defined by $\lambda(Y) = \mu(x_1)B_1 + \cdots + \mu(x_n)B_n$, is a semilinear mapping with respect to μ.
2. Show that every semilinear mapping with respect to μ is of the form given in Exercise 1 above.
3. Assume that λ is defined as in Exercise 1 above.
   a. Show that λ is one-to-one if and only if $B_1, \ldots, B_n$ are linearly independent.
   b. Prove that λ is one-to-one and onto if and only if $B_1, \ldots, B_n$ form a basis for V'.
4. If $\lambda: V \to V'$ is semilinear with respect to μ and λ is one-to-one and onto, prove that $\lambda^{-1}: V' \to V$ is a semilinear transformation with respect to the isomorphism $\mu^{-1}: k' \to k$.
*5. Assume $\lambda: V \to V'$ is a semilinear mapping with respect to μ and let U be a linear subspace of V.
   a. Prove that $\lambda(U)$ is a linear subspace of V'.
   b. If λ is one-to-one, prove that $\dim U = \dim \lambda(U)$.
   c. If λ is one-to-one and $A_1, \ldots, A_d$ are linearly independent vectors of V, prove that $\lambda(A_1), \ldots, \lambda(A_d)$ are linearly independent vectors of V.

In view of Exercise 4 above, we make the natural definition for non-singular semilinear transformations.

Definition 75.1.   A semilinear mapping $\lambda: V \to V'$ is called **nonsingular** if λ has an inverse.

We are interested in the nonsingular semilinear mappings of a vector space (V, k) onto itself. In this case, the isomorphism $\mu: k \to k$ can be any automorphism of k. Such mappings are naturally called **semilinear automorphisms** of V. We leave the proof of the following proposition to the reader.

Proposition 75.1.   The semilinear automorphisms of a vector space form a group *Sem* (under the composition of functions).

We now discuss the connection between semilinear and semiaffine transformations. For this, we choose a point $c \in X$ and consider the left vector space $X(c)$ over k (see Section 14, The Tangent Space $X(c)$). We know that $f: X(c) \to V$ and $g = f^{-1}: V \to X(c)$ are linear isomorphisms where

$$f(x) = \overrightarrow{c, x} \quad \text{for} \quad x \in X$$

and

$$g(A) = Ac \quad \text{for} \quad A \in V.$$

Consequently, if $\lambda: V \to V$ is a semilinear automorphism of V, we obtain a semilinear automorphism

$$L(c, \lambda): X(c) \to X(c)$$

by defining

$$L(c, \lambda)(x) = g(\lambda(f(x))) = (\lambda(\overrightarrow{c, x}))c.$$

The reader should observe that $L(c, \lambda)$ is simply the semilinear automorphism λ of V taken under f (equivalently, under g) from the vector space V to the vector space $X(c)$. This is the reason there can be no doubt that $L(c, \lambda)$ is a semilinear automorphism of $X(c)$, relative to the same automorphism μ of k as λ. In particular, $L(c, \lambda)$ is a linear automorphism of the tangent space $X(c)$ if and only if λ is a linear automorphism of V.

It is our purpose to show that $L(c, \lambda)$ is a semiaffine transformation of X which is affine if and only if λ is linear. We first prove:

Proposition 75.2.   Let $S(x, U)$ be an affine subspace of X. Then

$$L(c, \lambda)(S(x, U)) = S(y, \lambda(U))$$

where $y = L(c, \lambda)(x)$. Consequently, the affine subspaces $S(x, U)$ and $L(c, \lambda)(S(x, U))$ have the same dimension.

Proof. Select $A \in U$. Then $Ax \in S(x, U)$ and $L(c, \lambda)(Ax) = (\lambda(c, \overrightarrow{Ax}))c$. Since $\overrightarrow{c, Ax} = \overrightarrow{c, x} + A$, $\lambda(\overrightarrow{c, Ax}) = \lambda(\overrightarrow{c, x}) + \lambda(A)$ and we conclude that

$$\lambda(\overrightarrow{c, Ax}) = \lambda(A)[(\lambda(\overrightarrow{c, x}))c] = \lambda(A)y.$$

As A runs through U, $\lambda(A)$ runs through $\lambda(U)$. Moreover, $\lambda(U)$ is a linear subspace of V of the same dimension as U (Exercise 74.5). Done.

The main result now follows easily.

Proposition 76.1.   Let $n \geq 2$. Then $L(c, \lambda)$ is a semiaffine transformation of X which leaves the point c fixed.

Proof. $L(c, \lambda)$ is a semilinear automorphism of $X(c)$ and hence $L(c, \lambda)$ is a one-to-one mapping of X onto itself. By the previous proposition, $L(c, \lambda)$ maps each affine subspace of X onto an affine subspace of the same dimension. Hence $L(c, \lambda)$ is semiaffine (Proposition 66.1). Finally,

$$L(c, \lambda)(c) = \lambda(\overrightarrow{c, c})c = \mathbf{0}c = c.$$

Done.

Once we have defined semiaffine transformations for lines, it will be clear that Proposition 76.1 also holds for $n = 1$.

Proposition 76.1 enables us to construct semiaffine transformations of X galore, most of which are not dilations. All we have to do is choose an arbitrary semilinear automorphism λ of V and a point $c \in X$; the resulting $L(c, \lambda)$ is a semiaffine transformation of X.

Exercise

6. Let $r \in k$ and consider the function $\lambda: V \to V$ defined by $\lambda(A) = rA$.
   a. If dim $V \geq 1$, prove that λ is a linear transformation of V into itself if and only if r belongs to the center of k.
   b. If $r \neq 0$, prove that λ is a semilinear automorphism of V with respect to the inner automorphism μ of k defined by $\mu(t) = rtr^{-1}$ for all $t \in k$.
   c. If $r \neq 0$ and $c \in X$, prove that $L(c, \lambda)$ is the magnification $M(c, r)$.

The fact that $L(c, \lambda)$ leaves the point c fixed shows that $L(c, \lambda)$ is not the most general kind of semiaffine transformation. For example, translations different from 1_X have no fixed points, but are both affine and semiaffine transformations. We now bring out a property of $L(c, \lambda)$ which, surprisingly enough, will turn out to be a general property of all semiaffine transformations.

Exercise

*7. Put $L = L(c, \lambda)$ and prove that, if $x, y \in X$,

$$\overrightarrow{L(x), L(y)} = \lambda(\overrightarrow{x, y}).$$

(Hint: Use Exercise 8.10.)

Let $\lambda: V \to V$ be a semilinear automorphism of V with respect to the automorphism μ of k. We choose points $x \neq y$ and $p \neq q$ in X such that the line segments (x, y) and (p, q) are parallel. If L denotes $L(c, \lambda)$, the ratios

$$\frac{(p, q)}{(x, y)} \quad \text{and} \quad \frac{(L(p), L(q))}{(L(x), L(y))}$$

are known since L is semiaffine. If L is affine, these ratios are equal. If L is only semiaffine, we have never observed any relationship between these ratios. How blind we have been!

Proposition 77.1.   Let $\lambda: V \to V$ be a semilinear automorphism of V with respect to the automorphism μ of k and let $L = L(c, \lambda)$ be the associated semiaffine transformation. If the line segments (x, y) and (p, q) are parallel, then

$$\frac{(L(p), L(q))}{(L(x), L(y))} = \mu\left(\frac{(p, q)}{(x, y)}\right).$$

It follows that L is affine if and only if λ is linear.

Proof. Put $(p, q)/(x, y) = s$. Then $\overrightarrow{p, q} = s(\overrightarrow{x, y})$ and hence $\lambda(\overrightarrow{p, q}) = \mu(s)\lambda(\overrightarrow{x, y})$. According to Exercise 77.7,

$$\overrightarrow{L(p), L(q)} = \lambda(\overrightarrow{p, q}) = \mu(s)\lambda(\overrightarrow{x, y}) = \mu(s)\overrightarrow{L(x), L(y)}$$

and this says that

$$\frac{(L(p), L(q))}{(L(x), L(y))} = \mu(s).$$

We conclude that L is affine if and only if $\mu(s) = s$ for all $s \in k$ which is the same as to say that μ is the identity automorphism of k; that is, λ is linear. Done.

17. PARALLELOGRAMS

In the previous section, we saw how to proceed from a given semilinear automorphism λ of V and a point c to the semiaffine transformation $L(c, \lambda)$ of X. Now we want to go the other way. We want to associate a semilinear automorphism of V to every semiaffine transformation. This is much harder. The way is prepared by some remarks on parallelograms.

Definition 78.1. Assume that four distinct points x_1, x_2, x_3, x_4 and four distinct lines l_1, l_2, l_3, l_4 of X are such that

1. $l_1 \parallel l_3$ and $l_2 \parallel l_4$.
2. $\{x_1\} = l_4 \cap l_1$, $\{x_2\} = l_1 \cap l_2$, $\{x_3\} = l_2 \cap l_3$, $\{x_4\} = l_3 \cap l_4$.

The points x_1, x_2, x_3, x_4 are then called the **vertices of the parallelogram with sides** $\{x_1, x_2\}$, $\{x_2, x_3\}$, $\{x_3, x_4\}$, and $\{x_4, x_1\}$. (See Figure 78.1.)

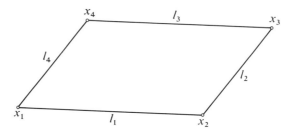

Figure 78.1

Observe that the sides of the parallelogram are nonoriented line segments.

Exercises

1. Let x_1, x_2, x_3, x_4 and l_1, l_2, l_3, l_4 be as given in Definition 78.1. Which pairs of sides should be called opposite and which pairs adjacent sides of the parallelogram? Which pairs of vertices should be called opposite and which pairs adjacent vertices of the parallelogram?

Define the diagonals of the parallelogram. (Diagonals are also non-oriented line segments.)

2. Consider the four points and the six lines of the plane over the field $k = \mathbf{Z}_2$ (Exercise 66.1).

   a. Prove that these four points are the vertices of three different parallelograms, depending on which four line segments are used as sides.

   b. Prove that, in each of these parallelograms, the two diagonals are parallel to one another. Surprised? We know from high school geometry that, in the real affine plane, two diagonals always intersect in a point inside the parallelogram. This exercise shows this is false when $k = \mathbf{Z}_2$.

3. Prove that semiaffine transformations map parallelograms onto parallelograms, preserving the relation of opposite and adjacent for sides and vertices.

4. Let l and m be two lines of X which have precisely one point x in common. Assume $y \in l$, $z \in m$ and $y \neq x$, $z \neq x$. Prove that the vectors $\overrightarrow{y, x}$ and $\overrightarrow{z, x}$ are linearly independent.

The lengths of opposite sides of a parallelogram can be compared even in affine geometry because the opposite sides are parallel line segments. The following proposition says that opposite sides of a parallelogram have the same length.

Proposition 79.1. Let x_1, x_2, x_3, x_4 and l_1, l_2, l_3, l_4 satisfy the conditions of Definition 78.1. Then

$$\overrightarrow{x_1, x_2} = \overrightarrow{x_4, x_3} \quad \text{and} \quad \overrightarrow{x_1, x_4} = \overrightarrow{x_2, x_3}.$$

Proof. Since $l_1 \parallel l_2$, $\overrightarrow{x_1, x_2} = s(\overrightarrow{x_4, x_3})$ and, since $l_2 \parallel l_4$, $\overrightarrow{x_1, x_4} = t(\overrightarrow{x_2, x_4})$. We must prove that $s = t = 1$. We observe that

$$\overrightarrow{x_1, x_2} + \overrightarrow{x_2, x_3} = \overrightarrow{x_1, x_3} = \overrightarrow{x_1, x_4} + \overrightarrow{x_4, x_3}$$

and hence

$$s(\overrightarrow{x_4, x_3}) + (\overrightarrow{x_2, x_3}) = t(\overrightarrow{x_2, x_3}) + \overrightarrow{x_4, x_3}.$$

It follows that

$$(s - 1)(\overrightarrow{x_4, x_3}) + (1 - t)(\overrightarrow{x_2, x_3}) = \mathbf{0}.$$

The vectors $\overrightarrow{x_4, x_3}$ and $\overrightarrow{x_2, x_3}$ are linearly independent; hence $s = t = 1$. Done.

Exercises

5. Let $x, y \in X$ and $A, B \in V$. Prove that $Ax = By$ if and only if $A = \overrightarrow{x, y} + B$.

6. Let x_1, x_2, x_3, x_4 and l_1, l_2, l_3, l_4 satisfy the conditions of Definition 78.1. We denote the corresponding parallelogram by P. The diagonals of P are $\{x_1, x_3\}$ and $\{x_2, x_4\}$. The diagonal $\{x_1, x_3\}$ is carried by the line $m_1 = S(x_1, \langle A \rangle)$ where $A = \overrightarrow{x_1, x_3}$; the diagonal $\{x_2, x_4\}$ is carried by the line $m_2 = S(x_2, \langle B \rangle)$, where $B = \overrightarrow{x_2, x_4}$.

   a. If the characteristic of k is not two, prove that $(\frac{1}{2}A)x_1 = (\frac{1}{2}B)x_2$ and that this point is the unique point of intersection of lines m_1 and m_2. Hence, in the real affine plane, the lines which carry the diagonals of a parallelogram intersect because the characteristic of $\mathbf{R}$ is not two.

   b. Assume the characteristic of k is not two and denote the point of intersection of the lines m_1 and m_2 by y. Prove that

   $$\overrightarrow{x_1, y} = \tfrac{1}{2}(\overrightarrow{x_1, x_3}) \qquad \text{and} \qquad \overrightarrow{x_2, y} = \tfrac{1}{2}(\overrightarrow{x_2, x_4}).$$

   c. Assume k is ordered. (Then the characteristic of k is 0 and $0 < \frac{1}{2} < 1$.) Prove that the point y of intersection of the lines m_1 and m_2 lies between the vertices x_1 and x_3 and also between the vertices x_2 and x_4. (Hence, in the real affine plane, the point of intersection of the carriers of the diagonals of a parallelogram lies inside the parallelogram because $\mathbf{R}$ is ordered.)

   Remark. If k is ordered, one means by the points of a line segment $\{x, y\}$ all the points between x and y. We denote this pointset by $[x, y]$. The sides and diagonals of the parallelogram P are then regarded as the pointsets $[x_1, x_2], [x_2, x_3], [x_3, x_4], [x_4, x_1], [x_1, x_3]$, and $[x_2, x_4]$. The results of the present exercise can now be expressed by saying that, over an ordered division ring, the diagonals of a parallelogram intersect at their common midpoint and this midpoint lies inside the parallelogram. This recovers the usual language and does away with the clumsy term "carrier of a diagonal."

7. Let $x_1, x_2 \in X$ and $A \in V$ where $A \neq \mathbf{0}$ and $x_2 \notin x_1 \vee Ax_1$. Prove that the points $x_1, x_2, x_3 = Ax_2, x_4 = Ax_1$ and the lines $l_1 = x_1 \vee x_2$,

$l_2 = x_2 \vee x_3$, $l_3 = x_3 \vee x_4$, $l_4 = x_4 \vee x_1$ form a parallelogram with $\{x_1, Ax_1\}$ and $\{x_2, Ax_2\}$ as opposite sides. Draw an appropriate diagram.

18. FROM SEMIAFFINE TO SEMILINEAR

We assume in this whole section that $n \geq 2$ and choose a semiaffine transformation σ of X. The purpose of this section is to construct a semilinear automorphism λ of V with the property that σ is equal to the semiaffine transformation $L(c, \lambda)$ multiplied by a translation. This is the content of Theorem 86.1. Its proof is long; the main difficulty lies in showing that the constructed λ is, indeed, a semilinear automorphism. In order to avoid a mile-long proof, we have made the various steps of the proof into propositions which are of interest for their own sake. This is particularly true of Proposition 84.1. All of this section up to Exercise 86.1 can be regarded as the proof of Theorem 86.1.

The first task is the construction of λ, that is, defining the action of λ on the vectors of V and the automorphism μ of k.

Proposition 81.1.   Let $A \in V$. The vector $\overrightarrow{\sigma(x), \sigma(Ax)}$ does not depend on the point $x \in X$.

Proof. If $A = 0$,

$$\overrightarrow{\sigma(x), \sigma(0x)} = \overrightarrow{\sigma(x), \sigma(x)} = 0$$

for all points $x \in X$. We, henceforth, assume $A \neq 0$ and choose points x and y in X. We must prove that

$$\overrightarrow{\sigma(x), \sigma(Ax)} = \overrightarrow{\sigma(y), \sigma(Ay)}.$$

Case 1. $y \notin x \vee Ax$. Then the four points x, y, Ay, Ax form a parallelogram with $x \vee Ax$ and $y \vee Ay$ as opposite sides. Hence the four points $\sigma(x), \sigma(y), \sigma(Ay), \sigma(Ax)$ form a parallelogram with $\sigma(x) \vee \sigma(Ax)$ and $\sigma(y) \vee \sigma(Ay)$ as opposite sides. According to Proposition 79.1,

$$\overrightarrow{\sigma(x), \sigma(Ax)} = \overrightarrow{\sigma(y), \sigma(Ay)}.$$

Case 2. $y \in x \vee Ax$. Since $n \geq 2$, there is a point $z \notin x \vee Ax$ from which it follows that $y \notin z \vee Ax$. (See Figure 82.1.) By Case 1,

$$\overrightarrow{\sigma(x), \sigma(Ax)} = \overrightarrow{\sigma(z), \sigma(Az)} = \overrightarrow{\sigma(y), \sigma(Ay)}.$$

Done.

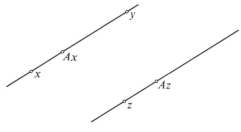

Figure 82.1

The above proposition tells us that the correspondence $A \to \overrightarrow{\sigma(x), \sigma(Ax)}$ is a well-defined function $\lambda: V \to V$; we now show that λ is a semilinear automorphism of V.

Proposition 82.1. For all vectors $A, B \in V$,

$$\lambda(A + B) = \lambda(A) + \lambda(B).$$

Furthermore, λ is a one-to-one mapping from V onto V.

Proof. 1. $\lambda(A + B) = \lambda(A) + \lambda(B)$. We choose a point $x \in X$ and use x to compute the vector $\lambda(A)$; the point Ax is then used to compute the vector $\lambda(B)$. This gives

$$\lambda(A) = \overrightarrow{\sigma(x), \sigma(Ax)} \qquad \text{and} \qquad \lambda(B) = \overrightarrow{\sigma(Ax), \sigma(B(Ax))}.$$

Consequently,

$$\lambda(A) + \lambda(B) = \overrightarrow{\sigma(x), \sigma(B(Ax))}$$

which equals $\lambda(A + B)$ since $B(Ax) = (A + B)x$.

2. λ is one-to-one. We know from Part 1 that λ is a homomorphism of the additive group of V into itself; hence it is sufficient to show that the kernel of λ is $\{0\}$. For this, we choose $A \in V$ and assume that $\lambda(A) = 0$. If $x \in X$, then $\lambda(A) = \overrightarrow{\sigma(x), \sigma(Ax)} = \mathbf{0}$ and hence $\sigma(x) = \sigma(Ax)$. But σ is one-to-one, so we conclude that $x = Ax$ or, equivalently, $A = \mathbf{0}$.

3. λ is onto. Select $B \in V$. We must find a vector A such that $\lambda(A) = B$. We choose a point $x \in X$, let $y = \sigma^{-1}(B\sigma(x))$ and put $A = \overrightarrow{x, y}$. Then $Ax = \sigma^{-1}(B\sigma(x))$ and hence

$$\lambda(A) = \overrightarrow{\sigma(x), \sigma(Ax)} = \overrightarrow{\sigma(x), B\sigma(x)} = B.$$

Done.

The above proposition says that $\lambda: V \to V$ is an automorphism of the additive group of V onto itself. However, nothing is said about the behavior of λ with respect to the product tA of a scalar t and a vector A. Consequently, we cannot claim that we have proved that λ sends every one-dimensional subspace of V onto a one-dimensional subspace. This is the essence of the next proposition.

Proposition 83.1.   Let $A \in V$ where $A \neq 0$. Then $\lambda(\langle A \rangle) = \langle \lambda(A) \rangle$. If $c \in X$ and l is the line $S(c, \langle A \rangle)$, then $\sigma(l) = S(\sigma(c), \langle \lambda(A) \rangle)$.

Proof. The line l is mapped onto the line $\sigma(l) = S(\sigma(c), U)$ where U is a one-dimensional linear subspace of V.

1. $U = \langle \lambda(A) \rangle$. The points c and Ac are distinct points of l and hence $\sigma(c)$ and $\sigma(Ac)$ are distinct points of $\sigma(l)$. Consequently,

$$U = \langle \overrightarrow{\sigma(c), \sigma(Ac)} \rangle = \langle \lambda(A) \rangle.$$

We now show that the direction space U of the line $\sigma(l)$ is also equal to $\lambda(\langle A \rangle)$.

2. $\lambda(\langle A \rangle) \subset U$. We choose $t \in k$ and prove that $\lambda(tA) \in U$. If $t = 0$, $\lambda(0A) = \lambda(0) = 0$ since λ is an additive automorphism of V. Thus we may assume that $t \neq 0$, in which case $\langle tA \rangle = \langle A \rangle$ and $l = S(c, \langle tA \rangle)$. By Part 1, the direction space U of $\sigma(l)$ is equal to $\langle \lambda(tA) \rangle$ and hence $\lambda(tA) \in U$.

3. $U \subset \lambda(\langle A \rangle)$. We select $B \in U$ and have to find a scalar t such that $B = \lambda(tA)$. Since $B\sigma(c) \in S(\sigma(c), U) = \sigma(l)$, there is a $t \in k$ such that $B\sigma(c) = \sigma((tA)c)$. We conclude that

$$\langle \overrightarrow{\sigma(c), \sigma((tA)c)} \rangle = \overrightarrow{\sigma(c), B\sigma(c)} = B.$$

We have shown that $U = \langle \lambda(A) \rangle = \lambda(\langle A \rangle)$. Done.

We have learned something about the behavior of λ with respect to the product tA. Nevertheless, we cannot claim to have proved that λ sends linearly independent vectors of V onto linearly independent vectors even though λ is an automorphism of the additive group of V. We now prove, however, that λ preserves the independence of two vectors.

Proposition 83.2.   If A and B are linearly independent vectors of V, the vectors $\lambda(A)$ and $\lambda(B)$ are also linearly independent.

Proof. Choose a point c in X and consider the lines $l = S(c, \langle A \rangle)$ and $m = S(c, \langle B \rangle)$. Since A and B are linearly independent, $l \cap m = \{c\}$; therefore, $\sigma(l) \cap \sigma(m) = \{\sigma(c)\}$ because σ is one-to-one. We know from Proposition 83.1 that $\sigma(l) = S(\sigma(c), \langle \lambda(A) \rangle)$ and $\sigma(m) = S(\sigma(c), \langle \lambda(B) \rangle)$. Thus the vectors $\lambda(A)$ and $\lambda(B)$ are necessarily linearly independent. Done.

Thus far, we have shown that λ is an automorphism of the additive group of V onto itself which sends one-dimensional subspaces of V onto one-dimensional subspaces and preserves the linear independence of two vectors. The fact that λ is a semilinear automorphism of V is a consequence of the following proposition which requires that dim $V \geq 2$.

Proposition 84.1. A function $f: V \to V$ is a semilinear automorphism of V if and only if it has the following three properties:

1. f is an automorphism of the additive group of V onto itself.
2. f sends one-dimensional subspaces of V onto one-dimensional subspaces.
3. If A and B are linearly independent vectors of V, the vectors $f(A)$ and $f(B)$ are also linearly independent.

Proof. We have already seen that every semilinear automorphism of V satisfies the three stated properties (Exercise 74.5). Conversely, assume that the function $f: V \to V$ has the above three properties. Because of Property 1, there only remains the task of constructing an automorphism μ of k such that $f(tA), = \mu(t)f(A)$ for $t \in k$ and $A \in V$.

We choose a nonzero vector $A \in V$. By Property 1, $f(A) \neq 0$ and, by Property 2, $f(\langle A \rangle) = \langle f(A) \rangle$. Consequently, if $t \in k$, $f(tA) = \mu_A(t)f(A)$ where $\mu_A(t) \in k$. Let us show that the function $t \to \mu_A(t)$ of k into k does not depend on the vector A. Select a second nonzero vector B in V; we must show that $\mu_A(t) = \mu_B(t)$ for all scalars t.

Case 1. A and B are linearly independent. Then $A + B \neq 0$ and we compute $f(t(A + B))$ in two different ways, namely,

$$f(t(A + B)) = \mu_{A+B}(t)f(A + B) = \mu_{A+B}(t)f(A) + \mu_{A+B}(t)f(B)$$

and

$$f(t(A + B)) = f(tA + tB) = f(tA) + f(tB) = \mu_A(t)f(A) + \mu_B(t)f(B).$$

By Property 3, $f(A)$ and $f(B)$ are linearly independent vectors of V and hence $\mu_A(t) = \mu_{A+B}(t) = \mu_B(t)$.

Case 2. A and B are linearly dependent. Since dim $V \geq 2$, there is a vector C in V which does not lie on the one-dimensional subspace $\langle A \rangle$.

Then both pairs of vectors A, C and B, C consist of linearly independent vectors. Hence, by Case 1, $\mu_A(t) = \mu_C(t) = \mu_B(t)$.

We conclude that the function $t \rightarrow \mu(t)$, where $\mu(t) = \mu_A(t)$, for any nonzero vector A of V is a well-defined function $\mu: k \rightarrow k$. Let us now prove that μ is an automorphism of k.

Suppose s, $t \in k$ and let us prove that $\mu(s + t) = \mu(s) + \mu(t)$. We compute $f((s + t)A)$ in two different ways, namely,

$$f((s + t)A) = \mu(s + t)f(A)$$

and

$$((s + t)A) = f(sA + tA) = f(sA) + f(tA)$$
$$= \mu(s)f(A) + \mu(t)f(A) = (\mu(s) + \mu(t))f(A).$$

Since $f(A) \neq \mathbf{0}$ by Property 1, we conclude that $\mu(s + t) = \mu(s) + \mu(t)$.

Next we show $\mu(st) = \mu(s)\mu(t)$ for s, $t \in k$. Since $\mu(s + t) = \mu(s) + \mu(t)$, it follows that $\mu(0) = 0$. Hence if $t = 0$, both $\mu(st)$ and $\mu(s)\mu(t)$ are zero. If $t \neq 0$, $tA \neq \mathbf{0}$ and we compute $f(stA)$ again in two ways, using the fact that $tA \neq \mathbf{0}$ in the second computation.

$$f(stA) = \mu(st)f(A)$$
$$f(stA) = f(s(tA)) = \mu(s)f(tA) = \mu(s)\mu(t)f(A).$$

Since $f(A) \neq \mathbf{0}$, $\mu(st) = \mu(s)\mu(t)$.

Thus far, we have shown that μ is a homomorphism from k into k. We must yet prove that μ is one-to-one and onto. Since μ is an additive homomorphism of k into k, we prove that μ is one-to-one by showing its kernel is $\{0\}$. Assume $\mu(t) = 0$; this implies $f(tA) = \mu(t)f(A) = \mathbf{0}$. By Property 1, f is an additive homomorphism of V; we conclude that $tA = \mathbf{0}$ and hence $t = 0$ (because $A \neq \mathbf{0}$). To show that μ is onto, we select $t \in k$ and must exhibit an $s \in k$ such that $\mu(s) = t$. Since $f(A) \neq \mathbf{0}$, we know from Property 2 that $f(\langle A \rangle) = \langle f(A) \rangle$. Hence there exists a scalar s such that $f(sA) = tf(A)$. But $f(sA) = \mu(s)f(A)$; we conclude that $\mu(s) = t$. This completes the proof that μ is an automorphism.

Finally, we must prove that f is semilinear with respect to μ; that is, $f(tB) = \mu(t)f(B)$ for all $t \in k$ and $B \in V$. If $B \neq \mathbf{0}$, this is true by definition of μ. If $B = \mathbf{0}$, $tB = \mathbf{0}$ and both $f(tB)$ and $f(B)$ are $\mathbf{0}$ by Property 1; all the more then is $\mu(t)f(B) = \mathbf{0}$. Done.

We have associated a semilinear automorphism λ of V with the semi-affine transformation σ of X. We now state the structure of σ in terms of λ. Let c be a point of X and denote the vector $\overrightarrow{c, \sigma(c)}$ by A. We recall that $L(c, \lambda)$ is a semiaffine transformation and that T_A is an affine transformation of X. Hence the product $T_A L(c, \lambda)$ is certainly semiaffine.

Theorem 86.1. $\sigma = T_A L(c, \lambda)$.

Proof. Select $x \in X$. Then

$$L(c, \lambda)(x) = (\lambda(\overrightarrow{c, x}))c$$

where

$$\lambda(\overrightarrow{c, x}) = \overrightarrow{\sigma(c), \sigma((\overrightarrow{c, x})c)} = \overrightarrow{\sigma(c), \sigma(x)}.$$

Consequently,

$$T_A L(c, \lambda)(x) = T_A((\overrightarrow{\sigma(c), \sigma(x)})c) = (\overrightarrow{\sigma(c), \sigma(x)})Ac = (\overrightarrow{\sigma(c), \sigma(x)})\sigma(c) = \sigma(x).$$

Done.

The above theorem gives us a good description of the semiaffine transformations of X. For a fixed choice $c \in X$, each semiaffine transformation is a semilinear automorphism of X viewed as a vector space with origin c (this is the $L(c, \lambda)$) followed by a translation.

Exercise

1. Prove that the representation of σ given in Theorem 86.1 is uniquely determined by the point c of X. Precisely, if $A, B \in V$ and λ, λ' are semilinear automorphisms of V such that $T_A L(c, \lambda) = \sigma = T_B L(c, \lambda')$, prove that $A = B$ and $\lambda = \lambda'$.

Let (p, q) and (x, y) be parallel line segments of X. The ratios

$$\frac{(p, q)}{(x, y)} \quad \text{and} \quad \frac{(\sigma(p), \sigma(q))}{(\sigma(x), \sigma(y))}$$

are known, but we have no law which relates them unless σ is an affine transformation. Such a law is now given. The semilinear automorphism of V associated with σ is again denoted by λ; the automorphism associated with λ is again denoted by μ.

Theorem 86.2.
$$\frac{(\sigma(p), \sigma(q))}{(\sigma(x), \sigma(y))} = \mu\left(\frac{(p, q)}{(x, y)}\right).$$

Proof. Choose a point c in X and put $\sigma = T_A L(c, \lambda)$ where $A = \overrightarrow{c, \sigma(c)}$. We write L for $L(c, \lambda)$ and obtain

$$\frac{(\sigma(p), \sigma(q))}{(\sigma(x), \sigma(y))} = \frac{(T_A L(p), T_A L(q))}{(T_A L(x), T_A L(y))} = \frac{(L(p), L(q))}{(L(x), L(y))}$$

since T_A is an affine transformation. Furthermore,

$$\frac{(L(p), L(q))}{(L(x), L(y))} = \mu\left(\frac{(p,q)}{(x,y)}\right),$$

by Proposition 77.1 which gives the desired equality. Done.

The relationship between affine transformations of X and linear automorphisms of V is given by

Corollary 87.1.   The semiaffine transformation σ of X is affine if and only if its associated semilinear automorphism λ is a linear automorphism of V.

Proof. According to the above theorem, σ is affine if and only if

$$\frac{(p,q)}{(x,y)} = \mu\left(\frac{(p,q)}{(x,y)}\right)$$

for every pair of parallel line segments (p,q) and (x,y) of X. This implies that μ is the identity automorphism of k which is the same as to say that λ is linear. Done.

Exercises

2. Let k be ordered. Prove that every affine transformation σ of X preserves betweenness. Precisely, if the point z lies between the points x and y, prove that $\sigma(z)$ lies between the points $\sigma(x)$ and $\sigma(y)$.
3. Let $k = \mathbf{Q}(\sqrt{2})$ where $\mathbf{Q}$ is the field of rational numbers. (This means $k = \{a + b\sqrt{2}\,|\,a, b \in \mathbf{Q}\}$ with addition and multiplication defined as follows:

$$(a + b\sqrt{2}) + (c + d\sqrt{2}) = (a + c) + (b + d)\sqrt{2},$$
$$(a + b\sqrt{2})(c + d\sqrt{2}) = (ac + 2bd) + (ad + bc)\sqrt{2}.)$$

Consider the natural ordering of k as a subfield of $\mathbf{R}$. Let $\mu: k \to k$ be the automorphism of k which maps $a + b\sqrt{2}$ onto $a - b\sqrt{2}$.
   a. Prove that μ does not preserve the ordering of k; that is, find $s, t \in k$ such that $s < t$ but $\mu(t) < \mu(s)$.
   b. Assume σ is a semiaffine transformation of X whose associated

semilinear automorphism λ corresponds to the automorphism μ of k. Prove that σ does not preserve betweenness.

4. Assume k is ordered. Let σ be a semiaffine transformation of X whose associated semilinear automorphism λ corresponds to an automorphism of k which is again denoted by μ. If μ preserves the order of k, prove that σ preserves betweenness in X. (Note: We had proved in Exercise 51.9 that every dilation preserves betweenness because, in this case, μ is an inner automorphism of k which, of course, preserves the order of k.)

*5. Let σ be a semiaffine transformation of X with a corresponding semilinear transformation λ. Prove that $\lambda(\overrightarrow{x, y}) = \overrightarrow{\sigma(x), \sigma(y)}$ for $x, y \in X$.

6. We proved, with some pain in Exercise 69.7, that a semiaffine transformation σ preserves independence of points. Now give a painless proof of the same fact by putting $\sigma = T_A L(c, \lambda)$ and combining Exercise 36.3 with Exercise 5 above.

19. SEMIAFFINE TRANSFORMATIONS OF LINES

In this section, X is a line, that is, $n = 1$. We have never defined semiaffine transformations for this case. The definition of affine transformation of a line (Definition 71.1) together with material of the previous section guides us to the correct definition.

Definition 88.1.   A **semiaffine transformation** is a permutation σ of X for which there exists an automorphism μ of k satisfying:

If $x \neq y$ and $p \neq q$ are points of X, then

$$\frac{(\sigma(p), \sigma(q))}{(\sigma(x), \sigma(y))} = \mu\left(\frac{(p, q)}{(x, y)}\right).$$

Exercises

1. Assume that λ is a semilinear automorphism of V with respect to the automorphism μ of k. If $c \in X$, prove that $L(c, \lambda)$ is a semiaffine transformation of X with respect to the same automorphism μ of k and $L(c, \lambda)$ leaves c fixed.

2. Assume that σ is a semiaffine transformation of X and $c \in X$. Find a vector $A \in V$ and a semilinear automorphism λ of V such that

$\sigma = T_A L(c, \lambda)$. (Hint: Use the definition of semiaffine to prove that the vector $\sigma(x), \overrightarrow{\sigma(Bx)}$ does not depend on the point x.)

3. Prove that the vector A and the semilinear automorphism λ of Exercise 2 above are uniquely determined by σ and c.

4. a. Prove that the affine transformations of X are semiaffine.
   b. Prove that a semiaffine transformation of X is affine if and only if the λ of Exercise 2 above is a linear automorphism of V.

5. Prove that dilations of X are semiaffine.

6. If k has no automorphisms except for the identity automorphism, prove that the semiaffine transformations are the same as the affine transformations.

7. If $k = \mathbf{R}$ or $k = \mathbf{Q}$, prove that the semiaffine transformations are the same as the affine transformations.

8. If k has only two elements, prove that there are precisely two semi-affine transformations and both of these are also affine transformations.

9. The concepts of affine and semiaffine transformations can also be developed for different affine spaces. Suppose that (X, V, k) and (X', V', k') are affine spaces where V' is a vector space over a division ring k'.
   a. Define a semiaffine transformation from X onto X'. (Hint: Such a transformation is a function from X onto X' which satisfies conditions almost identical to those of Definitions 65.1 or 88.1.)
   b. Assume that $k' = k$ and define an affine transformation from X onto X'. Observe that, if $k' \neq k$, the notion of affine transformation does not make sense.
   c. Assume that $k' = k$. Two affine spaces are called **isomorphic** if there exists an affine transformation from one onto the other. Prove that two affine spaces are isomorphic if and only if they have the same dimension. (Hint: X and X' have the same dimension if and only if there exists a linear isomorphism λ from V onto V'. Given λ, choose points $c \in X$ and $c' \in X'$ and study the mapping $u: X \to X'$ defined by

$$u(x) = (\lambda(\overrightarrow{c, x}))c' \qquad \text{for all} \qquad x \in X.$$

Similarly, let $u: X \to X'$ be an affine transformation. Choose a point c in X and study the mapping $\lambda: V \to V'$ defined by

$$\lambda A = \overrightarrow{u(c), u(Ac)} \qquad \text{for all} \qquad A \in V.)$$

   d. As much as possible, extend the theory of semiaffine and affine transformations of X to semiaffine and affine transformations from X to X'.

From this point on, when we speak of semiaffine transformations and no remark is made about the dimension, it is assumed that $n \geq 1$.

20. INTERRELATION AMONG THE GROUPS ACTING ON X AND ON V

The group Sa of semiaffine transformations, the group Af of affine transformations and the group Tr of translations all act on X. Of course, Tr is a subgroup of Af and Af is a subgroup of Sa.

The group of semilinear automorphisms of V acts on V and we denote this group by Sem. The group of the linear automorphisms of V also acts on V and it is customary to denote this group by $GL(n, k)$ and to call it the " general linear group " of dimension n over the division ring k. Of course, $GL(n, k)$ is a subgroup of Sem. The homomorphisms which go from the groups acting on X onto the groups acting on V will now be studied. We recall that a homomorphism from a group G *onto* a group H is called an epimorphism.

In Section 18, we were able to associate a semilinear automorphism λ of V with each semiaffine transformation σ of X. It is now more convenient to denote this λ by $\bar{\sigma}$; the mapping $\sigma \to \bar{\sigma}$ from Sa to Sem is denoted by ϕ.

Proposition 90.1. The mapping $\phi \colon Sa \to Sem$ is an epimorphism which has the group Tr of translations as kernel. Consequently, Tr is a commutative normal subgroup of Sa and $Sa/Tr \simeq Sem$.

Proof. 1. ϕ is a homomorphism. We choose $\sigma, \tau \in Sa$ and have to show that $\overline{\sigma\tau} = \bar{\sigma}\bar{\tau}$. If $x \in X$ and $A \in V$, $\overline{\sigma\tau}$ is defined by

$$\overline{\sigma\tau}(A) = \overrightarrow{(\sigma\tau)(x), (\sigma\tau)(Ax)} = \overrightarrow{\sigma(\tau(x)), \sigma(\tau(Ax))}.$$

Furthermore, $\bar{\tau}(A) = \overrightarrow{\tau(x), \tau(Ax)}$ from which we conclude that $(\bar{\tau}(A))\tau(x) = \tau(Ax)$ and hence $\sigma[(\bar{\tau}(A))\tau(x)] = \sigma(\tau(Ax))$. We now use the point $\tau(x)$ to compute $\bar{\sigma}(\bar{\tau}(A))$ and obtain

$$\bar{\sigma}(\bar{\tau}(A)) = \overrightarrow{\sigma(\tau(x)), \sigma[\bar{\tau}(A)\tau(x)]} = \overrightarrow{\sigma(\tau(x)), \sigma(\tau(Ax))}.$$

This shows that $\overline{\sigma\tau}(A) = \bar{\sigma}(\bar{\tau}(A))$ and hence $\overline{\sigma\tau} = \bar{\sigma}\,\bar{\tau}$.

2. ϕ maps Sa onto Sem. Let $\lambda \in Sem$. We choose $c \in X$ and put $\sigma = L(c, \lambda) \in Sa$. In order to show that $\lambda = \bar{\sigma}$, we select a vector $A \in V$ and have

$\bar{\sigma}(A) = \overrightarrow{\sigma(c), \sigma(Ac)}$. Furthermore, $\sigma(c) = L(c, \lambda)(c) = c$, and

$$\sigma(Ac) = L(c, \lambda)(Ac) = (\lambda(\overrightarrow{c, Ac}))c = \lambda(A)c.$$

It follows that

$$\bar{\sigma}(A) = \overrightarrow{c, \lambda(A)c} = \lambda(A)$$

which shows that $\bar{\sigma} = \lambda$.

3. The kernel K of ϕ is the group Tr. If $A \in V$, we first show that the translation T_A belongs to K, equivalently, that $\bar{T}_A = 1_V$ (the identity mapping of V). If $B \in V$ and $x \in X$,

$$\bar{T}_A(B) = \overrightarrow{T_A(x), T_A(Bx)} = \overrightarrow{Ax, B(Ax)} = B.$$

This shows that $\bar{T}_A = 1_V$ and hence $Tr \subset K$.

Conversely, let $\sigma \in K$. We choose a point $x \in X$ and put $A = \overrightarrow{x, \sigma(x)}$. In order to show that $\sigma = T_A$, we choose a second point $y \in X$ and prove that $\sigma(y) = T_A(y)$, equivalently, that $\overrightarrow{y, \sigma(y)} = A$. Let $B = \overrightarrow{x, y}$; see Figure 91.1.

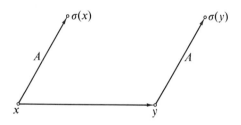

Figure 91.1

Since $\bar{\sigma} = 1_V$,

$$B = \bar{\sigma}(B) = \overrightarrow{\sigma(x), \sigma(Bx)} = \overrightarrow{\sigma(x), \sigma(y)}.$$

Furthermore,

$$\overrightarrow{x, \sigma(y)} = \overrightarrow{x, \sigma(x)} + \overrightarrow{\sigma(x), \sigma(y)} = A + B$$

and

$$\overrightarrow{x, \sigma(y)} = \overrightarrow{x, y} + \overrightarrow{y, \sigma(y)} = B + \overrightarrow{y, \sigma(y)}.$$

We conclude that $A + B = B + \overrightarrow{y, \sigma(y)}$ and hence $A = \overrightarrow{y, \sigma(y)}$. Therefore, $\sigma = T_A$ which shows that $K \subset Tr$.

Thus far, we have shown that ϕ is an epimorphism with kernel Tr; it now follows from general principles of group theory that Tr is a normal subgroup of Sa and $Sa/Tr \simeq Sem$. We have previously established that Tr is commutative. Done.

We denote the restriction of the epimorphism ϕ to Af by ϕ_0 and show in the next proposition that ϕ_0 is an epimorphism from Af onto $GL(n, k)$. The maps ϕ and ϕ_0 are pictured in Figure 92.1. It is customary that arrows with double arrowheads indicate epimorphisms and that vertical lines indicate inclusions of the lower group in the upper group.

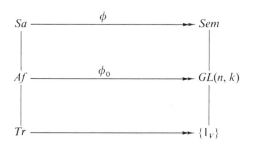

Figure 92.1

Since the group Tr is normal in Sa, the translation group is certainly normal in the smaller group Af. This fact is also contained in the following proposition.

Proposition 92.1. ϕ_0 is an epimorphism from the group Af onto the group $GL(n, k)$ which has the group Tr of translations as kernel. Consequently, Tr is a commutative normal subgroup of Af and $Af/Tr \simeq GL(n, k)$.

Proof. According to Corollary 87.1, if $\sigma \in Sa$, $\phi(\sigma) \in GL(n, k)$ if and only if $\sigma \in Af$. This is the same as saying that $Af = \phi^{-1}(GL(n, k))$. Since ϕ_0 is the restriction of ϕ to Af, we conclude that ϕ_0 is an epimorphism from Af onto $GL(n, k)$ with kernel Tr. We already know that Tr is a commutative group. Finally, the isomorphism $Af/Tr \simeq GL(n, k)$ and the fact that Tr is a normal subgroup of Af follow from the general principles of group theory. Done.

Exercises

1. Let $\lambda \in Sem$. Then there is an automorphism μ of k such that $\lambda(tA) = \mu(t)\lambda(A)$ for all $A \in V$ and $t \in k$. Prove that μ is uniquely determined by λ. This requires the convention that dim $V \neq 0$ which was made at the end of Section 19.
2. Prove that the automorphisms of k form a group (under the composition of functions).

3. Denote the group of automorphisms of k by *Aut k*. By Exercise 1
   above there is a mapping
   $$h: Sem \rightarrow Aut\ k$$
   which associates to each $\lambda \in Sem$ the automorphism μ of k with the
   property that $\lambda(tA) = \mu(t)\lambda(A)$ for all $A \in V$ and $t \in k$.
   a. Prove that $h: Sem \rightarrow Aut\ k$ is an epimorphism.
   b. Prove that $GL(n, k)$ is the kernel of h.
   c. Prove that $GL(n, k)$ is a normal subgroup of *Sem*.
   d. Prove that *Af* is a normal subgroup of *Sa*. (Hint: Consider Figure
      92.1; since $GL(n, k)$, is normal in *Sem* and $Af = \phi^{-1}(GL(n, k))$,
      the fact that *Af* is normal in *Sa* follows from a general principle of
      group theory.)
   e. Observe that, in Figure 92.1, every group is normal in every larger
      group.
*4. a. Assume $\phi: G \rightarrow G'$ is an epimorphism of groups. If G is commuta-
      tive, prove that G' is commutative.
   b. If $n \geq 2$, prove that the groups *Sa* and *Af* are not commutative.
      (Hint: From linear algebra, we know that, for $n \geq 2$, $GL(n, k)$ is
      not commutative.)

21. DETERMINATION OF AFFINE TRANSFORMATIONS BY INDEPENDENT POINTS AND BY COORDINATES

A linear transformation of the n-dimensional vector space V is completely
determined by its action on n linearly independent vectors of V. An analogous
statement holds for the affine transformations of our affine space (X, V, k).

Proposition 93.1.   Let $x_0, \ldots, x_n$ and $y_0, \ldots, y_n$ be two sets of $n + 1$
linearly independent points of X. Then there exists a unique affine transformation σ of X with the property that $\sigma(x_i) = y_i$ for $i = 0, 1, \ldots, n$.

Proof. 1.   *Uniqueness.* Suppose that σ is an affine transformation which sends
x_i to y_i for $i = 0, \ldots, n$. Let λ be the linear automorphism of V which corresponds to σ and put $A = \overrightarrow{x_0, y_0}$. Then $\sigma = T_A L(x_0, \lambda)$ (Theorem 86.1).
Hence we only have to prove that λ is uniquely determined by the two sets of
independent points $x_0, \ldots, x_n$ and $y_0, \ldots, y_n$. We know that the n vectors
$\overrightarrow{x_0, x_1}, \ldots, \overrightarrow{x_0, x_n}$ are linearly independent because $x_0, \ldots, x_n$ are independent points (Exercise 36.3); for the same reason, the n vectors $\overrightarrow{y_0, y_1}, \ldots,$

$\overrightarrow{y_0,y_n}$ are linearly independent. By Exercise 88.5, $\lambda(\overrightarrow{x_0,x_i}) = \overrightarrow{y_0,y_i}$; consequently, λ is the unique linear transformation of V which sends the n linearly independent vectors $\overrightarrow{x_0,x_1}, \ldots, \overrightarrow{x_0,x_n}$ onto, respectively, the independent vectors $\overrightarrow{y_0,y_1}, \ldots, \overrightarrow{y_0,y_n}$.

2. *Existence.* Let A and λ be defined as in the uniqueness part of the proof. There only remains to be shown that $T_A L(x_0, \lambda)(x_i) = y_i$ for $i = 0, \ldots, n$. Clearly,

$$L(x_0, \lambda)(x_i) = (\lambda(\overrightarrow{x_0,x_i}))x_0 = (\overrightarrow{y_0,y_i})x_0$$

and hence

$$T_A L(x_0, \lambda)(x_i) = A((\overrightarrow{y_0,y_i})x_0) = \overrightarrow{y_0,y_i}(Ax_0) = (\overrightarrow{y_0,y_i})y_0 = y_i.$$

Done.

An affine transformation σ sends $n + 1$ independent points onto $n + 1$ independent points (Exercise 69.7). As a consequence of the previous proposition, an affine transformation is completely known by its action on $n + 1$ independent points $x_0, \ldots, x_n$ of X. In other words, if σ and τ are affine transformations such that $\sigma(x_i) = \tau(x_i)$ for $i = 0, \ldots, n$, then $\sigma = \tau$. Since the identity function 1_X of X is an affine transformation leaving all the points fixed, we have:

Corollary 94.1. If an affine transformation σ of X leaves $n + 1$ independent points of X fixed, then $\sigma = 1_X$.

Exercises

1. Let $x_0, \ldots, x_n$ and $y_0, \ldots, y_n$ be two sets of $n + 1$ independent points of X and let μ be an automorphism of k. Prove that there exists a unique semiaffine transformation σ of X such that $\sigma(x_i) = y_i$ for $i = 0, \ldots, n$ and the semilinear automorphism of V associated with σ corresponds to the automorphism μ of k.

2. Assume that k has an automorphism μ which is not the identity automorphism. Let $x_0, \ldots, x_n$ be $n + 1$ independent points of X. Prove that there exists a semiaffine transformation of X which leaves the points $x_0, \ldots, x_n$ fixed but is different from 1_X.

3. Let $x_1, \ldots, x_d$ and $y_1, \ldots, y_d$ be two sets of d independent points of X and hence $1 \le d \le n + 1$.
   a. If μ is an automorphism of k, prove that there is at least one semiaffine transformation σ of X such that $\sigma(x_i) = y_i$ for $i = 1, \ldots, d$ and the semilinear automorphism of V associated with σ corresponds to the automorphism μ of k.
   b. Prove that there exists at least one affine transformation σ of X such that $\sigma(x_i) = y_i$ for $i = 1, \ldots, d$.
   c. If $d < n + 1$ and X is not the line over the field with two elements, prove that the σ's in a and b are not unique.

We now derive the equations of an affine transformation σ in terms of a coordinate system $c, A_1, \ldots, A_n$ of X. For this, we put $\sigma = T_B L(c, \lambda)$ where $B = c, \sigma(c)$ and λ is the linear automorphism of V associated with σ. If x is a point of X, the coordinates $(x_1, \ldots, x_n)$ of x are, by definition, the coordinates of the vector $\overrightarrow{c, x}$ which means that $\overrightarrow{c, x} = x_1 A_1 + \cdots + x_n A_n$. The notation $x = (x_1, \ldots, x_n)$ is used. We put

$$\lambda(A_i) = a_{1i} A_1 + \cdots + a_{ni} A_n$$

for $i = 1, \ldots, n$ and observe that

$$\lambda(\overrightarrow{c, x}) = \left(\sum_{i=1}^{n} x_i a_{1i} \right) A_1 + \cdots + \left(\sum_{i=1}^{n} x_i a_{ni} \right) A_n.$$

Hence the coordinates of the point $L(c, \lambda)(x) = (\lambda(\overrightarrow{c, x}))c$ are

$$\left(\sum_{i=1}^{n} x_i a_{1i}, \ldots, \sum_{i=1}^{n} x_i a_{ni} \right).$$

We now compute the coordinates of the point

$$y = \sigma(x) = T_B(c, \lambda)(x) = T_B[((\lambda(\overrightarrow{c, x}))c] = (B + \lambda(\overrightarrow{c, x}))c.$$

These coordinates are, by definition, the coordinates of the vector $B + \lambda(\overrightarrow{c, x})$ and hence if $B = (b_1, \ldots, b_n)$ and $y = (y_1, \ldots, y_n)$,

$$y_1 = x_1 a_{11} + x_2 a_{12} + \cdots + x_n a_{1n} + b_1$$
$$\vdots$$
$$y_n = x_1 a_{n1} + x_2 a_{n2} + \cdots + x_n a_{nn} + b_n.$$

In the above system of equations, the "constants" $b_1, \ldots, b_n$ are the coordinates of the point $\sigma(c)$. (See Figure 96.1.) We see that the coordinates $(y_1, \ldots, y_n)$ of the point $y = \sigma(x)$ are expressed in terms of the coordinates $(x_1, \ldots, x_n)$ of the point x by means of "right" linear equations. If k is not

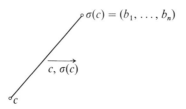

Figure 96.1

commutative, the coefficients a_{ij} must be written on the right side of the variables $x_1, \ldots, x_n$. If k is commutative, these coefficients may be written on either side of the variables.

The right linear equations which express the coordinates of the point $y = \sigma(x)$ in terms of the coordinates of the point x are the most general kind of right linear equations. Any system of n such equations, subject only to the condition that the $n \times n$ matrix (a_{ij}) represent a linear automorphism λ of V, describes the affine transformation $T_B L(c, \lambda)$ of X where B is the vector with coordinates $(b_1, \ldots, b_n)$. If k is commutative, the condition that the matrix (a_{ij}) represents a linear automorphism of V is equivalent to saying that the determinant of (a_{ij}) is not zero. If k is not commutative, the same is true, but the notion of determinant is now more complicated. (See Artin [2], pp. 151–158.)

Example 1

Let $n = 2$ and c, A_1, A_2 be a coordinate system for X. Consider the points $p_0 = (-1, 1), p_1 = (-2, 3), p_2 = (0, 3), q_0 = (1, 2), q_1 = (1, 3)$ and $q_2 = (5, 1)$. We ask the reader to verify that the two sets of points $\{p_0, p_1, p_2\}$ and $\{q_0, q_1, q_2\}$ are independent. Hence there is a unique affine transformation σ

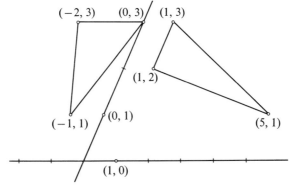

Figure 96.2

of X such that $\sigma(p_i) = q_i$ for $i = 1, 2$. The situation is pictured in Figure 96.2. What are the equations of σ? There is really not much else we can do but solve the three systems of equations

$$1 = -a_{11} + a_{12} + b_1$$
$$2 = -a_{21} + a_{22} + b_2$$

$$1 = -2a_{11} + 3a_{12} + b_1$$
$$3 = -2a_{21} + 3a_{22} + b_2$$

$$5 = 3a_{12} + b_1$$
$$1 = 3a_{22} + b_2.$$

The solution is $a_{11} = 2$, $a_{12} = 1$, $a_{21} = -1$, $a_{22} = 0$, $b_1 = 2$, and $b_2 = 1$. Hence the equations of σ are

$$y_1 = x_1 2 + x_2 + 2$$
$$y_2 = -x_1 + 1.$$

How did we solve the above system of six equations in six unknowns? Of course, by the high school method of elimination of variables. Whenever a system of linear equations is simple enough to be solved by hand, the high school method is by far the quickest. For instance, this very efficient method is much faster than the use of Kramer's rule or of triangular matrices. If the coefficients a_{ij} and the constants b_i of the system are real numbers and are so complicated that the system cannot be solved any longer by the high school method, computing machines must be used. The whole problem of solving equations then becomes more one of analysis than of algebra.

Exercises

4. Let $n = 3$ and c, A_1, A_2, A_3 be a coordinate system for X. Consider the points

$$p_0 = (1, 0, 1), \quad p_1 = (0, 1, 1), \quad p_2 = (1, 0, 0), \quad p_3 = (-1, 1, 2)$$
$$q_0 = (1, 2, 1), \quad q_1 = (-1, 2, 0), \quad q_3 = (0, 3, 4), \quad q_4 = (1, -2, 5).$$

a. Prove that the two sets $\{p_0, p_1, p_2, p_3\}$ and $\{q_0, q_1, q_2, q_3\}$ both consist of four independent points.

b. Find the equations of the affine transformation σ with the property that $\sigma(p_i) = q_i$ for $i = 0, 1, 2, 3$.

5. Let k be commutative and $c, A_1, \ldots, A_n$ be a coordinate system of X. From the equations of an affine transformation, it is clear that a complete classification of all affine transformations comes down to writing the nonsingular $n \times n$ matrices with elements in k as products

of matrices of "simplest possible form." These simplest matrices are the so-called elementary matrices (see MacLane and Birkhoff, [16], p. 280). If this program is carried out for the real affine plane, one comes to the conclusion that affine transformations of the plane are products of translations, reflections, shears, compressions, and elongations. We urge the reader to study such a development (see MacLane and Birkhoff, [16], pp. 279–284).

22. THE THEOREM OF DESARGUES

We assume in this whole section that $n \geq 2$. A triangle xyz is determined by three noncollinear points x, y, and z of X which are called the vertices of the triangle; the sides of the triangle are the line segments $\{x, y\}$, $\{y, z\}$ and $\{z, y\}$. If xyz and $x'y'z'$ are two triangles, there always exists an affine transformation σ of X such that $\sigma(x) = x'$, $\sigma(y) = y'$ and $\sigma(z) = z'$. This follows from Exercise 95.3 since three points are noncollinear if and only if they are independent. We also know that σ is not unique unless $n = 2$.

As we shall explain later on, the theorem of Desargues really belongs in projective geometry and not in affine geometry. The "affine part" of the theorem is concerned with the question of when there exists a dilation which sends the vertices of triangle xyz onto those of the triangle $x'y'z'$. Since a dilation sends every line onto a parallel line, it is obvious that such a dilation can only exist if the sides of the two triangles are parallel in pairs; see Figure 98.1. (Most figures in this section have been drawn for the plane even though we are not assuming that $n = 2$.)

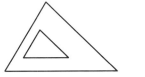

Figure 98.1

The following proposition states that this condition is also sufficient in order that the triangles be interrelated by a dilation. The symbol $\parallel$ will be used for parallel line segments as well as for parallel lines.

Proposition 98.1. Let xyz and $x'y'z'$ be two triangles. There exists a dilation D such that

$$D(x) = x', \quad D(y) = y', \quad D(z) = z'$$

if and only if

$$\{x, y\} \parallel \{x', y'\}, \qquad \{y, z\} \parallel \{y', z'\}, \qquad \{z, x\} \parallel \{z', x'\}.$$

If the sides satisfy this condition, the dilation D is unique.

Proof. If a dilation with the required properties exists, it is obvious that the sides of the triangle satisfy the stated condition. Now assume that the sides satisfy this condition.

Since $\{x, y\} \parallel \{x', y'\}$, there exists a unique dilation D such that $D(x) = x'$ and $D(y) = y'$. We must yet prove $D(z) = z'$. Because $\{z\} = (y \vee z) \cap (z \vee x)$ and D is one-to-one, we have $\{D(z)\} = D(y \vee z) \cap D(z \vee x)$. Furthermore, $D(y \vee z)$ is the line through y' which is parallel to $y \vee z$ or, equivalently, $D(y \vee z) = y' \vee z'$. Similarly, $D(z \vee x) = z' \vee x'$ and we conclude that

$$\{D(z)\} = (y' \vee z') \cap (z' \vee x') = \{z'\}.$$

This shows that $D(z) = z'$. Done.

When two triangles xyz and $x'y'z'$ are related by a dilation D as in the above proposition, the lines $x \vee x'$, $y \vee y'$, and $z \vee z'$ enjoy a special property.

Proposition 99.1. Let xyz and $x'y'z'$ be two triangles which have no vertex in common. Suppose there exists a dilation D with the property that $D(x) = x'$, $D(y) = y'$, and $D(z) = z'$. Then the three lines $x \vee x'$, $y \vee y'$, and $z \vee z'$ are either parallel or have a point in common.

Proof. The three lines $x \vee x'$, $y \vee y'$, and $z \vee z'$ are traces of D (Exercise 44.12). (We recall that a line l is a trace of D if $D(l) = l$.) Hence, if D is a translation, these three lines are parallel and, if D is a magnification, they pass through the center of the magnification (Exercise 46.16). Done.

The following surprising corollary of Propositions 98.1 and 99.1 needs no proof. It constitutes the affine part of Desargues' theorem.

Corollary 99.1 (the affine part of the theorem of Desargues). Let xyz and $x'y'z'$ be two triangles which have no vertex in common. Assume that $\{x, y\} \parallel \{x' \ y'\}$, $\{y, z\} \parallel \{y', z'\}$, and $\{z, x\} \parallel \{z', x'\}$. Then the three lines $x \vee x'$, $y \vee y'$, and $z \vee z'$ are either parallel or have a point in common.

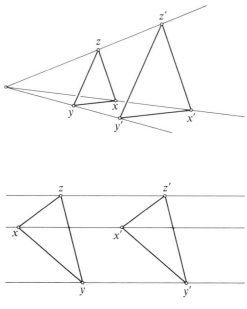

Figure 100.1

Figure 100.1 illustrates the theorem of Desargues.

Figure 100.2 shows that the converses of Proposition 99.1 and Corollary 99.1 are false.

The only way to obtain a good feeling for the surprise element in the theorem of Desargues is by drawing pairs of triangles whose sides are

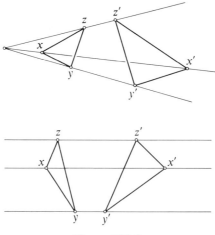

Figure 100.2

parallel in pairs, but which are in as unlikely positions as possible. Then one completes the drawing and observes that in each case the theorem of Desargues holds. Figure 101.1 shows two such pictures.

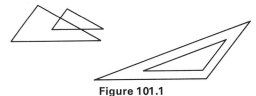

Figure 101.1

Exercises

1. Figure 101.2 displays two parallelograms whose sides are parallel in pairs, but the diagonal $\{x, z\}$ is not parallel to the diagonal $\{x', z'\}$.

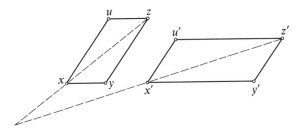

Figure 101.2

a. Prove that there is no dilation which maps the vertices of parallelogram $xyzu$ onto those of parallelogram $x'y'z'u'$. Conclude that the theorem of Desargues does not hold for quadrilaterals in the plane.

b. Draw some pictures which show that the theorem of Desargues does not hold for plane polygons with more than three sides. We shall see in the next exercise that this can be done because the vertices of such a polygon are dependent. (The figure below refers to Exercise 2.)

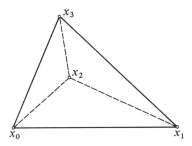

Figure 101.3

2. A **simplex** of X is determined by $n + 1$ independent points $x_0, \ldots, x_n$ of X which are called the **vertices of the simplex**; the **sides of the simplex** are the $\binom{n+1}{2}$ line segments $\{x_i, x_j\}$ for $0 \le i < j \le n$. In the plane, the triangles are the simplexes and, when $n = 3$, the tetrahedrons are the simplexes (see Figure 101.3). Prove that the theorem of Desargues holds for simplexes. Precisely, formulate and prove Propositions 98.1 and 99.1 and Corollary 99.1 for simplexes.

3. Let a, b, c be distinct lines of X which are either parallel or meet in a point p. Let x, x' be points on a, y, y' be points on b and z, z' be points on c which are distinct from p if the lines a, b, and c meet. Assume, furthermore, that $x \vee y \parallel x' \vee y'$ and $y \vee z \parallel y' \vee z'$. (See Figure 102.1.)

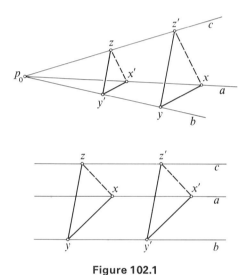

Figure 102.1

Prove that $x \vee z \parallel x' \vee z'$. (This is Artin's version of Desargues' theorem, tailored to the geometric axioms which he uses for the affine plane; see Artin, [2], Theorem 2.15, p. 70.) However, we are not assuming that $n = 2$, even though the pictures have been drawn for that case.

Side Remark on the Projective Plane. This remark is meant only for readers who have some knowledge of the projective plane. Let xyz and $x'y'z'$ be two triangles in the projective plane P. We assume that these triangles have no vertex in common and, dually, that the six lines $a = x \vee y$, $b = y \vee z$, $c = z \vee x$, $a' = x' \vee y'$, $b' = y' \vee z'$ and $c' = z' \vee x'$ are mutually distinct. We recall that, in P, two distinct lines always have a unique point in common. The theorem of Desargues for the projective plane states that the following two conditions are equivalent.

1. The three points of intersection $a \cap a'$, $b \cap b'$ and $c \cap c'$ lie on a line.
2. The three lines $x \vee x'$, $y \vee y'$ and $z \vee z'$ have a point in common.

We indicate how this famous theorem follows immediately from Corollary 99.1 by general principles of projective geometry. If Condition 1 holds, we consider the affine plane for which the (projective) line containing the three points is the line at infinity. In this plane, the "corresponding" sides of the two triangles are parallel and this corollary states precisely that Condition 2 holds. By dualizing the just proven implication $1 \Rightarrow 2$, we obtain the implication $2 \Rightarrow 1$; therefore, the two conditions are equivalent.

In the two pictures of Figure 100.2, interpreted in the projective plane, Condition 2 is satisfied. The reader should complete these pictures and convince himself that Condition 1 is also satisfied.

By now, the reader has surely seen that the conclusion of Corollary 99.1 is nothing but the affine formulation of Condition 2. The reader should also write the affine formulation of Condition 1 and, thereby, obtain a most remarkable if and only if statement for the affine plane. It contains much more than Corollary 99.1. This concludes our remarks on the projective plane.

Gérard Desargues (1593–1662) was a French architect and scientist who lived in Lyon. He was about 200 years ahead of his time when, in 1639, he published some of the fundamental concepts of projective geometry; his theorem of Desargues was published in 1648. The actual birth of projective geometry is usually considered to have occurred in 1822 when the French engineer of Napoleon's armies, Jean Victor Poncelet (1788–1868), published his *Traité des propriétés projectives des figures*, [19]. (See D. J. Struik, *A Concise History of Mathematics*, [21].)

23. THE THEOREM OF PAPPUS

We assume throughout this section that $n = 2$. The theorem of Pappus deals with hexagons. Nowadays, the interest in the theorem does not lie in hexagons, but in the fact that this theorem holds if and only if the division ring k is a field. Pappus, who lived around 300 or 400 A.D. in Alexandria, never dreamed about noncommutative division rings. His geometry was based exclusively on the real numbers and there his theorem holds. As in the case of the theorem of Desargues, the theorem of Pappus basically belongs in projective geometry. The reasons for this will be explained toward the end of the section, but first we shall discuss the affine part of the theorem.

A **hexagon** $xx'yy'zz'$ is determined by an *ordered* set of distinct points x, x', y, y', z, z' of X which are called the **vertices of the hexagon**. The **sides of the hexagon** are the six line segments $\{x, x'\}$, $\{x', y\}$, $\{y, y'\}$, $\{y', z\}$, $\{z, z'\}$, and $\{z', x\}$. Observe that the sides of the hexagon are defined in terms of the ordering of the vertices and that no restriction has been placed on the relative position of the six vertices; they may all lie on a line. The three pairs of opposite sides of the hexagon are $\{x, x'\}$ and $\{y', z\}$, $\{x', y\}$ and $\{z', z\}$, $\{y, y'\}$ and $\{z', x\}$. In Figure 104.1 the three pairs of opposite sides are designated by corresponding shades of grey.

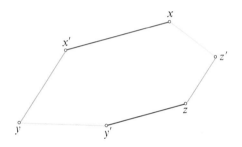

Figure 104.1

In the hexagon of Figure 104.1, the pair of opposite sides in black and the pair of opposite sides in medium grey are parallel, but the pair of opposite sides in light grey are not parallel. The affine part of the theorem of Pappus states that this cannot happen in degenerate hexagons and we shall see that this theorem is correct if and only if k is a field. We call the hexagon

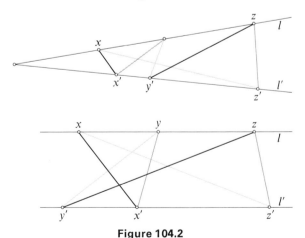

Figure 104.2

$xx'yy'zz'$ **degenerate** if both sets of points x, y, z and x', y', z' lie on lines. Although we do not exclude the case where these two lines coincide and hence the six vertices are collinear, such high degeneracy is of no interest. Figure 104.2 depicts two degenerate hexagons with the three pairs of opposite sides again colored black, grey, and light grey. The reader should draw several more degenerate hexagons for himself, always pointing out the pairs of opposite sides.

We now formulate the affine part of Pappus' theorem and assert its equivalence with the commutativity of k. Observe that, if one restricts oneself to affine geometry over fields, as in Pappus' time, the theorem reduces to Statement 1.

Theorem 105.1. The following two statements are equivalent.

1. **(the affine part of the theorem of Pappus)** If two of the pairs of opposite sides of a degenerate hexagon consist of parallel line segments, the third pair also consists of parallel line segments.
2. k is a field.

Before giving the proof, we draw two degenerate hexagons (Figures 105.1 and 106.1), each having the property that two pairs of opposite sides consist of parallel line segments (the black and the grey pair). Observe that

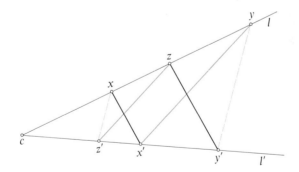

Figure 105.1

the light dashed pair of opposite sides are also parallel. Of course, the reader should repeat this experiment many times and observe in each case that the "remaining" pair of opposite sides are parallel. Mathematics is a very experimental science!

Proof of Theorem 105.1. Assume that k is a field and let us prove Statement 1. Assume $xx'yy'zz'$ is a degenerate hexagon with the property that

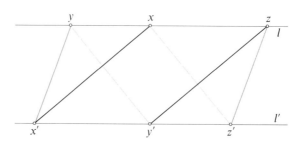

Figure 106.1

$\{x, x'\} \parallel \{y', z\}$ and $\{x', y\} \parallel \{z, z'\}$; see Figures 105.1 and 106.1. We must prove that $\{y, y'\} \parallel \{z', x\}$. We denote the line which carries the points x, y, z by l and the line which carries the points x', y', z' by l'. If $l = l'$, it is trivial that $\{y, y'\} \parallel \{z', x\}$; hence we assume that $l \neq l'$.

Case 1. l and l' are not parallel. Since $n = 2$, l and l' have a unique point c in common and we have the situation represented by Figure 105.1.

We first prove that none of the six vertices coincides with the point c. Suppose $y = c$. Then $x' \vee y = l'$ and, since $\{x', y\} \parallel \{z, z'\}$, it follows that $l' \parallel z \vee z'$ and hence $l' = z \vee z'$. This implies both c and z lie on l and l' and, therefore, $l = l'$, contrary to assumption. One argues similarly for the other vertices.

Since $c \neq x$ and $c \neq z$, there exists a magnification $M(c, r)$ such that $M(c, r)(x) = z$. For similar reasons, there exists a magnification $M(c, s)$ such that $M(c, s)(z') = x'$. Let us prove that $M(c, sr)(x) = y$ and $M(c, rs)(z') = y'$. Since $\{x', y\} \parallel \{z, z'\}$ and l is a trace of $M(c, s)$, the point $M(c, s)(z)$ lies on both l and $x' \vee y$; hence $M(c, s)(z) = y$. (We leave it for the reader to show that $l \cap (x' \vee y) = \{y\}$, something which is suggested, but not proved, by the dangerous Figure 105.1.) Consequently,

$$y = M(c, s)(M(c, r)(x)) = M(c, sr)(x).$$

One shows, in the same way, that

$$y' = M(c, r)(M(c, s)(z')) = M(c, rs)(z').$$

We now use, for the first time, the fact that k is a field. Therefore, $sr = rs$ and hence the magnifications $M(c, rs)$ and $M(c, sr)$ are one and the same. This magnification maps the point z' on the point y' and x on y which shows that $\{y, y'\} \parallel \{z', x\}$.

Case 2. l and l' are parallel. The reasoning is entirely similar to the one given in Case 1, except that the magnifications $M(c, r)$ and $M(c, s)$ are replaced by the translations $T_{\overrightarrow{x,z}}$ and $T_{\overrightarrow{z',x'}}$. We leave the details to the reader. It is interesting to observe that, since translations commute even when k is noncommutative, the conclusion $\{y, y'\} \parallel \{z', x\}$ is now correct *even if k is not commutative*. This completes the proof that Statement 2 implies Statement 1.

Assume that Statement 1 holds and let us prove that k is commutative. We choose $r, s \in k$ and show that $rs = sr$. This equality holds even if k is not commutative as long as $r = 0$ or $s = 0$ or $rs = 1$. Thus we assume that $r \neq 0$, $s \neq 0$ and $rs \neq 1$ (which implies that $sr \neq 1$). Since $n = 2$, X contains two lines l and l' which are not parallel and we denote their unique point of intersection by c. We choose a point $x \neq c$ on l and a point $z' \neq c$ on l'. Finally, we let $M(c, r)(x) = z$ and $M(c, s)(z') = x'$ (see Figure 107.1).

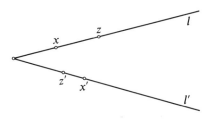

Figure 107.1

We first show that $z \in l$, $z \neq c$, $z \neq x$ and, similarly, that $x' \in l'$, $x' \neq c$, $x' \neq z'$. The point z lies on l because l is a trace of $M(c, r)$ while $x \neq c$ implies that $z \neq c$; finally, $z \neq x$ because $r \neq 1$. The same argument shows that $x' \in l'$, $x' \neq c$ and $x' \neq z'$.

Next we put $M(c, s)(z) = y$ and $M(c, r)(x') = y'$.

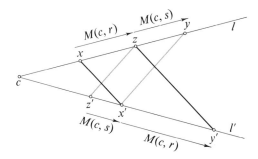

Figure 107.2

Now $y \in l$ because l is a trace of $M(c, s)$ while $z \neq c$ implies that $y \neq c$. The fact that $y \neq x$ follows from $rs \neq 1$ and $s \neq 1$ implies that $y \neq z$. The same argument shows that $y' \in l'$ and y' is not c, z', or x'.

We consider the degenerate hexagon $xx'yy'zz'$. We know that

$\{x, x'\} \parallel \{y', z\}$ because $M(c, r)$ is a magnification and $\{x', y\} \parallel \{z, z'\}$ because $M(c, s)$ is a magnification. Then, by Statement 1, we conclude that $\{y, y'\} \parallel \{z', x\}$.

Let us prove that

$$M(c, rs)(x) = M(c, sr)(x) = y.$$

By definition,

$$M(c, sr)(x) = M(c, s)(M(c, r)(x)) = M(c, s)(z) = y.$$

In the same way, one proves that $M(c, rs)(z') = y'$. We now consider the equality $M(c, rs)(x) = y$. For this, we observe that $\{x\} = (z' \vee x) \cap l$ and hence $M(c, rs)(x)$ is the point of intersection of l with the line l'' which passes through the point $M(c, rs)(z') = y'$ and is parallel to $z' \vee x$. Since $\{y, y'\} \parallel \{z', x\}$, $l = y \vee y'$, and $M(c, rs)(x) = y$.

The magnifications $M(c, rs)$ and $M(c, sr)$ have the same action on the two distinct points x and c; therefore, $M(c, rs) = M(c, sr)$ and hence $rs = sr$. Done.

Exercises

In the following exercises, $xx'yy'zz'$ stands for a degenerate hexagon where $\{x, x'\} \parallel \{y', z\}$ and $\{x', y\} \parallel \{z, z'\}$. The line which carries the points x, y, and z is denoted by l and the line which carries the points x', y', and z' by l'.

1. If $l \parallel l'$, prove that $\{y, y'\} \parallel \{z', x\}$ even if k is not commutative. Note that this is the same as giving the proof for Case 2 on page 106.
2. Assume that the lines $z \vee z'$ and $x \vee x'$ intersect at the point a and the lines $x' \vee y$ and $y' \vee z$ intersect at b. See Figure 108.1. If k is a field,

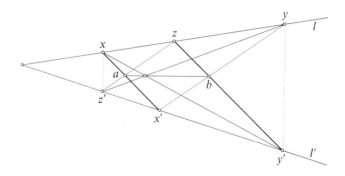

Figure 108.1

prove that the three lines $a \vee b$, $x \vee y'$ and $z' \vee y$ have a point in common or are parallel (whether or not $l \parallel l'$). If $l \parallel l'$, prove that the same result holds even if k is not commutative.

3. Assume that the lines (in Figure 109.1) $x \vee x'$ and $y \vee y'$ intersect at a point p and the lines $y' \vee z$ and $z' \vee x$ intersect at q. If k is a field, prove that the three lines $p \vee q$, $x' \vee z$ and $y \vee z'$ have a point in common

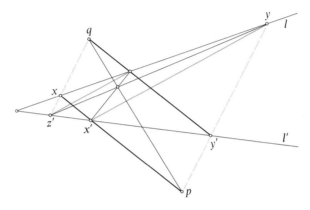

Figure 109.1

or are parallel (whether or not $l \parallel l'$). If $l \parallel l'$, prove that the same result holds even if k is not commutative.

We refer to Statement 1 of Theorem 105.1 as the theorem of Pappus. There is a very interesting unsolved problem connected with this theorem in the case that the affine plane X contains only finitely many points. Of course, this happens if and only if k is finite. Since Wedderburn's theorem states that finite division rings are fields, we have the following corollary of Theorem 105.1.

Corollary 109.1. If an affine plane contains only finitely many points, the theorem of Pappus holds.

The above corollary is a geometric formulation of Wedderburn's theorem. There exists no simple geometric proof of this corollary which avoids the use of Wedderburn's theorem. Such a proof would, of course, constitute a simple geometric proof of Wedderburn's theorem.

Side Remark on Associativity. The multiplication in a division ring is associative by definition. If we had based our geometry on nonassociative rings, we could have asked whether there is a geometric theorem which is equivalent to associativity of multiplication just as Pappus' theorem is equivalent to the commutativity of multiplication. It is fascinating that the theorem of Desargues plays this role; see R. H. Bruck, "Recent Advances in the Foundations of Euclidean Geometry," *The American Mathematical Monthly*, Vol. 62, No. 7, No. 4 of the Slaught Memorial Papers, pp. 2–17, [6].

Side Remark on the Projective Plane. This remark is meant only for readers with some knowledge of the projective plane P. Let $xx'yy'zz'$ be a degenerate hexagon in P where x, y, z lie on l and x', y', z' lie on l'. We assume $l \neq l'$ and put

$$\{p\} = (x \vee x') \cap (y' \vee z), \qquad \{q\} = (x' \vee y) \cap (z \vee z'),$$
$$\{r\} = (y \vee y') \cap (z' \vee x).$$

Hence the points p, q, and r are the points of intersection of the three pairs of lines which are determined by the three pairs of opposite sides of the hexagon.

The theorem of Pappus for the projective plane states that the points p, q, and r lie on a line. This theorem is correct if and only if k is a field and we indicate how this follows immediately from Theorem 105.1 by general principles of projective geometry. Let k be a field and denote the (projective) line through the points p and q by l''. In order to show that $r \in l''$, consider the affine plane for which l'' is the line at infinity. In this plane, the hexagon has the property that $\{x, x'\} \parallel \{y', z\}$ and $\{x', y\} \parallel \{z, z'\}$. By Theorem 105.1, one concludes that $\{y, y'\} \parallel \{z', x\}$ which is the same as saying that $r \in l''$. Conversely, suppose that the theorem of Pappus holds for the projective plane. In order to show that k is a field, remove any (projective) line from this plane. In the resulting affine plane, Statement 1 of Theorem 105.1 holds and hence k is a field.

The theorem of Pappus for the projective plane can, of course, be formulated for the affine plane. The resulting theorem for the affine plane is very surprising and contains much more than Statement 1 of Theorem 105.1. Of course, this affine theorem will hold if and only if k is a field. We urge the reader to carry out the investigation. This concludes our remarks on the projective plane.

Pappus wrote the last of the great mathematical works of the school at Alexandria. His *Mathematical Synagogues* (synagogue comes from the Greek word for collection) is the only source for a great mass of the Greek geometry of antiquity. The depth of his *Synagogues* is incredible. The seventh book contains a generalization of the theorem of Desargues on triangles to arbitrary plane polygons. If Poncelet in his *Traité* [19](Vol. I, Sec. 168, p. 68) had not credited the theorem on triangles to Desargues, it would probably have been called another Pappus' theorem. Although Pappus gives many historical notes in his *Synagogues*, it is hard to tell how old this theorem was already in his time. (See Hk De Vries, *Beknopt Leerboek Der Projectieve Meetkunde*, pp. 57–58, [9].)

chapter 2
metric vector spaces

Introduction. Affine geometry is the study of incidence and parallelism. We now want to introduce a "measuring rod" into affine space in order to study such metric concepts as distance, orthogonality, similarity, congruence, etc. As we shall see, there are many measuring rods that can be imposed on the same affine space. Although such a rod allows us to introduce new concepts into affine geometry, it is important to realize that the underlying affine structure of the space is in no way affected. In particular, even when the measuring rod is non-Euclidean, the fifth parallel axiom remains valid.

Of course, the measuring rod for our affine space (X, V, k) must be defined axiomatically and this will be done by choosing a symmetric bilinear function for the vector space V. Hence it will first be necessary to make a thorough study of vector spaces which have been provided with a symmetric bilinear function. Such vector spaces are called "metric vector spaces" and they are the subject of the present chapter. Once such notions as length of a vector, orthogonality of vectors, etc., have been studied in V, the transition to the space X is quite easy and will be carried out in Chapter 3.

24. INNER PRODUCTS

In the remainder of the book, k stands for a field with characteristic different from two. Hence k is commutative and every one of its elements may be divided by two. As before, V is an n-dimensional vector space over k where n is finite. Since k is commutative, scalars may be written either on the left or on the right of the vectors.

A **symmetric bilinear function** is a function from $V \times V$ into k which satisfies five axioms. The image of the ordered pair $(A, B) \in V \times V$ under the function is designated by AB. Using, again, capital letters to denote vectors and lower case letters to denote scalars, the five axioms are:

1. $A(B + C) = AB + AC.$
2. $(A + B)C = AC + BC.$
3. $(tA)B = t(AB).$
4. $A(tB) = t(AB).$
5. $AB = BA.$

Axioms 2 and 3 say that function $(A, B) \rightarrow AB$ is linear in the first variable; Axioms 1 and 4 assert that the function is linear in the second variable. Hence Axioms 1 through 4 say that the function is bilinear. Finally, Axiom 5 states that the function is symmetric.

A symmetric bilinear function is also called an **inner product** or a **scalar product** for V. A vector space, provided with an inner product, is called a **metric vector space**, a **vector space with metric** or even a **geometry**. The inner product is the measuring rod for V and the metric geometry one obtains depends on the choice of this inner product.

It is very important to adopt the geometric attitude toward metric vector spaces. This is done by taking our pictures and language from Euclidean geometry. First of all, the vectors of V are viewed as arrows with the same initial point, namely, the origin $\mathbf{0}$ of V. Given two nonzero vectors A and B, we also view the angle $\theta = \angle(A, B)$ formed by them; see Figure 114.1. We visualize this angle even though we do not make a formal definition of it; the reason is that this can be done only if the field k is ordered (see page 420).

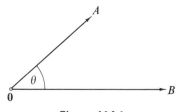

Figure 114.1

If A and B are nonzero vectors in Euclidean n-space, the cosine of $\angle(A, B)$ is defined as usual and the well known Euclidean formula is

$$AB = (\text{length } A)(\text{length } B) \cos \angle(A, B).$$

This is why, in an arbitrary metric vector space, one thinks of the scalar AB as something like the cosine of the angle formed by the vectors A and B even though this is correct in Euclidean geometry only if the two vectors have unit length. We do not attempt to use the Euclidean formula to define the notion of cosine for an arbitrary metric vector space because it is usually not possible to define the length of a vector as it is done in Euclidean space. Let us see why this is true.

If $A \in V$, we write A^2 for AA. In Euclidean geometry, $A^2 = (\text{length } A)^2$; more precisely, the length of A is the nonnegative square root of the real number A^2. It now becomes clear why it is usually not possible to define the length of A in an arbitrary metric vector space. The only reasonable definition for this length would be $\sqrt{A^2}$, but this does not work since A^2 may not even have a square root in k. This can even happen when k is the field of real numbers since in non-Euclidean geometries over the real numbers A^2 may very well be negative. Thus one thinks of the scalar A^2 as something like the square of the length of the vector A, but one does not make a formal definition of the length of a vector.

However, we do make a formal definition of orthogonality. The definition is based on the fact that two vectors A and B in Euclidean n-space are orthogonal if and only if $AB = 0$ (see the Euclidean formula).

Definition 115.1. Two vectors A and B of a metric vector space are **orthogonal (perpendicular)** if $AB = 0$. Notation: $A \perp B$.

The definition does not exclude the possibility that one or both of the vectors A, B is equal to the origin **0**. Since $A0 = 0A = 0$ for all $A \in V$ (see Exercise 116.2), the origin is orthogonal to all vectors of V.

We now extend the concept of orthogonality to linear subspaces of V.

Definition 115.2. Two linear subspaces U and W of V are **orthogonal (perpendicular)** if $A \perp B$ for all $A \in U$ and $B \in W$. Notation: $U \perp W$.

It should be pointed out that this concept of orthogonal (perpendicular) subspaces does not agree with the classical concept of perpendicular planes in Euclidean three-space E_3. We are accustomed to calling two planes U and

W of E_3 perpendicular if U contains a line l which is perpendicular to W; see Figure 116.1. This does not mean that every vector in the plane U is ortho-

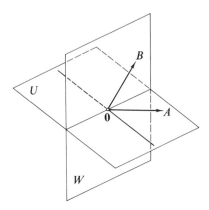

Figure 116.1

gonal to every vector in the plane W. The vectors $A \in U$ and $B \in W$ in Figure 116.1 are such examples. They are, clearly, not orthogonal. We shall see in Exercise 152.10b how the classical concept of perpendicular planes in E_3 can be expressed by using " orthogonal complements."

Our first example of a metric vector space is the **null space** whose inner product is defined by $AB = 0$ for all A, B in V. Null spaces are of no interest by themselves, but we shall see that they may occur as subspaces of geometries (metric vector spaces) which are themselves not null spaces; it is important to know what kind of null spaces lie embedded in a given geometry.

We recall that, throughout this book, **Q** denotes *the field of rational numbers*, **R** *the field of real numbers*, and **C** *the field of complex numbers*.

Exercises

1. Prove that a null space is a metric vector space; that is, show that the function from $V \times V \rightarrow k$ defined by $(A, B) \rightarrow 0$ is bilinear and symmetric.

*2. If V is a metric vector space, prove that $A0 = 0A = 0$ for all $A \in V$.

*3. Assume V is a metric vector space.

a. If U is a linear subspace of V, prove that

$$U^* = \{A \mid A \in V, \quad AB = 0 \text{ for all } B \in U\}$$

is a linear subspace of V. This subspace U^* is called the *orthogonal complement* of U (see Definition 148.1).

b. If U and W are linear subspaces of V, prove that

$$U \perp W \Leftrightarrow W \subset U^* \Leftrightarrow U \subset W^*.$$

*4. Let $A_1, \ldots, A_n$ be a coordinate system for the n-dimensional vector space V. If $B, C \in V$ and

$$B = b_1 A_1 + \cdots + b_n A_n, \qquad C = c_1 A_1 + \cdots + c_n A_n,$$

define $BC = b_1 c_1 + \cdots + b_n c_n$.

  a. Prove that the function $(B, C) \to BC$ is an inner product for V. (This is a special case of Exercise 5 below.) If $k = \mathbf{R}$, the resulting geometry is called *n-dimensional Euclidean geometry*.
  b. Prove that $\mathbf{0}$ is the only vector which is orthogonal to all vectors of V.
  c. If $k = \mathbf{R}$, prove that $A^2 = 0$ if and only if $A = \mathbf{0}$. Prove also that this is false if k is the field of complex numbers and $n \geq 2$.

*5. Let $A_1, \ldots, A_n$ be a coordinate system for the n-dimensional vector space V. Choose a symmetric bilinear form

$$\sum_{i,j=1}^{n} g_{ij} x_i y_j \in k[x_1, \ldots, x_n, y_1, \ldots, y_n];$$

"symmetric" means that $g_{ij} = g_{ji}$ for all $i, j = 1, \ldots, n$. If $B, C \in V$ and $B = b_1 A_1 + \cdots + b_n A_n$, $C = c_1 A_1 + \cdots + c_n A_n$, define

$$BC = \sum_{i, j=1}^{n} g_{ij} b_i c_j.$$

Prove that the function $(B, C) \to BC$ is an inner product for V. (We shall see in the next section that all inner products of V can be obtained in this way.)

Let $k = \mathbf{R}$ and V be n-dimensional Euclidean geometry described in Exercise 117.4. This language is not quite defensible since true Euclidean space is the metric affine space $(X, V, \mathbf{R})$ to which the metric vector space V gives rise (see Chapter 3). However, the step from V to $(X, V, \mathbf{R})$ is so small that this convenient "abuse of language" is justified. We commit the same abuse of language with regard to other metric vector spaces we shall encounter.

In Euclidean geometry, the only vector orthogonal to all vectors is the origin (Exercise 117.4b). This is not true in general; in the null space, every vector is orthogonal to all of V. This observation leads to the following definition.

Definition 117.1. A metric vector space is called **nonsingular** if the origin is the only vector which is orthogonal to all vectors.

The following exercises develop the fact that in a metric vector space the theorem of Pythagoras and, more generally, the law of cosines hold. Hence these laws are not restricted to Euclidean geometry, but hold in all geometries which are based on an inner product.

Exercises

In the following exercises, V is a metric vector space and A, B are arbitrary vectors in V.

6. a. Prove that $(A + B)^2 = A^2 + 2AB + B^2$.
   b. Prove that $(A - B)^2 = A^2 - 2AB + B^2$.
   c. Prove that $A \perp B$ if and only if $(A + B)^2 = A^2 + B^2$, equivalently, if and only if $(A - B)^2 = A^2 - B^2$.
*7. Interpret A^2 as "the square of the length of the vector A" and consider Figure 118.1.

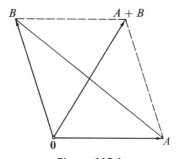

Figure 118.1

a. State why Part c of the above exercise expresses the theorem of Pythagoras for both triangles $0AB$ and $0A(A + B)$. Note also that the formula

$$(A - B)^2 = A^2 - 2AB + B^2$$

expresses the law of cosines for triangle $0AB$. The term $-2AB$ is the correction term which has to be added to the theorem of Pythagoras if the triangle is not a right triangle. In Euclidean geometry, one can go further and put

$$-2AB = -2(\text{length } A)(\text{length } B) \cos \angle (A, B).$$

b. Prove the "parallelogram law." This says that the sum of the squares of the lengths of the four sides of a parallelogram is equal to the sum of the squares of the lengths of the two diagonals.

c. Prove that the diagonals of the parallelogram $0A(A + B)B$ are orthogonal if and only if $A^2 = B^2$. Interpret this result as: The diagonals of a parallelogram are orthogonal if and only if the parallelogram is a rhombus.

The fact that the law of cosines follows immediately from the axioms of the inner product allows us to interpret these axioms geometrically as being the algebraic formulation of the law of cosines. Hence, by basing all our metric geometry on the five axioms of the inner product, we are using the law of cosines as the principal axiom of metric geometry. This is quite a feather in the cap of Pythagoras since the law of cosines is nothing but the theorem of Pythagoras for arbitrary triangles.

We finish this section with a proposition which depends heavily on the fact that the characteristic of k is different from two. Suppose that V is a metric vector space, but that we only know the value of A^2 for each vector $A \in V$. Can we compute the inner product AB for all vectors $A \neq B$? The answer is affirmative, but only because char $k \neq 2$.

Proposition 119.1.   Let V be a metric vector space. The function $A \to A^2$ from V into k determines the inner product completely.

Proof. Select $A, B \in V$; then
$$(A + B)^2 = A^2 + 2AB + B^2$$
and hence, since char $k \neq 2$,
$$AB = \tfrac{1}{2}((A + B)^2 - A^2 - B^2).$$
The right-hand side of the last equality is completely determined by the function $A \to A^2$; thus AB is also determined. Done.

Exercises

*8. Assume that the metric space V is not the null space, that is, there are vectors $A, B \in V$ such that $AB \neq 0$. Prove that there exists a vector $C \in V$ such that $C^2 \neq 0$.

9. Prove that the zero-dimensional vector space $\{0\}$ is the only non-singular null space.

25. INNER PRODUCTS IN TERMS OF COORDINATES

We assume, in the remainder of this chapter, that our n-dimensional vector space V over k has been provided with a metric.

Whenever a coordinate system is chosen for V, it is understood that $n \geq 1$ because the zero-dimensional vector space has no bases and hence no coordinate systems. As in Chapter 1, a coordinate system means an ordered basis.

Let $A_1, \ldots, A_n$ be a coordinate system for V. The $n \times n$ symmetric matrix $G = (g_{ij})$ where $g_{ij} = A_i A_j$ is completely determined by the inner product of V and the given coordinate system. Of course, it is immaterial whether one talks about the symmetric matrix (g_{ij}) or the symmetric bilinear form

$$\sum_{i,j=1}^n g_{ij} x_i y_j \in k[x_1, \ldots, x_n, y_1, \ldots, y_n],$$

as either one determines the other uniquely.

Conversely, the knowledge of the matrix G and the fixed coordinate system $A_1, \ldots, A_n$ enables us to compute the inner product BC of any two vectors B and C. We simply put

$$B = b_1 A_1 + \cdots + b_n A_n, \qquad C = c_1 A_1 + \cdots + c_n A_n$$

and conclude from the five axioms of the inner product that

$$BC = \sum_{i,j=1}^n g_{ij} b_i c_j.$$

In other words, BC is computed by substituting the coordinates of the vectors B and C in the symmetric bilinear form

$$\sum_{i,j=1}^n g_{ij} x_i y_j.$$

Using the result of Exercise 117.5, it follows easily that

Proposition 120.1.   A coordinate system of V gives rise to a one-to-one correspondence between the inner products of V and the $n \times n$ symmetric matrices with entries in k, equivalently, between the inner products of V and the symmetric bilinear forms of $k[x_1, \ldots, x_n, y_1, \ldots, y_n]$.

For the above result, the characteristic of k was actually immaterial. For what is coming, it is crucial that the characteristic is different from two. The square B^2 of the vector $B = b_1 A_1 + \cdots + b_n A_n$ is equal to

$$B^2 = \sum_{i,j=1}^n g_{ij} b_i b_j.$$

Thus B^2 is computed by substituting the coordinates of B in the quadratic form

$$f = \sum_{i,j=1}^n g_{ij} x_i x_j \in k[x_1, \ldots, x_n].$$

Clearly, f is obtained from the bilinear form

$$\sum_{i,j=1}^{n} g_{ij} x_i y_j$$

by replacing y_j by x_j. Since the inner product is completely determined by the function $A \to A^2$ (Proposition 119.1), it follows that the quadratic form f determines the inner product and hence the symmetric matrix G. The rule for computing the g_{ij}s from f is obtained by observing that

$$f = \sum_{i=1}^{n} g_{ii} x_i^2 + 2 \sum_{1 \le i < j \le n} g_{ij} x_i x_j.$$

Consequently, g_{ii} is the coefficient of x_i^2 for $i = 1, \ldots, n$ and g_{ij} is one-half the coefficient of $x_i x_j$ for $1 \le i < j \le n$. We have proved:

Proposition 121.1. A coordinate system of V gives rise to a one-to-one correspondence between the inner products of V and the quadratic forms of $k[x_1, \ldots, x_n]$.

Example 1

Let $n = 3$ and consider the quadratic form

$$f = 5x_1^2 - 8x_2^2 + 7x_3^2 + 37x_1x_2 + 18x_1x_3 - 11x_2x_3.$$

The symmetric matrix corresponding to f is

$$G = \begin{pmatrix} 5 & \frac{37}{2} & 9 \\ \frac{37}{2} & -8 & -\frac{11}{2} \\ 9 & -\frac{11}{2} & 7 \end{pmatrix}.$$

Of course, G determines an inner product for V only after a coordinate system has been chosen.

Exercises

Let $A_1, \ldots, A_n$ be a coordinate system for the vector space V.

1. Find the symmetric $n \times n$ matrix G which makes V into the null space.
2. Let $k = \mathbf{R}$ and determine the matrix G in such a way that V is an n-dimensional Euclidean space. Also, write the corresponding symmetric bilinear form and quadratic form.

3. Assume $n = 2$ and $G = \left(\begin{smallmatrix} 0 & 1 \\ 1 & 0 \end{smallmatrix}\right)$. Find all vectors which are orthogonal to A_1; to A_2; to $A_1 + A_2$; to $A_1 - A_2$.

Propositions 120.1 and 121.1 enable us to understand a major difference between the algebra of the past century and that of the present one. Quadratic forms used to be studied, say 50 to 100 years ago, by direct manipulation on the quadratic polynomials, or, equivalently, on the symmetric matrices. Now, one interprets the quadratic form as a metric vector space and studies the form by studying the geometry of that space. The foundations of the geometric theory of quadratic forms were laid as late as 1936 by Ernst Witt, "Theorie der quadratischen Formen in beliebegen Korpern," *Journal für die reine und angewandte Mathematik*, Vol. 176, Oct. 1936, pp. 31–44, [23]. This geometrization of the theory of quadratic forms is typical of the geometrization which practically all of modern mathematics has undergone in the present century. Artin used to say that "In spite of modern boasting," mathematics of the present century is much more geometrical than that of the past century.

We return to our metric vector space V with its coordinate system $A_1, \ldots, A_n$ and symmetric matrix G. The following criterion for nonsingularity is often convenient.

Proposition 122.1.   The metric vector space V is nonsingular if and only if $\det G \neq 0$.

Proof. If $B = b_1 A_1 + \cdots + b_n A_n$ and $C = c_1 A_1 + \cdots + c_n A_n$,

$$BC = \sum_{i,j=1}^{n} g_{ij} b_i c_j = (b_1, \ldots, b_n) G \begin{pmatrix} c_1 \\ \vdots \\ c_n \end{pmatrix}.$$

Hence a vector is orthogonal to B if and only if its coordinates $(x_1, \ldots, x_n)$ satisfy the equation

$$(b_1, \ldots, b_n) G \begin{pmatrix} x_1 \\ \vdots \\ x_n \end{pmatrix} = 0.$$

It follows that B is orthogonal to all vectors of V if and only if

$$(b_1, \ldots, b_n) G = (0, \ldots, 0).$$

The matrix G has a nonzero left solution $(b_1, \ldots, b_n)$ if and only if $\det G = 0$.
Done.

When analyzing a geometry, it is always important to know whether there are nonzero "isotropic" vectors.

Definition 123.1. A vector $A \in V$ is called **isotropic** or a **null vector** if $A^2 = 0$.

The origin $\mathbf{0}$ is always isotropic. Even if the geometry is nonsingular, there may be many nonzero isotropic vectors. This is seen by observing that a vector is isotropic if and only if its coordinates satisfy the equation

$$\sum_{i,j=1}^{n} g_{ij} x_i x_j = 0.$$

If V is not the null space, not all g_{ij}s are zero and this is the equation of an $(n - 1)$-dimensional cone with $\mathbf{0}$ as vertex. The term cone refers to the fact that, if the coordinates of a vector A satisfy the equation, the coordinates of every vector tA also satisfy the equation. In other words, the cone is made up of lines through $\mathbf{0}$, the so-called *generators of the cone*. Adopting the language of physics, where a large part of linear algebra originates, we refer to the cone of isotropic vectors as the **light cone**. The light cone may consist of the origin alone, as in the case of Euclidean geometry, or it may consist of two lines, as in the case of the Lorentz plane (see page 126). If V is the null space, the light cone is all of V.

26. CHANGE OF COORDINATE SYSTEM

Let $A_1, \ldots, A_n$ and $B_1, \ldots, B_n$ be coordinate systems for our metric vector space V. We denote the matrix of the metric relative to the coordinate system $A_1, \ldots, A_n$ by $G = (A_i A_j)$ and the matrix relative to the coordinate system $B_1, \ldots, B_n$ by $G' = (B_i B_j)$. How are these matrices related?

Since $A_1, \ldots, A_n$ and $B_1, \ldots, B_n$ are coordinate systems of the same vector space, there exists a nonsingular $n \times n$ matrix P such that

$$(B_1, \ldots, B_n) = (A_1, \ldots, A_n)P.$$

Observe that, on the right side of this matrix equality, there occurs the product of an $1 \times n$ matrix whose entries are vectors with the $n \times n$ matrix P whose entries are scalars. The axioms for scalar multiplication guarantee that the usual rules of matrix multiplication remain valid when such matrices are multiplied. For instance, taking transposes on both sides of the above matrix equality, we obtain

$$\begin{pmatrix} B_1 \\ \vdots \\ B_n \end{pmatrix} = P^{\mathrm{T}} \begin{pmatrix} A_1 \\ \vdots \\ A_n \end{pmatrix}$$

where P^{T} denotes the transpose of P. The five axioms for the inner product also guarantee that the rules of matrix multiplication hold when matrices are multiplied whose entries are vectors. For example,

$$(A_1, \ldots, A_n) \begin{pmatrix} A_1 \\ \vdots \\ A_n \end{pmatrix} = A_1{}^2 + \cdots + A_n{}^2$$

and

$$\begin{pmatrix} A_1 \\ \vdots \\ A_n \end{pmatrix} (A_1, \ldots, A_n) = G.$$

Using these remarks, it takes only a line to give the relationship between G and G'.

Proposition 124.1. The matrices of the metric relative to different coordinate systems are related by the formula

$$G' = P^{\mathrm{T}} G P$$

where P is a nonsingular $n \times n$ matrix.

Proof.

$$G' = \begin{pmatrix} B_1 \\ \vdots \\ B_n \end{pmatrix} (B_1, \ldots, B_n) = P^{\mathrm{T}} \begin{pmatrix} A_1 \\ \vdots \\ A_n \end{pmatrix} (A_1, \ldots, A_n) P = P^{\mathrm{T}} G P.$$

Done.

Two $n \times n$ matrices G and G' are called **congruent** if there exists a non-singular matrix P such that $G' = P^{\mathrm{T}} G P$. We conclude from Proposition 124.1 that two $n \times n$ matrices are congruent if and only if they represent the same metric of V relative to appropriate coordinate systems. Once the notion of isometric vector spaces has been introduced (Section 27), it will be clear that the study of metric vector spaces is equivalent to the study of symmetric matrices under the equivalence relation of congruence. However, the direct study of the geometry of vector spaces is infinitely more instructive, beautiful, and simple than the study of the congruence relation of symmetric matrices. We shall make use of the congruence relation only to define the discriminant of a geometry.

If G and G' represent the metric of V with respect to appropriate coordinate systems, then $G' = P^TGP$ and

$$\det G' = (\det P^T)(\det G)(\det P) = (\det G)(\det P)^2$$

where $\det P \neq 0$. Assume first that V is nonsingular. Then $\det G$ and $\det G'$ belong to k^* where k^* denotes the multiplicative group of nonzero elements of k (Proposition 122.1). Since k^* is commutative, the set of its squares

$$k^{*2} = \{c^2 \,|\, c \in k^*\}$$

is a subgroup of k^* (see Exercise 130.3) and we can form the quotient group k^*/k^{*2}. The equality

$$\det G' = (\det G)(\det P)^2$$

shows that the coset $(\det G)k^{*2}$ in k^*/k^{*2} does not depend on the coordinate system, that is,

$$(\det G)k^{*2} = (\det G')k^{*2}.$$

Definition 125.1.   If V is nonsingular, the coset $(\det G)k^{*2}$ is called the **discriminant** of V. Notation: disc V.

The discriminant is the first invariant of a metric vector space we encounter. "Invariant" should mean a quantity which remains unchanged under isometries; since we have not yet said what an isometry is, let invariant simply mean a quantity which is intrinsically determined by the geometry of V.

We shall usually refer to the determinant $\det G$ itself as the discriminant of V. This is rash abuse of language since the discriminant is the coset $(\det G)k^{*2}$ in k^*/k^{*2}. Nevertheless, this little crime is convenient, and as long as the abuse is clearly realized, it will not cause confusion.

If V is singular, $\det G = 0$ for all coordinate systems. This follows either from Proposition 122.1 or from the equality $\det G' = (\det G)(\det P)^2$. In this case, we say that the discriminant of V is zero. The zero-dimensional vector space $\{0\}$ has no coordinate systems and hence no discriminant.

The discriminant is a very weak invariant of a metric vector space in the sense that it does not tell much about the geometry of the space. For example, if $k = \mathbf{R}$, $\mathbf{R}^*/\mathbf{R}^{*2}$ contains only two elements, namely, the coset $\mathbf{R}^*$ of positive real numbers and the coset $(-1)\mathbf{R}^*$ of negative real numbers. These two cosets have 1 and -1 as representatives. If V is a nonsingular space over $\mathbf{R}$, its discriminant is completely determined by the sign of det G. It is certainly useful to know det G is positive or negative, but, as expected, this cannot give much information about the underlying geometry. On the other hand, discriminants can often be used to distinguish one metric from another.

We now consider some examples of metric vector spaces. They have been chosen for their interest in physics and geometry and will be referred to often. In all examples, it is assumed that V is an n-dimensional vector space for which a coordinate system $A_1, \ldots, A_n$ has been chosen and the metric is determined by choosing the matrix $G = (g_{ij})$ where $g_{ij} = A_i A_j$. Finally, the abuse of language introduced in our discussion of the discriminant will occur in all examples.

Example 1

n-dimensional Euclidean space. We know from Exercise 117.4 that n-dimensional Euclidean space is obtained by letting $k = \mathbf{R}$ and choosing G to be the unit matrix

$$G = \begin{pmatrix} 1 & & \bigcirc \\ & \ddots & \\ \bigcirc & & 1 \end{pmatrix}.$$

The discriminant is 1 and hence this geometry is nonsingular. The symmetric bilinear form is $x_1 y_1 + \cdots + x_n y_n$ and the quadratic form is $x_1^2 + \cdots + x_n^2$. There are no nonzero isotropic vectors and vectors $A_1, \ldots, A_n$ of the coordinate system are mutually orthogonal.

Example 2

The Lorentz plane. Here $n = 2$, $k = \mathbf{R}$ and $G = \begin{pmatrix} 1 & 0 \\ 0 & -1 \end{pmatrix}$. The Lorentz plane is nonsingular since the discriminant is -1. The symmetric bilinear form is $x_1 y_1 - x_2 y_2$ and the quadratic form is $x_1^2 - x_2^2$. The coordinates of the isotropic vectors are the solutions to the equation

$$x_1^2 - x_2^2 = (x_1 + x_2)(x_1 - x_2) = 0.$$

Thus, the light cone consists of the two lines

$$U = \langle A_1 + A_2 \rangle \quad \text{and} \quad W = \langle A_1 - A_2 \rangle.$$

(When $Y_1, \ldots, Y_m \in V$, $\langle Y_1, \ldots, Y_m \rangle$ denotes the linear subspace spanned by $Y_1, \ldots, Y_m$.) The picture below of the Lorentz plane uses the fact that $A_1 \perp A_2$.

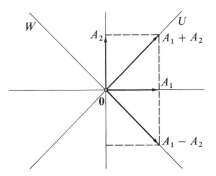

Figure 127.1

The Lorentz plane L and the Euclidean plane E_2 provide us with a good example how different metrics can be imposed on the same vector space. The discriminant of E_2 is 1 while the discriminant of L is -1. Since -1 is not a square in $\mathbf{R}$, the geometries are certainly distinct; since every nonzero vector in E_2 has positive square, it is not possible to find a coordinate system for the Euclidean plane such that the matrix of the Euclidean metric becomes $\begin{pmatrix} 1 & 0 \\ 0 & -1 \end{pmatrix}$. We shall see in the next section that E_2 and L are even "inequivalent," something which is to be expected because E_2 has no nonzero isotropic vectors while L has infinitely many of them.

H. Lorentz (1853–1928) was one of the great Dutch physicists. His "Lorentz transformation" was fundamental in Einstein's development of relativity theory and he (Lorentz) received the Nobel prize in 1902.

Example 3

Minkowski space. Here $n = 4$, $k = \mathbf{R}$, and

$$G = \begin{pmatrix} 1 & 0 & 0 & 0 \\ 0 & 1 & 0 & 0 \\ 0 & 0 & 1 & 0 \\ 0 & 0 & 0 & -1 \end{pmatrix}.$$

This discriminant is -1 and hence Minkowski space is a nonsingular four-space. The symmetric bilinear form is $x_1 y_1 + x_2 y_2 + x_3 y_3 - x_4 y_4$ and the quadratic form is $x_1^2 + x_2^2 + x_3^2 - x_4^2$. The three-dimensional light cone has the equation

$$x_1^2 + x_2^2 + x_3^2 - x_4^2 = 0$$

and, therefore, there are infinitely many isotropic vectors. The vectors A_1, A_2, A_3, A_4 of the coordinate system are mutually orthogonal and

$A_4{}^2 = -1$. We leave to the reader the discussion that Minkowski space and Euclidean four-space are distinct geometries and that any reasonable notion of equivalence for metric vector spaces should show that these geometries are inequivalent.

Hermann Minkowski (1864–1909) was a famous German mathematician who had Einstein as a student. When Minkowski developed his geometry in order to express the laws of relativity, Einstein said, "Since the mathematicians have invaded the theory of relativity, I do not understand it myself any more." For the importance of Minkowski space for relativity theory and the related work of Lorentz, see P. G. Bergmann, *An Introduction to the Theory of Relativity*, [5].

Example 4

Negative Euclidean space of dimension *n*. In this example, n is arbitrary, $k = \mathbf{R}$, and G is the negative of the unit matrix, that is,

$$G = \begin{pmatrix} -1 & & \bigcirc \\ & \ddots & \\ \bigcirc & & -1 \end{pmatrix}.$$

Negative Euclidean space is nonsingular since its discriminant is $(-1)^n$. The symmetric bilinear form is $-x_1 y_1 - \cdots - x_n y_n$ and the quadratic form is $-x_1{}^2 - \cdots - x_n{}^2$. There are no nonzero isotropic vectors and the vectors $A_1, \ldots, A_n$ of the coordinate system are mutually orthogonal.

When n is odd, the discriminants show that Euclidean n-space and negative Euclidean n-space are different geometries. If n is even, both geometries have discriminant 1. Nevertheless, these geometries are obviously distinct since in Euclidean space all nonzero vectors have a positive square while in negative Euclidean space all nonzero vectors have a negative square. This also indicates that these two geometries will certainly be inequivalent.

Example 5

Let k be an algebraically closed field and

$$G = \begin{pmatrix} 1 & & \bigcirc \\ & \ddots & \\ \bigcirc & & 1 \end{pmatrix}.$$

(Readers who are not familiar with algebraically closed fields should restrict themselves to the field $\mathbf{C}$ of complex numbers whenever algebraically closed fields are mentioned in this book.) The discriminant is 1, the symmetric bilinear form is $x_1 y_1 + \cdots + x_n y_n$ and the quadratic form is $x_1^2 + \cdots + x_n^2$. The geometry is nonsingular and the vectors $A_1, \ldots, A_n$ are mutually orthogonal. The $(n-1)$-dimensional light cone has equation $x_1^2 + \cdots + x_n^2 = 0$; consequently, if $n \geq 2$, there are infinitely many isotropic vectors.

Since G is the unit matrix, it is sometimes claimed that this geometry is the analogue of Euclidean geometry for the field k. However, this is a misconception as the field **C** of complex numbers already shows. The true analogue of Euclidean geometry, in the case of the field **C**, is unitary geometry which is not even based on a symmetric bilinear form, but on a Hermitian form (see R. Godement, *Algebra*, Section 36, [12]). Returning to the case of an algebraically closed field, we shall later see that for $n = 2$, the present geometry is a generalization of the Lorentz plane. The precise structure of the present geometry for arbitrary n will become clear after we have discussed "Artinian geometry" (Section 35, Artinian Spaces). However, we want to warn the reader now not to think of the present geometry as a generalization of Euclidean geometry. This concludes our discussion of this example.

Exercises

1. Assume $n = 2$ and that A_1, A_2 is a coordinate system for V. Let the metric on V be given by the matrix $G = \begin{pmatrix} 0 & 1 \\ 1 & 0 \end{pmatrix}$ relative to this coordinate system. Find a new coordinate system B_1, B_2 for V such that the same metric relative to the new coordinate system is given by the matrix $G' = \begin{pmatrix} 1 & 0 \\ 0 & -1 \end{pmatrix}$. Hence, if $k = \mathbf{R}$, the matrix G defines the Lorentz plane. (Hint: Let $B_1 = aA_1 + bA_2$ and $B_2 = cA_1 + dA_2$; then find appropriate values for a, b, c, and d.)

*2. Let $n = 2$ and A_1, A_2 be a coordinate system for V. Relative to this coordinate system, assume that the metric is given by the matrix $G = \begin{pmatrix} 1 & 0 \\ 0 & -1 \end{pmatrix}$. (Hence, if $k = \mathbf{R}$, the matrix G defines the Lorentz plane.)

   a. Prove that the vectors $A_1 + A_2$ and $A_1 - A_2$ are not orthogonal. Hence the lines U and W of Figure 127.1 in the Lorentz plane are not orthogonal, even though they appear so in the figure. How dangerous those pictures!

   b. Find the matrix of the metric relative to the coordinate system $A_1 + A_2$, $A_1 - A_2$.

   c. Find scalars a_1, a_2, b_1, b_2 such that the vectors

   $$M = a_1 A_1 + a_2 A_2 \quad \text{and} \quad N = b_1 A_1 + b_2 A_2$$

   form a coordinate system for V and such that the matrix of the given metric becomes $\begin{pmatrix} 0 & 1 \\ 1 & 0 \end{pmatrix}$.

    d. Find the symmetric bilinear form and the quadratic form of V relative to the coordinate system M, N found in Part c.

    e. Conclude from Part d that, if $t \in k$, there is a $B \in V$ such that $B^2 = t$. In words: The metric represents all scalars. Observe how different the Lorentz plane is from the Euclidean plane. The Lorentz plane represents all scalars while the Euclidean plane represents all nonnegative scalars.

*3. a. If G is a commutative group, prove that the set of squares $\{g^2 \mid g \in G\}$ is a subgroup of G. This is the reason that the set of nonzero squares k^{*2} of the field k is a group.

    b. If G is not commutative, the set of squares may or may not be a group. For instance, the symmetric group S_3 is not commutative. Prove, however, that the set of squares is the alternating group A_3.

4. a. If k is algebraically closed, prove that k^*/k^{*2} consists of one element. (For those readers who are not familiar with algebraically closed fields, prove that $\mathbf{C}^*/\mathbf{C}^{*2}$ consists of one element.)

    b. For the field $\mathbf{Q}$ of rational numbers, prove that $\mathbf{Q}^*/\mathbf{Q}^{*2}$ is an infinite group. (Hint: Prove that the set $\{p\mathbf{Q}^{*2} \mid p$ a prime number$\}$ is an infinite subset of $\mathbf{Q}^*/\mathbf{Q}^{*2}$.)

5. Assume k is finite. Exercises a through c which follow depend on the fact that char $k \neq 2$.

    a. Prove that k contains p^h elements where p is an odd prime and h is a positive integer.

    b. Prove that the group k^{*2} of nonzero squares of k is a cyclic group of order $\frac{1}{2}(p^h - 1)$. (Hint: Use the fact that a finite subgroup of the multiplicative group of a field is always cyclic; see E. Artin, *Galois Theory*, Notre Dame Mathematical Lectures, Number 2, p. 49, [1].) In order to compute the order of k^{*2}, find the kernel of the epimorphism $c \to c^2$ from k^* onto k^{*2}. (Epimorphism means homomorphism onto.)

    c. Prove that k^*/k^{*2} has two elements.

    d. If k is a finite field of characteristic 2, how should Parts a through c be reworded?

6. Let $k = \mathbf{Z}_5$. Assume $n = 2$ and that A_1, A_2 is a coordinate system for V. Relative to this coordinate system, let metrics for V be given by the matrices

$$\begin{pmatrix} 3 & 2 \\ 2 & 3 \end{pmatrix}, \quad \begin{pmatrix} 1 & 2 \\ 2 & 3 \end{pmatrix}, \quad \begin{pmatrix} 3 & 2 \\ 2 & 4 \end{pmatrix}.$$

    a. Determine which of the three metrics are nonsingular.

    b. Prove that the three metrics have distinct discriminants and hence are distinct metrics.

*7. We say that "*every element of k has a square root*" if, for each $c \in k$, there exists a $d \in k$ such that $d^2 = c$.

    a. If k is finite, prove that not every element of k has a square root. (This is correct only because char $k \neq 2$.)

    b. Prove that every element of an algebraically closed field has a square root.

    c. If k is a finite field of characteristic 2, prove that every element of k has a square root.

*8. Assume that every element of k has a square root.

    a. Prove that all nonsingular geometries have discriminant 1.

    b. Prove that every nonsingular geometry represents all scalars. (Hint: Choose $A \in V$ such that $A^2 = a \in k^*$ (Exercise 119.8).)

*9. Assume $n = 3$ and A_1, A_2, A_3 is a coordinate system for V. Relative to this coordinate system, let the metric of V be given by the matrix

$$\begin{pmatrix} 1 & 0 & 0 \\ 0 & 1 & 0 \\ 0 & 0 & -1 \end{pmatrix}.$$

The metric is clearly nonsingular and has discriminant -1.

    a. Find the symmetric bilinear form and the quadratic form of the metric relative to the given coordinate system.

    b. If k is **R** or **C**, prove that V has infinitely many null lines. (It is sufficient to show this for **R**.) Draw an accurate picture of the light cone for the case that $k = \mathbf{R}$.

    c. Find three linearly independent isotropic vectors B_1, B_2, B_3 in V. Give the matrix, the symmetric bilinear form, and the quadratic form of V relative to the coordinate system B_1, B_2, B_3. Compute the discriminant of V, again using this new coordinate system.

27. ISOMETRIES

In general, two mathematical structures are called "equivalent" or *isomorphic* if there exists a one-to-one mapping from one of the structures onto the other one which preserves all intrinsic structure. Such a mapping is called an *isomorphism* and hence two structures are isomorphic if there exists an isomorphism between them. If the two mathematical structures are metric vector spaces V and W over the field k, the intrinsic structure consists of the vector space structure of V and W together with the two inner products, one for V and one for W. In this situation, an isomorphism is called an *isometry* and isomorphic metric vector spaces are called isometric. The explicit definitions are now given.

Definition 132.1.   An **isometry** from V to W is a function $\sigma: V \to W$ satisfying

1. σ is one-to-one and onto.
2. σ is a linear transformation.
3. $AB = (\sigma A)(\sigma B)$ for all $A, B \in V$.

Two metric vector spaces V and W are called **isometric** if there exists an isometry from one onto the other. Notation: $V \simeq W$.

Condition 1 states that V and W are isomorphic as sets. Conditions 1 and 2 together state that σ is a nonsingular linear transformation from V onto W and hence V and W are isomorphic as vector spaces. In Condition 3, AB denotes the inner product in V while $(\sigma A)(\sigma B)$ denotes the inner product in W; this condition states that σ preserves the metric structure of the two spaces. It is always understood that isometric vector spaces are defined over the same field.

Observe that we have denoted the image of a vector $A \in V$ by σA without using parentheses. This will cause no confusion and, as will become apparent, it makes the exposition much easier to read in many places.

A Remark on Terminology. We would have preferred to say "congruence" instead of isometry since congruence has always stood for isomorphism in metric geometry. Actually, our first choice would have been simply to say isomorphism rather than take part in the proliferation of terminology which besets mathematics. Instead of using only the term isomorphism, mathematicians invent a new term each time an isomorphism occurs. For example, the concept of isomorphism is called:

homeomorphism in topology;
diffeomorphism in differential geometry;
one-to-one correspondence in set theory;
nonsingular linear transformation onto in linear algebra;
congruence in metric geometry;
isometry in the theory of metric vector spaces.

Thank goodness, in group theory an isomorphism is still called an isomorphism!

Exercise

1. Assume $\sigma: V \to W$ is an isometry. Since σ is one-to-one and onto, σ has an inverse $\gamma: W \to V$. Prove that γ is also an isometry. Consequently, if $V \simeq W$, there is an isometry from V to W and an isometry from W to V.

It is fortunate that the condition $AB = (\sigma A)(\sigma B)$ does not have to be checked for all vectors of V, but only for those of a coordinate system. We usually say **linear isomorphism** instead of nonsingular linear transformation onto. A **linear automorphism** is a linear isomorphism from a vector space onto itself.

Proposition 133.1. Let $\sigma: V \to W$ be a linear isomorphism from the metric vector space V onto the metric vector space W. Assume that $A_1, \ldots, A_n$ is a coordinate system for V. Then σ is an isometry if and only if $A_i A_j = (\sigma A_i)(\sigma A_j)$ for $i, j = 1, \ldots, n$.

Proof. The "only if" part is obvious. In order to prove the "if" part, we assume that $A_i A_j = (\sigma A_i)(\sigma A_j)$ for $i, j = 1, \ldots, n$. Conditions 1 and 2 of Definition 132.1 are satisfied since σ is a linear isomorphism. Consequently, there only remains to be shown that, if $X, Y \in V$, then $XY = (\sigma X)(\sigma Y)$. For this, we put

$$X = \sum_{i=1}^{n} x_i A_i \quad \text{and} \quad Y = \sum_{j=1}^{n} y_j A_j.$$

Then

$$(\sigma X)(\sigma Y) = \left(\sum_{i=1}^{n} x_i \sigma A_i \right) \left(\sum_{j=1}^{n} y_j \sigma A_j \right)$$

$$= \sum_{i,j=1}^{n} x_i y_j (\sigma A_i)(\sigma A_j)$$

$$= \sum_{i,j=1}^{n} x_i y_j A_i A_j$$

$$= XY.$$

Done.

Exercises

2. Let $\sigma: V \to W$ be a linear isomorphism between the metric vector spaces V and W. Assume S is a subset of V which spans V. (This means that every vector of V can be written as a finite linear combination of vectors in S.) Prove that σ is an isometry if and only if $AB = (\sigma A)(\sigma B)$ for all $A, B \in S$.

3. Prove that the metric vector spaces V and W are isometric if and only if there are coordinate systems for them relative to which the metrics

have the same symmetric matrix (and hence the same symmetric bilinear form and the same quadratic form).

*4. One says that a *scalar c is represented* by a metric vector space if there exists a vector A in the space such that $A^2 = c$. Prove that isometric vector spaces represent the same scalars. We shall see in Exercise 162.2 that the converse is false: Two metric vector spaces can represent the same set of scalars and still not be isometric.

5. We know that a metric vector space gives rise to an equivalence class of congruent symmetric matrices, one matrix for each coordinate system of the space. Prove that two metric vector spaces are isometric if and only if they give rise to the same class of matrices. This shows that the problem of classifying metric vector spaces under the relation of isometry is the same as that of classifying symmetric matrices under the congruence relation. Both problems are totally unsolved; some remarks relative to the situation are made at the end of this section.

In the definition of isometry, it may very well happen that $V = W$. In that case, the function $\sigma: V \to V$ of that definition is a linear automorphism of V, satisfying $AB = (\sigma A)(\sigma B)$ for all $A, B \in V$. The next exercise gives a method for constructing nonisometric geometries for the same vector space.

Exercise

6. Let G be a symmetric $n \times n$ matrix with entries in k. For each coordinate system of V, G makes V into a metric vector space. Prove that all these metric vector spaces are isometric.

For an arbitrary field k, there are no decisive criteria which state when two given metric vector spaces are isometric. Of course, isometric spaces must represent the same scalars (Exercise 134.4) and this often provides an easy check. A little further help comes from

Proposition 134.1. Isometric metric vector spaces have the same discriminant.

Proof. If V and W are isometric vector spaces, they have coordinate systems relative to which the metrics have the same symmetric matrix, therefore, certainly the same discriminant. Done.

Exercises

*7. Consider the Euclidean plane, the Lorentz plane, and the negative Euclidean plane. Prove that no two of these planes are isometric. Observe, however, that the Euclidean and negative Euclidean plane have the same discriminant.

8. Consider Euclidean four-space, Minkowski space, and negative Euclidean four-space. Prove that no two of these three metric vector spaces are isometric. We shall see in Example 161.2 that there are altogether five nonsingular, nonisometric metrics for real four-space.

9. Which two of the spaces mentioned in the previous exercise have the same discriminant?

In the next four exercises, we study metric lines. That is, V is a metric vector space of dimension one over k.

10. Prove that two metric lines are isometric if and only if they have the same discriminant.

*11. Prove that there is a one-to-one correspondence between the elements of the group $k*/k*^2$ and the set of classes of nonsingular isometric lines. By abuse of language (counting each class of nonsingular isometric lines as one), we say there are just as many lines as there are elements in the group $k*/k*^2$.

12. Prove that a line is singular if and only if it is a null line. Again, by abuse of language, we say there is only one singular line.

*13. We continue to count all isometric lines as one.
    a. Prove that there are infinitely many nonisometric lines over **Q**.
    b. Prove that the null line, the Euclidean line, and the negative Euclidean line are the only nonisometric lines over **R**.
    c. Prove that there are two nonisometric lines over a field all of whose elements have a square root. (Hint: Use Exercise 131.8.)
    d. If k is finite, prove that there are three nonisometric lines over k. (Hint: Use Exercise 130.5c).

When $n \geq 2$, the classification problem of metric vector spaces over an arbitrary field is completely unsolved. One would like to associate a set of invariants to a metric vector space with the property that two spaces are isometric if and only if they have the same invariants. Here, invariant simply means a manageable quantity such as the discriminant which does not change under isometries. We now mention some of the special fields for which the classification problem has been solved.

1. *k is a field all of whose elements have a square root.* These fields were discussed in Exercises 130.7 and 131.8; we recall that all algebraically closed

fields have this property. We shall prove (Proposition 157.1) that all non-singular geometries of the same dimension are isometric.

2. $k = \mathbf{R}$. This problem is much harder, but has been completely solved by James Joseph Sylvester (1844–1897). It turns out that there are precisely $n + 1$ nonsingular, nonisometric geometries when dim $V = n$. It will be shown (Corollary 178.1) that this result holds for every ordered field in which all positive elements have a square root. Sylvester was one of the leading English mathematicians of the nineteenth century. He had great influence on American mathematics, having taught at both the University of Virginia (1841–1842) and Johns Hopkins University (1877–1883). He founded the *American Journal of Mathematics* and was one of the first to establish graduate work in mathematics at American universities.

3. $k = \mathbf{Q}$, or, *more generally, k is an algebraic number field.* The problem is again solved but needs advanced algebraic number theory, namely, class field theory, for its solution. (See O. T. O'Meara, *Introduction to Quadratic Forms,* [18].)

4. k is a finite field. The problem is again completely solved, this time by algebra not any harder than occurs in this book. For each dimension, there are now precisely two nonsingular, nonisometric geometries. (See Artin [2], pp. 143–148.)

The solution of the classification problem, in the case that k is an algebraic number field, shows clearly that presentday mathematics cannot begin to tackle the problem for arbitrary fields.

Exercises

In the following exercises, $k = \mathbf{R}$, $n = 2$, and A_1, A_2 is a coordinate system of V. The metrics of V are determined by symmetric matrices relative to this coordinate system.

*14. If the matrices

$$G = \begin{pmatrix} 1 & \frac{3}{2} \\ \frac{3}{2} & 2 \end{pmatrix} \quad \text{and} \quad G' = \begin{pmatrix} -7 & \frac{3}{4} \\ \frac{3}{4} & -2 \end{pmatrix}$$

are used to define metrics for V, prove that the resulting planes are not isometric.

*15. Suppose that the metric of V is determined by the matrix

$$G = \begin{pmatrix} 3 & -1 \\ -1 & 2 \end{pmatrix}.$$

a. Prove that this metric is the Euclidean metric. (Warning: It is not sufficient to show that disc $V = 1$. One must find a new coordinate system B_1, B_2 for V relative to which the metric is given by the unit matrix.)

b. Find a 2×2 matrix P such that

$$\begin{pmatrix} 1 & 0 \\ 0 & 1 \end{pmatrix} = P^T G P.$$

*16. Assume that the metric of V is determined by the matrix

$$G = \begin{pmatrix} 3 & -1 \\ -1 & -2 \end{pmatrix}.$$

a. Prove that the resulting plane is the Lorentz plane. (Again, coordinate systems will have to be used. Later we shall be able to solve this problem much easier; see Exercise 167.15.)

b. Find a 2×2 matrix P such that

$$\begin{pmatrix} 1 & 0 \\ 0 & -1 \end{pmatrix} = P^T G P.$$

*17. Use the methods of the last two exercises to show that the matrices

$$G = \begin{pmatrix} 1 & \frac{3}{2} \\ \frac{3}{2} & 2 \end{pmatrix} \quad \text{and} \quad G' = \begin{pmatrix} -7 & \frac{3}{4} \\ \frac{3}{4} & -2 \end{pmatrix}$$

(of Exercise 14 above) determine, respectively, the Lorentz plane and the negative Euclidean plane.

Remark. After we have completed Section 31 on rectangular co-ordinate systems, this set of exercises will be much easier to work; in fact, they almost become trivial.

28. SUBSPACES

If U is a linear subspace of our metric vector space V, U is a metric vector space in its own right. The inner product of U is defined by: If A, $B \in U$, then AB is the inner product of A and B as vectors of V. According to the rules of logic, we are obliged to check that the five axioms for the inner product are satisfied for U, but this is a triviality. (This is where mathematical logic lets us down. It gives us no way to avoid trivial proofs.) We usually say subspace instead of linear subspace or metric linear subspace.

Example 1

Let V be the Lorentz plane. Assume that the coordinate system A_1, A_2 has been chosen in such a way that the metric is given by the matrix $\left(\begin{smallmatrix} 1 & 0 \\ 0 & -1 \end{smallmatrix}\right)$. Consider the one-dimensional linear subspaces (lines) $\langle A_1 \rangle, \langle A_2 \rangle, U = \langle A_1 + A_2 \rangle$, and $W = \langle A_1 - A_2 \rangle$; all are pictured in Figure 138.1. The vector

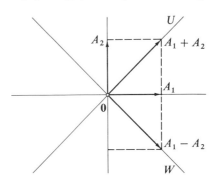

Figure 138.1

A_1 is a coordinate system for the line $\langle A_1 \rangle$ and the metric of $\langle A_1 \rangle$ is given by the 1×1 matrix $(A_1{}^2) = (1)$. Thus $\langle A_1 \rangle$ is the Euclidean line.

A_2 is a coordinate system for the line $\langle A_2 \rangle$ and the metric of $\langle A_2 \rangle$ is given by the 1×1 matrix $(A_1{}^2) = (-1)$. Hence $\langle A_2 \rangle$ is the negative Euclidean line.

Since

$$(A_1 + A_2)^2 = A_1{}^2 + 2A_1 A_2 + A_2{}^2 = 1 + 0 - 1 = 0,$$

we conclude that U is the null line. Analogously, W is the null line.

We know that there are only three nonisometric lines over $\mathbf{R}$, namely, the Euclidean line, the negative Euclidean line, and the null line (Exercise 135.13b). All three of these lines lie embedded in the Lorentz plane. Observe, in particular, that the Lorentz plane, although nonsingular, contains null lines.

Exercises

1. Prove that the Lorentz plane contains no null lines, except for the two lines $U = \langle A_1 + A_2 \rangle$ and $W = \langle A_1 - A_2 \rangle$ mentioned in Example 138.1. (Hint: Use Example 126.2.)
2. Let V be Euclidean n-space. Prove that every one-dimensional subspace of V (line) is the Euclidean line. (Actually, much more is true. In

Exercise 159.7b, we shall prove the following. Each m-dimensional subspace of V is Euclidean m-space for $m = 1, 2, \ldots, n$.)

3. Let V be negative Euclidean n-space. Prove that every line of V is the negative Euclidean line. (It is true that all subspaces of V are negative Euclidean spaces; see Exercise 166.11b.)

4. Let V be Minkowski space. Assume that the coordinate system A_1, A_2, A_3, A_4 has been chosen in such a way that the metric is given by the matrix

$$\begin{pmatrix} 1 & 0 & 0 & 0 \\ 0 & 1 & 0 & 0 \\ 0 & 0 & 1 & 0 \\ 0 & 0 & 0 & -1 \end{pmatrix}.$$

Put

$$
\begin{array}{ll}
U_1 = \langle A_1, A_2 \rangle & U_4 = \langle A_1, A_2, A_3 \rangle \\
U_2 = \langle A_1, A_4 \rangle & U_5 = \langle A_1, A_2, A_4 \rangle \\
U_3 = \langle A_2, A_4 \rangle & U_6 = \langle A_2, A_3, A_4 \rangle.
\end{array}
$$

a. For each of the above subspaces of V, find the matrix of the metric relative to the indicated coordinate system.

b. Identify each of the planes U_1, U_2, U_3. Conclude that U_2 and U_3 are isometric while U_1 is not isometric to either U_2 or U_3.

c. Prove that U_5 is isometric to U_6 while U_4 is not isometric to either U_5 or U_6. Which well known three-space is U_4?

*5. Let $k = \mathbf{R}, n = 2$, and assume the metric of V is given by the symmetric bilinear form $x_1 y_1$ relative to the coordinate system A_1, A_2.

a. Prove that $\langle A_1 \rangle$ is the Euclidean line and $\langle A_2 \rangle$ is the null line.

b. Prove that the lines $\langle A_1 \rangle$ and $\langle A_2 \rangle$ are orthogonal.

c. Prove that V is a singular plane which is the direct sum of a Euclidean line and an orthogonal null line.

Suppose that $\mathcal{W}$ is a finite dimensional vector space over k without a metric. To every subspace U of $\mathcal{W}$ is associated the quotient space $\mathcal{W}/U$. We now review from linear algebra how $\mathcal{W}/U$ becomes a vector space over k. Since U is a subgroup of the commutative group $\mathcal{W}$, every $A \in \mathcal{W}$ determines the coset

$$A + U = \{A + B \mid B \in U\}$$

of A modulo U. The commutativity of $\mathcal{W}$ makes it unnecessary to distinguish between left cosets and right cosets and we may write $U + A$ instead of $A + U$. These cosets are the vectors of the vector space $\mathcal{W}/U$. Addition of vectors and scalar multiplication are defined as follows: For $A, B \in \mathscr{W}$ and $t \in \mathscr{k}$,

$$(A + U) + (B + U) = (A + B) + U$$
$$t(A + U) = tA + U.$$

It must first be checked that the above definition makes sense. The danger lies in the fact that we are given the coset and not the vector A. Precisely, $A + U = A' + U$ if and only if $A - A' \in U$ which is a long way from $A = A'$. Once this danger has been overcome, it must be shown that the above definitions satisfy the axioms of a vector space. If the reader feels the need to review the notion of quotient space, we urge him to work out the details and do the next set of exercises. Of course, he should also visualize the vector space $\mathscr{W}/U$ correctly; the coset $A + U$ is obtained by translating the linear subspace U by the vector A and hence can easily be visualized. Figure 140.1 pictures the vector space $\mathscr{W}/U$ for dim $\mathscr{W} = 2$ and dim $U = 1$.

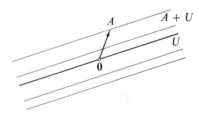

Figure 140.1

Exercises

In the following exercises, $\mathscr{W}$ is a vector space without a metric and U is a subspace of $\mathscr{W}$.

6. a. Prove that U is the origin of $\mathscr{W}/U$.
   b. Prove that $A + U$ is the origin of $\mathscr{W}/U$ if and only if $A \in U$.
   c. If $U = \{\mathbf{0}\}$, prove that the vector spaces $\mathscr{W}$ and $\mathscr{W}/U$ are identical. (For the purists: There is an abuse of set theory here.)
   d. If $U = \mathscr{W}$, prove that $\mathscr{W}/U$ consists of only the origin.
7. Let dim $\mathscr{W} = p$ and dim $U = q$.
   a. Prove that dim $\mathscr{W}/U = p - q$.
   b. If $p = 3$ and $q = 1$, how should $\mathscr{W}/U$ be visualized?
   c. If $p = 3$ and $q = 2$, how should $\mathscr{W}/U$ be visualized?

We now return to our metric vector space V. If U is a subspace of V, it is very tempting to define a metric for V/U by

$$(A + U)(B + U) = AB \qquad \text{for} \quad A, B \in V.$$

However, this definition simply does not make sense. If it did make sense and $A + U = A' + U$ and $B + U = B'' + U$, it would be necessary that $AB = A'B'$. However, all we know is that $A' = A + C$ and $B' = B + D$ where $C, D \in U$ and, therefore,

$$A'B' = AB + AD + CB + CD.$$

It is easy to give examples where $AD + CB + CD \neq 0$ (see Exercise 141.8). We come to the disappointing conclusion that the vector space V/U does not inherit a metric from V.

The previous sentence should be understood correctly. It is easy to give V/U many metrics; one simply chooses a coordinate system for V/U and a symmetric matrix. However, none of these metrics has any relation to the metric of V and this is what we mean by saying that V/U does not inherit a metric from V. We are forced to regard V/U as a vector space without a metric. There are exceptional subspaces of V whose quotient spaces inherit a metric from V and we discuss them in the next section.

Exercise

*8. Let V be three-dimensional Euclidean space and assume that the metric on V is given by the matrix

$$\begin{pmatrix} 1 & 0 & 0 \\ 0 & 1 & 0 \\ 0 & 0 & 1 \end{pmatrix}$$

relative to the coordinate system A_1, A_2, A_3. If

$$U = \langle A_1 + A_2, A_2 + A_3 \rangle,$$

show that V/U does not inherit the metric of V, that is, find vectors A, A', B, B' such that $A + U = A' + U$ and $B + U = B' + U$, but $AB \neq A'B'$.

29. THE RADICAL

To every subspace U of our metric vector space V corresponds the orthogonal complement U^* which is also a linear subspace. We may choose $U = V$; in this special case, the orthogonal complement of V is called the radical of V.

Definition 142.1.   The **radical** of V is the linear subspace

$$\{A \,|\, A \in V, \quad AB = 0 \quad \text{for all} \quad B \in V\}.$$

Notation: rad V.

Since rad V is a subspace of V, $\mathbf{0} \in$ rad V. It may very well happen that rad $V = \{\mathbf{0}\}$. In fact, it is an immediate consequence of the definition of non-singular spaces that rad $V = \{\mathbf{0}\}$ if and only if V is nonsingular. Since every vector A of rad V is orthogonal to all of V, A is certainly orthogonal to all vectors of rad V. Consequently, as a subspace of V, rad V *is a null space.* This shows, in particular, that all vectors of rad V are isotropic.

Example 1

In order to obtain a radical different from $\{\mathbf{0}\}$, we must consider a singular space. Let $n = 2$ and suppose that the metric of V is given by the bilinear form $x_1 y_1$ relative to the coordinate system A_1, A_2. We studied this plane in Exercise 139.5 in the case that $k = \mathbf{R}$. If

$$B = b_1 A_1 + b_2 A_2 \qquad \text{and} \qquad C = c_1 A_1 + c_2 A_2,$$

then we have $BC = b_1 c_1$. Consequently, $B \in$ rad V if and only if $b_1 c_1 = 0$ for all $c_1 \in k$; this is the same as saying that $b_1 = 0$. It follows that rad $V = \langle A_2 \rangle$. Since the discriminant of the line $\langle A_1 \rangle$ is 1, we see that V is the direct sum of its radical $\langle A_2 \rangle$ and the nonsingular line $\langle A_1 \rangle$; moreover, these two lines are orthogonal. We shall show in this section that every metric vector space is the direct sum of its radical and an orthogonal, nonsingular subspace.

Exercises

1. Prove that the Lorentz plane and Minkowski space have radical $\{\mathbf{0}\}$, but that both these spaces have null subspaces of positive dimension.
2. Let $n = 3$ and assume that the metric of V is given by the matrix

$$\begin{pmatrix} 1 & 0 & 0 \\ 0 & 1 & 0 \\ 0 & 0 & -1 \end{pmatrix}$$

   relative to some coordinate system. Prove that rad $V = \{\mathbf{0}\}$, but that V contains null subspaces of positive dimension.
3. Prove that Euclidean n-space and negative Euclidean n-space have radical $\{\mathbf{0}\}$ and contain no null subspaces of positive dimension.
4. Prove that every subspace of V is orthogonal to rad V.

5. Prove that two null spaces are isometric if and only if they have the same dimension.

*6. Let $\sigma: V \to W$ be an isometry from the metric vector space V onto the metric vector space W. Prove that $\sigma(\text{rad } V) = \text{rad } W$.

7. Prove that the radicals of isometric spaces have the same dimension and, therefore, the radicals are isometric.

The radical of V is one of the exceptional subspaces of V alluded to at the end of the previous section. The quotient space $V/\text{rad } V$ inherits a metric from V which is even nonsingular. If $A, B \in V$, we define the inner product of $A + \text{rad } V$ and $B + \text{rad } V$ of $V/\text{rad } V$ as AB.

Proposition 143.1.   Using the definition

$$(A + \text{rad } V)(B + \text{rad } V) = AB,$$

the quotient space $V/\text{rad } V$ becomes a nonsingular metric vector space.

Proof. To show that the definition makes sense, select $A, A', B, B' \in V$ in such a way that

$$A + \text{rad } V = A' + \text{rad } V \qquad \text{and} \qquad B + \text{rad } V = B' + \text{rad } V.$$

We must prove that $AB = A'B'$. Now $A' = A + C$ and $B' = B + D$ where $C, D \in \text{rad } V$. Therefore,

$$A'B' = AB + AD + CB + CD.$$

Since $C, D \in \text{rad } V$, they are orthogonal to all vectors of V and hence $AD + CB + CD = 0$; consequently, $AB = A'B'$. Knowing that the product $(A + \text{rad } V)(B + \text{rad } V) = AB$ is well defined, it must be shown that the five axioms for an inner product are satisfied. We leave this as an exercise for the reader.

Finally, we must show that the metric of $V/\text{rad } V$ is nonsingular. Assume that $A + \text{rad } V$ is orthogonal to all of $V/\text{rad } V$. We must prove that $A + \text{rad } V$ is the origin of $V/\text{rad } V$, that is, $A \in \text{rad } V$. For every $B \in V$,

$$(A + \text{rad } V)(B + \text{rad } V) = AB = 0$$

implies that $A \in \text{rad } V$. Done.

The metric of $V/\text{rad } V$ is completely determined by the metric of V. Hence to every metric vector space, there corresponds a unique nonsingular space, namely, $V/\text{rad } V$. The dimension of $V/\text{rad } V$ is called the **rank** of the metric vector space V.

Exercises

8. Let U be a subspace of rad V.
   a. Prove that V/U becomes a metric vector space under the definition
      $(A + U)(B + U) = AB$.
   b. Prove that V/U is nonsingular if and only if $U = \text{rad } V$.
9. Let U be a subspace of V. Prove that the product $(A + U)(B + U) = AB$ is well defined if and only if $U \subset \text{rad } V$. (The "if" part was done in the above exercise.)
10. If V is nonsingular, prove that the nonsingular space $V/\text{rad } V$ is identical to V.
*11. Let $\sigma: V \to W$ be an isometry. Prove that
   a. $V/\text{rad } V \simeq W/\text{rad } W$. (Hint: $\sigma(\text{rad } V) = \text{rad } W$ by Exercise 143.6.)
   b. If U is a subspace of V and $U \simeq V/\text{rad } V$, then $\sigma U \simeq W/\text{rad } W$.

In linear algebra, there is the notion of the direct sum $U_1 \oplus \cdots \oplus U_s$ of vector spaces $U_1, \ldots, U_s$. We recall that a vector A of $U_1 \oplus \cdots \oplus U_s$ can be written in one and only one way as a sum $A = A_1 + \cdots + A_s$ where $A_i \in U_i$ for $i = 1, \ldots, s$. For metric vector spaces, there is the stronger notion of orthogonal sum.

Definition 144.1. Let $U_1, \ldots, U_s$ be subspaces of the metric vector space V. Then V is called the **orthogonal sum** of $U_1, \ldots, U_s$ if the following two conditions are satisfied.

1. $V = U_1 \oplus \cdots \oplus U_s$.
2. $U_i \perp U_j$ for $i \neq j$.

Notation: $V = U_1 \oplus \cdots \oplus U_s$.

Let $V = U_1 \oplus \cdots \oplus U_s$. Each U_i is a metric vector space as a subspace of V. Furthermore, it is easy to see that the metric of V is completely determined by the metrics of $U_1, \ldots, U_s$. Namely, if $A, B \in V$, then $A = A_1 + \cdots + A_s$ and $B = B_1 + \cdots + B_s$ where $A_i, B_i \in U_i$ for $i = 1, \ldots, s$. Since $A_i B_j = 0$ for $i \neq j$, $AB = A_1 B_1 + \cdots + A_s B_s$. Here $A_i B_i$ is the inner product in the metric vector space U_i and hence the metrics of $U_1, \ldots, U_s$ determine the inner product AB completely.

Suppose that $\mathscr{W}$ is a vector space without a metric which is the direct sum of subspaces $U_1, \ldots, U_s$, that is, $\mathscr{W} = U_1 \oplus \cdots \oplus U_s$. If each U_i is a metric vector space, the above reasoning shows that there can be at most one metric for $\mathscr{W}$ satisfying

1. $\mathscr{W}$ induces on each U_i the given metric.
2. $\mathscr{W} = U_1 \oplus \cdots \oplus U_s$.

It is, furthermore, clear that such a metric always exists and is, in fact, described below.

If $A, B \in \mathcal{W}$, then

$$A = A_1 + \cdots + A_s \qquad \text{and} \qquad B = B_1 + \cdots + B_s$$

where $A_i, B_i \in U_i$. Define

$$AB = A_1 B_1 + \cdots + A_s B_s$$

where $A_i B_i$ is the inner product in the given metric of U_i.

We leave it to the reader to show that this definition satisfies the five axioms of the inner product and that the resulting metric for $\mathcal{W}$ has the two required properties.

Finally, let $U_1, \ldots, U_s$ be a totally arbitrary family of metric vector spaces. The spaces $U_1, \ldots, U_s$ may or may not be subspaces of the same larger space. If they are, they may have large intersections; it may even be that $U_1 = \cdots = U_s$. Anything goes! In this case, one means, by the orthogonal sum $U_1 \oplus \cdots \oplus U_s$ of these spaces, the direct sum $U_1 \oplus \cdots \oplus U_s$ with the metric described above.

Exercises

*12. Let $V = U \oplus N$ where U is a nonsingular space and N is a null space.
   a. Prove that $N = \text{rad } V$. (Hint: For the inclusion rad $V \subset N$, assume $A = B + C \in \text{rad } V$ where $B \in U$ and $C \in N$; then show that $B = \mathbf{0}$. The inverse inclusion is easy.)
   b. Prove that $U \simeq V/\text{rad } V$. (Hint: By Part a, we know that $V = U \oplus \text{rad } V$; consider the mapping $A \to A + \text{rad } V$ from U into $V/\text{rad } V$.)

13. Let W and W' be nonsingular spaces with N and N' null spaces. Assume that $W \oplus N \simeq W' \oplus N'$. Prove that $W \simeq W'$ and $N \simeq N'$.

*14. Let $V = U \oplus N$ where N is a null space and U is some subspace of V. Prove that $N = \text{rad } V$ if and only if U is nonsingular. (Hint: The "if" part was done in Exercise 12 above. For the "only if" part, show that $A \in \text{rad } U$ implies $A \in U \cap \text{rad } V = \{\mathbf{0}\}$.)

We now use the notions of radical and orthogonal sum to show that one may restrict oneself to nonsingular geometries in the study of metric vector spaces. If one knows everything about nonsingular spaces, one knows everything about all metric vector spaces. The basic reason is contained in the following proposition.

Proposition 146.1.   The metric vector space V is the orthogonal sum of its radical and a nonsingular space U where $U \simeq V/\mathrm{rad}\ V$.

Proof. Since rad V is a linear subspace of V, we know from linear algebra that there exists a linear subspace U of V such that $V = U \oplus \mathrm{rad}\ V$. Since rad V is orthogonal to all subspaces of V, $V = U \boxdot \mathrm{rad}\ V$; rad V is a null space and hence U is nonsingular (Exercise 145.14). Finally, the isometry $U \simeq V/\mathrm{rad}\ V$ was proved in Exercise 145.12. Done.

In general, the subspace U in the above proof is not unique. Nevertheless, its geometry is unique since $U \simeq V/\mathrm{rad}\ V$. Therefore, the geometry of V is completely determined by the geometry of the nonsingular space $V/\mathrm{rad}\ V$ and the dimension of rad V .This is the reason why we may restrict ourselves to nonsingular spaces when trying to classify metric vector spaces.

Exercises

15. Let $U_1, \ldots, U_s$ and $W_1, \ldots, W_s$ be metric vector spaces where $U_i \simeq W_i$ for $i = 1, \ldots, s$. Prove that $U_1 \boxdot \cdots \boxdot U_s \simeq W_1 \boxdot \cdots \boxdot W_s$.
16. Let V and W be metric vector spaces of the same dimension. If $V/\mathrm{rad}\ V \simeq W/\mathrm{rad}\ W$, prove that $V \simeq W$.
17. As always, dim $V = n$ and the rank of V is the dimension of $V/\mathrm{rad}\ V$.
    a. Prove that $n = \mathrm{rank}\ V + \dim(\mathrm{rad}\ V)$.
    b. Prove that V is nonsingular if and only if n is the rank of V.
18. Let $n = 2$ and assume that V has rank 1.
    a. Prove that rad V is a line.
    b. Prove that all lines of V different from rad V are nonsingular and isometric.
19. Let $V = U_1 \boxdot \cdots \boxdot U_s$. Prove that
    a. rad $V = (\mathrm{rad}\ U_1) \boxdot \cdots \boxdot (\mathrm{rad}\ U_s)$.
    b. $V/\mathrm{rad}\ V \simeq (U_1/\mathrm{rad}\ U_1) \boxdot \cdots \boxdot (U_s/\mathrm{rad}\ U_s)$.
    c. $\dim(\mathrm{rad}\ V) = \dim(\mathrm{rad}\ U_1) + \cdots + \dim(\mathrm{rad}\ U_s)$.
    d. rank $V = \mathrm{rank}\ U_1 + \cdots + \mathrm{rank}\ U_s$.
    e. V is nonsingular if and only if each U_i is nonsingular.
20. Let $V = U_1 \boxdot \cdots \boxdot U_s$. Choose a coordinate system $A_{i1}, \ldots, A_{in_i}$ for U_i, $i = 1, \ldots, s$ and consider the resulting coordinate system

$$A_{11}, \ldots, A_{1n_1}; \ \ldots; \ A_{s1}, \ldots, A_{sn_s}$$

for V.

a. Prove that the matrix of V relative to this coordinate system is decomposed into s diagonal blocks

$$\begin{pmatrix} G_1 & & & \text{O} \\ & G_2 & & \\ & & \ddots & \\ \text{O} & & & G_s \end{pmatrix}$$

where G_i is the matrix of U_i relative to the coordinate system $A_{i1}, \ldots, A_{in_i}$.

b. Prove that the discriminant of V is the product of the discriminants of each U_i; that is,

$$\text{disc } V = \prod_{i=1}^{s} \text{disc } U_i.$$

(Be honest in your proof; the discriminant is not the same as the determinant.)

21. Let r be the rank of V, $A_1, \ldots, A_n$ a coordinate system of V, and G the matrix of V relative to this coordinate system.

   a. Prove that there exists a nonsingular matrix P such that

   $$P^{\mathrm{T}}GP = \begin{pmatrix} G' & \text{O} \\ \text{O} & \text{O} \end{pmatrix}$$

   where G' is a nonsingular, symmetric $r \times r$ matrix.

   b. Prove that the rank of the matrix G is equal to the rank of V.

22. Let G be a symmetric $n \times n$ matrix of rank r. Prove that there exists a nonsingular $n \times n$ matrix P such that

   $$P^{\mathrm{T}}GP = \begin{pmatrix} G' & \text{O} \\ \text{O} & \text{O} \end{pmatrix}$$

   where G' is a nonsingular, symmetric $r \times r$ matrix.

30. ORTHOGONALITY

We now concentrate on nonsingular spaces. This does not mean that we forget all about singular geometries because a nonsingular space may well contain singular subspaces. The two null lines of the Lorentz plane and the infinitely many null lines of Minkowski space are specific examples. In general, a nonsingular space may contain many singular subspaces which are not null spaces. (See Exercises 151.4, 152.6, and 152.7.)

The possible occurrence of singular subspaces in a nonsingular space is the only thing one has to become accustomed to when thinking about non-Euclidean geometries. In Euclidean n-space, all subspaces are again Euclidean and hence nonsingular. This is not true, in general, and a large part of the geometry we develop is devoted to taking care of the singular subspaces of nonsingular geometries.

We begin by investigating the orthogonal complement U^* of a linear subspace U of V. As always, V is a metric vector space of dimension n.

Definition 148.1.   Let U be a subspace of V. The linear subspace $\{A \mid A \in V,\, AB = 0 \text{ for all } B \in U\}$ of V is called the **orthogonal complement** of U. Notation: U^*.

It was proved in Exercise 116.3 that U^* is a linear subspace of V. The following proposition gives the principal reason it is such an advantage to restrict oneself to nonsingular spaces.

Proposition 148.1.   Let V be nonsingular. If U is a subspace of V, $\dim U + \dim U^* = n$.

Proof. Let $A_1, \ldots, A_n$ be a coordinate system for V and G the matrix of V relative to this coordinate system. We assume that $\dim U = m$ and choose a coordinate system $B_1, \ldots, B_m$ for U. If

$$B_i = b_{i1} A_1 + \cdots + b_{in} A_n$$

for $i = 1, \ldots, m$, we know from linear algebra that the $m \times n$ matrix

$$B = \begin{pmatrix} b_{11} & \cdots & b_{1n} \\ \vdots & & \vdots \\ b_{m1} & \cdots & b_{mn} \end{pmatrix}$$

has rank m. Since V is nonsingular, $\det G \neq 0$ and hence the $m \times n$ matrix BG also has rank m.

A vector $X = x_1 A_1 + \cdots + x_n A_n$ belongs to U^* if and only if

$$(b_{i1}, \ldots, b_{in}) G \begin{pmatrix} x_1 \\ \vdots \\ x_n \end{pmatrix} = 0 \qquad \text{for} \quad i = 1, \ldots, m.$$

In other words, U^* is the solution set of the system of m homogeneous equations in n variables

$$BG \begin{pmatrix} x_1 \\ \vdots \\ x_n \end{pmatrix} = 0.$$

Since the matrix BG has rank m, this solution set has dimension $n - m$. Done.

Remark. There is a much more elegant proof of the above proposition which does not make use of coordinate systems. (See Artin [2], Theorem 3.5, p. 117.) We avoided this proof because it is based on a part of linear algebra which is not very well known and takes many pages to develop, namely, the theory of bilinear forms. The reader can find this theory in Artin [2], pp. 16–23. In his "Suggestions for the Use of this Book," Artin says (of pages 16–23), "The content of this paragraph is of such a fundamental importance for most of mathematics that every effort should be devoted to its mastery."

Proposition 148.1 expresses a geometric property of the orthogonal complement to which we are well accustomed from Euclidean geometry. For example, in the Euclidean plane the orthogonal complement of a line is a line while in Euclidean three-space, the orthogonal complement of a line is a plane. The next proposition is of the same nature. Both propositions are false if V is singular (see Exercise 152.8).

Proposition 149.1. Let V be nonsingular. If U is a subspace of V,

1. $U^{**} = U$.
2. $U \cap U^* = \text{rad } U = \text{rad } U^*$.
3. The following five statements are equivalent.
   - a. U is nonsingular.
   - b. U^* is nonsingular.
   - c. $U \cap U^* = \{0\}$.
   - d. $V = U + U^*$.
   - e. $V = U \oplus U^*$.

Proof. 1. It is obvious that $U \subset U^{**}$. By Proposition 148.1,

$$\dim U^{**} = n - \dim U^* = n - (n - \dim U) = \dim U.$$

Since $U \subset U^{**}$ and $\dim U = \dim U^{**}$, we conclude that $U = U^{**}$.

2. $U \cap U^*$ consists of all vectors of U which are orthogonal to all of U which is the same as saying that $U \cap U^* = \text{rad } U$. If we apply this same reasoning to U^*, we obtain $U^* \cap U^{**} = \text{rad } U^*$. But by 1, $U^{**} = U$ and hence rad $U = \text{rad } U^*$.

3. The equivalence of a, b, and c is an immediate consequence of 2. Furthermore,

$$\dim(U + U^*) + \dim(U \cap U^*) = \dim U + \dim U^* = n.$$

Since $U + U^* = V$ if and only if $\dim(U + U^*) = n$, the equivalence of c and d follows. Finally, e $\Rightarrow$ d trivially while d $\Rightarrow$ e follows from the fact that U^* is orthogonal to U and d $\Rightarrow$ c. Done.

In Euclidean n-space E_n, all subspaces U are nonsingular and hence $U \cap U^* = \{0\}$ and $E_n = U \oplus U^*$. This agrees with the usual picture we have in Euclidean three-space of a line U which is orthogonal to a plane U^*.

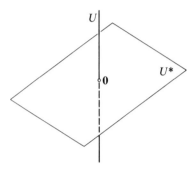

Figure 150.1

Exercises

1. A metric vector space is called **anisotropic** if $A^2 \neq 0$ for every nonzero vector A.
   a. Prove that V is anisotropic if and only if all its subspaces are nonsingular.
   b. If V is anisotropic and U is a subspace of V, prove that $U \cap U^* = \{0\}$ and $V = U \oplus U^*$. (Hence Figure 150.1 gives the right picture for all anisotropic spaces.)

*2. Let U be a subspace of V.
   a. Prove that rad $U \subset U^*$.
   b. If U is a null space, prove that $U \subset U^*$ and, when V is nonsingular, $U = $ rad U^*. (Thus Figure 150.1 is very dangerous when V is not anisotropic. It gives the right picture only if U is nonsingular.)

3. Let V be the Lorentz plane. Assume that a coordinate system A_1, A_2 has been chosen so that $\left(\begin{smallmatrix} 1 & 0 \\ 0 & -1 \end{smallmatrix}\right)$ is the matrix of V.
   a. If $U = \langle A_1 + A_2 \rangle$, prove that $U = U^*$. (Hint: Use Proposition 148.1 to compute $\dim U^*$; then use Exercise 2 above.)
   b. If $U = \langle A_1 - A_2 \rangle$, prove that $U = U^*$.
   c. Find the orthogonal complement of the line $\langle A \rangle$ where A has coordinates, respectively, $(1, 0)$, $(0, 1)$, $(1, 2)$.

*4. Let $n = 3$, $k = \mathbf{R}$, and the matrix of V be

$$\begin{pmatrix} 1 & 0 & 0 \\ 0 & 1 & 0 \\ 0 & 0 & -1 \end{pmatrix}$$

relative to the coordinate system A_1, A_2, A_3. This space was discussed in Exercise 131.9. There we saw that the light cone C of V has the equation $x_1^2 + x_2^2 - x_3^2 = 0$ and hence is a cone of revolution with $\langle A_3 \rangle$ as axis. Moreover, the null lines of V are the generators of C. See Figure 151.1. Let U be a null line of V.

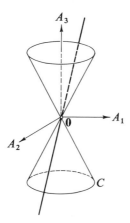

Figure 151.1

a. Prove that U^* is a singular plane whose radical is U. Consequently, U^* has rank 1 and is not a null plane. Here then is a nonsingular space which contains singular planes which are not null planes.
b. Prove that U is the only null line of U^*.
c. Prove that U^* is the tangent plane of C along the generator U. (Hint: This is a geometric consequence of Part b.) See Figure 151.2.

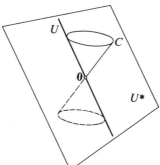

Figure 151.2

*5. Let U be a null line of Minkowski space. Prove that U^* is a singular three-space of rank 2 and hence is not a null space. Also prove that $U = \text{rad } U^*$.

*6. We now generalize Exercises 4 and 5 above to arbitrary dimension and any field k (char $k \neq 2$). Assume that V is nonsingular and contains a null line U.

a. Prove that $n \geq 2$.

b. Prove that U^* is an $(n-1)$-dimensional singular subspace of V with rank $n-2$ and rad $U^* = U$.

*7. An $(n-1)$-dimensional subspace of an n-dimensional vector space is called a **hyperplane**. Let V be nonsingular and not anisotropic and assume that $n \geq 3$. Prove that V contains singular hyperplanes which are not null spaces.

*8. Assume $n = 3$ and A_1, A_2, A_3 is a coordinate system of V such that

$$\begin{pmatrix} 1 & 0 & 0 \\ 0 & 0 & 0 \\ 0 & 0 & 0 \end{pmatrix}$$

is the matrix of V.

a. Prove that V is singular.

b. Find rad V.

c. Compute the rank of V.

d. Find $\langle A_3 \rangle^*$ and observe that Proposition 148.1 does not hold for singular spaces.

e. Find $\langle A_3 \rangle^{**}$ and observe that Proposition 149.1 does not hold for singular spaces.

9. We generalize Exercise 8 above to arbitrary dimension n. Let V be singular and U a nonzero subspace of rad V.

a. Prove that dim $U + \dim U^* = \dim U + n > n$.

b. Prove that $U^{**} = \text{rad } V$. Hence, if $U \neq \text{rad } V$, then $U^{**} \neq U$.

*10. Let U and W be linear subspaces of a nonsingular space V.

a. Prove that $U^* \perp W^* \Leftrightarrow U^* \subset W \Leftrightarrow W^* \subset U$. (Hint: Use Exercise 116.3b.)

b. If dim $V = 3$ and U, W are planes, prove that $U^* \perp W^*$ if and only if U (W) contains a line which is perpendicular to W (U). Thus two planes U and W of Euclidean three-space are perpendicular in the classical sense (see page 116) if and only if $U^* \perp W^*$.

31. RECTANGULAR COORDINATE SYSTEMS

The Euclidean plane has coordinate systems which consist of two ortho-gonal vectors. Similarly, Euclidean three-space has coordinate systems con-sisting of three mutually orthogonal vectors; see Figure 153.1. It is the purpose

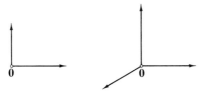

Figure 153.1

of this section to show that a metric vector space always possesses rectangular coordinate systems and draw some conclusions from this property.

Definition 153.1. A coordinate system $A_1, \ldots, A_n$ of V is called a **rectangular coordinate system** if $A_i \perp A_j$ for $i \neq j$.

If n linearly independent vectors of V are mutually orthogonal but not ordered, they are called an **orthogonal basis**. Since we always order our bases so that every vector obtains an n-tuple of coordinates $(x_1, \ldots, x_n)$, we shall stick to the term rectangular coordinate system.

Should one admit isotropic vectors in a rectangular coordinate system? The answer is that one has no control over this. The geometry of V decides how many isotropic vectors occur in a rectangular coordinate system.

Proposition 153.1. Let $A_1, \ldots, A_n$ be a rectangular coordinate system of V. Suppose that $A_i^2 \neq 0$ for $i = 1, \ldots, r$ and $A_i^2 = 0$ for $i = r + 1, \ldots, n$. Then

$$\langle A_1, \ldots, A_r \rangle \simeq V/\mathrm{rad}\ V \quad \text{and} \quad \langle A_{r+1}, \ldots, A_n \rangle = \mathrm{rad}\ V.$$

Proof. It is obvious that

$$V = \langle A_1, \ldots, A_r \rangle \oplus \langle A_{r+1}, \ldots, A_n \rangle.$$

The metric of $\langle A_1, \ldots, A_r \rangle$ is nonsingular since it is given by the matrix

$$\begin{pmatrix} A_1^2 & & \bigcirc \\ & \ddots & \\ \bigcirc & & A_r^2 \end{pmatrix}$$

and hence has the nonzero discriminant $A_1{}^2 \cdots A_r{}^2$. Similarly, $\langle A_{r+1}, \ldots, A_n \rangle$ is a null space since its matrix is the zero matrix. Since V is the orthogonal sum of the nonsingular space $\langle A_1, \ldots, A_r \rangle$ and the null space $\langle A_{r+1}, \ldots, A_n \rangle$, the desired conclusions follow (Exercise 145.12). Done.

We see from the above proposition that the number of isotropic vectors in a rectangular coordinate system of V is the dimension of rad V and that the number of nonisotropic vectors is the rank of V. In particular, there are no isotropic vectors in a rectangular coordinate system of a nonsingular space.

Before showing that a metric vector space always has a rectangular coordinate system, we prove that orthogonality of *nonisotropic vectors* implies that they are linearly independent. We are not assuming that V is nonsingular.

Proposition 154.1. If m nonisotropic vectors $A_1, \ldots, A_m$ of V are mutually orthogonal, they are linearly independent.

Proof. Suppose that $a_1 A_1 + \cdots + a_m A_n = 0$. Then for $1 \leq i \leq m$,

$$(a_1 A_1 + \cdots + a_m A_m)A_i = 0A_i = 0.$$

On the other hand,

$$(a_1 A_1 + \cdots + a_m A_m)A_i = a_i A_i{}^2.$$

Consequently, $a_i A_i{}^2 = 0$ and, since $A_i{}^2 \neq 0$, $a_i = 0$. Done.

Exercises

1. Let $A_1, \ldots, A_n$ be a rectangular coordinate system of V.
   a. Prove that the matrix of V is

   $$\begin{pmatrix} A_1{}^2 & & \bigcirc \\ & \ddots & \\ \bigcirc & & A_n{}^2 \end{pmatrix}.$$

   b. Prove that V is nonsingular if and only if none of the n vectors A_i is isotropic.
   c. Prove that $V = \langle A_1 \rangle \oplus \cdots \oplus \langle A_n \rangle$.

*2. Assume $V = U \oplus W$. If $A_1, \ldots, A_r$ is a rectangular coordinate system of U and $A_{r+1}, \ldots, A_n$ is a rectangular coordinate system of W, prove that $A_1, \ldots, A_n$ is a rectangular coordinate system of V.

3. Let $k = \mathbf{R}$. Write the standard matrix for each of the following geometries and show that the underlying coordinate system is rectangular.
   a. Euclidean n-space.
   b. Negative Euclidean space.
   c. The Lorentz plane.
   d. Minkowski space.
   e. The orthogonal sum of the Euclidean plane and the negative Euclidean line. What kind of a cone is the light cone of this space?
4. Use rectangular coordinate systems to prove that
   a. The Lorentz plane is the orthogonal sum of a Euclidean line and a negative Euclidean line.
   b. Minkowski space is the orthogonal sum of Euclidean three-space and a negative Euclidean line; of a Euclidean plane and a Lorentz plane; of the space of Exercise 3e above and a Euclidean line; of three Euclidean lines and a negative Euclidean line.

We now prove the main theorem of this section. Since a zero-dimensional vector space has no coordinate systems, we must assume $n \geq 1$.

Theorem 155.1.   If $n \geq 1$, V has a rectangular coordinate system.

Proof. Case 1. V is nonsingular. If $n = 1$, every nonzero vector of V is a rectangular coordinate system. We assume dim $V = n$ and make the induction hypothesis that the theorem holds for all nonsingular spaces of dimension less than n. Since V is nonsingular, there is a nonisotropic vector A_1 in V and the line $\langle A_1 \rangle$ is nonsingular. Consequently,

$$V = \langle A_1 \rangle \oplus \langle A_1 \rangle^*$$

where $\langle A_1 \rangle^*$ is a nonsingular space of dimension $n - 1$. By the induction hypothesis, $\langle A_1 \rangle^*$ has a rectangular coordinate system $A_2, \ldots, A_n$; then $A_1, \ldots, A_n$ is a rectangular coordinate system of V (Exercise 154.2).
   Case 2. V is singular. We put

$$V = U \oplus \text{rad } V$$

where U is nonsingular and rad $V \neq \{0\}$. By Case 1, U has a rectangular coordinate system $A_1, \ldots, A_r$. Every coordinate system $A_{r+1}, \ldots, A_n$ of the null space rad V is a rectangular coordinate system. The vectors $A_1, \ldots, A_n$ form a rectangular coordinate system of V. Done.

Exercises

5. If $A_1, \ldots, A_n$ is a rectangular coordinate system of V and $a_1, \ldots, a_n$ are nonzero scalars, prove that $a_1 A_1, \ldots, a_n A_n$ is also a rectangular coordinate system.

6. Let G be an $n \times n$ symmetric matrix of rank r. Prove that there exists an $n \times n$ nonsingular matrix P such that

$$
P^{\mathsf{T}} G P = \begin{pmatrix} a_1 & & & & & \\ & \ddots & & & \bigcirc & \\ & & a_r & & & \\ & & & 0 & & \\ & \bigcirc & & & \ddots & \\ & & & & & 0 \end{pmatrix}
$$

where $a_i \neq 0$, $i = 1, \ldots, r$. (Hint: Choose a coordinate system $A_1, \ldots, A_n$ for V and give V the metric corresponding to G. Then choose a rectangular coordinate system.)

Although Theorem 155.1 is extremely useful, it is not very deep. Nevertheless, the fact that every metric vector space has a rectangular coordinate system is not trivial. Our geometric intuition tells us that no great geometry can spring from just the existence of rectangular coordinate systems. Here, our geometric intuition serves us better than our algebraic intuition. The existence of rectangular coordinate systems is equivalent to the theorem which states that every symmetric matrix is equivalent to a diagonal matrix. Hence, in order to classify symmetric matrices under the congruence relation, it is only necessary to decide when two diagonal matrices are congruent. This appears to be a tremendous simplification of the problem of classifying symmetric matrices. But it is not! Sylvester's solution of the classification problem of metric vector spaces over the real numbers (Section 34) and the difficult solution of the same problem over the rational numbers (see O'Meara [18]) show this very clearly. Artin once said in class, "The reduction of symmetric matrices to diagonal form hasn't begun to scratch the classification problem."

In the next two sections, we shall show how far one can go in classifying metric vector spaces by use of rectangular coordinate systems in the case of two special kinds of fields.

32. CLASSIFICATION OF SPACES OVER FIELDS WHOSE ELEMENTS HAVE SQUARE ROOTS

We recall that every algebraically closed field, and hence the field of complex numbers, has the property that every element has a square root.

Proposition 157.1.   If every element of k has a square root, two metric vector spaces over k are isometric if and only if they have the same dimension and rank. If $n \geq 1$, there exists a rectangular coordinate system of V such that the matrix of V becomes

$$\begin{pmatrix} 1 & & & & \bigcirc \\ & \ddots & & & \\ & & 1 & & \\ & & & 0 & \\ & & & & \ddots \\ \bigcirc & & & & 0 \end{pmatrix}.$$

The number of ones is the rank of V.

Proof. Case 1. V is nonsingular. Let $A_1, \ldots, A_n$ be a rectangular coordinate system of V. We put $A_i^2 = a_i$ for $i = 1, \ldots, n$; since V is nonsingular, all $a_i \neq 0$. Consider the new rectangular coordinate system

$$\frac{1}{\sqrt{a_1}} A_1, \ldots, \frac{1}{\sqrt{a_n}} A_n.$$

Since $[(1/\sqrt{a_i})A_i]^2 = (1/a_i)A_i^2 = 1$, the matrix of V, with respect to the new coordinate system, is

$$\begin{pmatrix} 1 & & \bigcirc \\ & \ddots & \\ \bigcirc & & 1 \end{pmatrix}.$$

This shows, in particular, that two nonsingular spaces are isometric if and only if they have the same dimension.

Case 2. V is singular. We put $V = U \perp \mathrm{rad}\, V$ where $U \simeq V/\mathrm{rad}\, V$. The nonsingular space U is either $\{0\}$ or has a rectangular coordinate system $A_1, \ldots, A_r$, where r is the rank of V and $A_i^2 = 1$ for $i = 1, \ldots, r$ (Case 1). If

$A_{r+1}, \ldots, A_n$ is any coordinate system of rad V, the matrix of V relative to the coordinate system $A_1, \ldots, A_n$ is

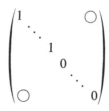

where r is the number of ones. Consequently, two spaces are isometric if and only if they have the same dimension and rank. Done.

A rectangular coordinate system $A_1, \ldots, A_n$ with the property that $A_1{}^2 = \cdots = A_n{}^2 = 1$ is called an **orthonormal coordinate system** or an **orthonormal basis** if the vectors are not ordered. It is rare that a metric vector space has an orthonormal coordinate system for, when this happens, the discriminant of the space is obviously 1. Even when the discriminant is 1, a space seldom has an orthonormal coordinate system (see Exercise 158.3). However, if every element of k has a square root, we see, from Proposition 157.1, that every nonsingular space of dimension at least one has an orthonormal coordinate system.

Exercises

1. Prove that two isometric spaces always have the same rank and dimension.
2. Prove that the Lorentz plane and Minkowski space have no orthonormal coordinate system.
*3. Prove that the negative Euclidean plane has no orthonormal coordinate system even though its discriminant is 1.
4. Let $k = \mathbf{Z}_5$ and $n = 2$. Assume that the metric of V is given by $\left(\begin{smallmatrix} 1 & 2 \\ 2 & 2 \end{smallmatrix}\right)$ relative to some coordinate system. Prove that V has no orthonormal coordinate system.
*5. Assume that V represents only squares, that is, A^2 is a square in k for every $A \in V$. This clearly happens when every element of k has a square root. It may also happen when every element of k does not have a square root. For example, Euclidean n-space represents only squares even though the negative real numbers have no square root in $\mathbf{R}$.

a. If $n \geq 1$ and V represents only squares, prove that V has a coordinate system relative to which the matrix of V is

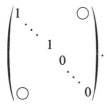

b. Prove that two spaces which represent only squares are isometric if and only if they have the same dimension and rank.

*6. In Example 126.1, Euclidean n-space E_n was introduced as an n-dimensional metric vector space over **R** with an orthonormal coordinate system. Prove that E_n may also be defined as an n-dimensional metric vector space over **R** which is positive definite, meaning that $A^2 > 0$ for all $A \neq \mathbf{0}$. This also makes the zero-dimensional vector space $\{\mathbf{0}\}$ Euclidean as one wants. (Hint: Apply Exercise 158.5. Observe that the step from an orthonormal coordinate system to positive definite is trivial.)

*7. Assume k is an ordered field and V is positive definite, that is, $A^2 > 0$ for all $A \neq \mathbf{0}$.

a. Prove that any subspace of V is also positive definite. Thus the property of being positive definite is inherited by all subspaces of V.

b. Prove that every m-dimensional subspace of Euclidean n-space is a Euclidean m-space.

33. CLASSIFICATION OF SPACES OVER ORDERED FIELDS WHOSE POSITIVE ELEMENTS HAVE SQUARE ROOTS

We assume for this whole section that k is ordered and that every positive element of k has a square root. The field **R** has this property; **Q** does not have this property (why not?) even though it is ordered. The field **C** of complex numbers cannot be ordered. Actually, **R** belongs to a large class of fields, all of which have the above property, namely, the "real closed fields" (see S. Lang, *Algebra*, pp. 271–282, [15]). Readers who are not acquainted with ordered fields should restrict themselves to **R**.

Proposition 159.1.   Assume that V is nonsingular and $n \geq 1$. Then V has a rectangular coordinate system $A_1, \ldots, A_n$ where $A_1{}^2 = \cdots = A_r{}^2 = 1$ and $A_{r+1}^2 = \cdots = A_n{}^2 = -1$ for some r, $0 \leq r \leq n$.

Proof. Let $A_1, \ldots, A_n$ be a rectangular coordinate system of V. Since V is nonsingular, we can put $A_i^2 = a_i$ where $a_i \neq 0$. Assume that the vectors $A_1, \ldots, A_n$ have been ordered so that $a_i > 0$ for $i = 1, \ldots, r$ and $a_j < 0$ for $j = r + 1, \ldots, n$. Since every positive element of k has a square root, there are elements $c_1, \ldots, c_n \in k$ such that $c_i^2 = a_i$ for $i = 1, \ldots, r$ and $c_j^2 = -a_j$ for $j = r + 1, \ldots, n$ ($c_i = \sqrt{a_i}$ and $c_j = \sqrt{-a_j}$). Consider the new rectangular coordinate system

$$c_1^{-1}A_1, \ldots, c_n^{-1}A_n.$$

Then $(c_i^{-1}A_i)^2 = a_i^{-1}A_i^2 = 1$ for $i = 1, \ldots, r$ and $(c_j^{-1}A_j)^2 = -a_j^{-1}A_j^2 = -1$ for $j = r + 1, \ldots, n$. Done.

Clearly, the matrix of V relative to the new coordinate system in the above proposition is

$$\begin{pmatrix} 1 & & & & & & \bigcirc \\ & \ddots & & & & & \\ & & 1 & & & & \\ & & & -1 & & & \\ & & & & \ddots & & \\ \bigcirc & & & & & & -1 \end{pmatrix}$$

with r ones and $n - r$ negative ones on the main diagonal. There are only $n + 1$ matrices of this type, one for each $r = 0, \ldots, n$. For example, when $n = 3$, the four matrices are

$$\begin{pmatrix} 1 & 0 & 0 \\ 0 & 1 & 0 \\ 0 & 0 & 1 \end{pmatrix}, \quad \begin{pmatrix} 1 & 0 & 0 \\ 0 & 1 & 0 \\ 0 & 0 & -1 \end{pmatrix}, \quad \begin{pmatrix} 1 & 0 & 0 \\ 0 & -1 & 0 \\ 0 & 0 & -1 \end{pmatrix}, \quad \begin{pmatrix} -1 & 0 & 0 \\ 0 & -1 & 0 \\ 0 & 0 & -1 \end{pmatrix}.$$

This already shows that there are at most $n + 1$ nonisometric, nonsingular spaces of dimension n. Sylvester's theory shows that no two of these $n + 1$ matrices represent isometric geometries (Corollary 178.1) and hence there are precisely $n + 1$ nonisometric, nonsingular spaces of dimension n.

Assume $k = \mathbf{R}$ and let us look again at some familiar spaces now that we have Proposition 159.1. There are precisely two nonisometric, nonsingular lines over $\mathbf{R}$, the Euclidean line and the negative Euclidean line. This result was actually obtained prior to Proposition 159.1 (Exercise 135.11). We now consider the cases when $n = 2$ and $n = 4$. The three-dimensional case is the subject of Exercise 162.1. It is convenient to use the term *real space* when talking about a metric vector space over $\mathbf{R}$.

Example 1

Assume $k = \mathbf{R}$ and $n = 2$. Every nonsingular real plane has a coordinate system such that its matrix is one of the following.

$$\begin{pmatrix} 1 & 0 \\ 0 & 1 \end{pmatrix}, \quad \begin{pmatrix} 1 & 0 \\ 0 & -1 \end{pmatrix}, \quad \begin{pmatrix} -1 & 0 \\ 0 & -1 \end{pmatrix}.$$

For as low a dimension as two, it is not difficult to show, without using Sylvester's theory, that the corresponding three planes are not isometric and this was done in Exercise 135.7. Hence we know that *the Euclidean plane, the Lorentz plane, and the negative Euclidean plane are the only nonisometric, nonsingular real planes.*

Example 2

Let $k = \mathbf{R}$ and $n = 4$. Every nonsingular real four-space has a coordinate system such that its matrix is one of the following.

$$G_1 = \begin{pmatrix} 1 & 0 & 0 & 0 \\ 0 & 1 & 0 & 0 \\ 0 & 0 & 1 & 0 \\ 0 & 0 & 0 & 1 \end{pmatrix}, \quad G_2 = \begin{pmatrix} 1 & 0 & 0 & 0 \\ 0 & 1 & 0 & 0 \\ 0 & 0 & 1 & 0 \\ 0 & 0 & 0 & -1 \end{pmatrix},$$

$$G_3 = \begin{pmatrix} 1 & 0 & 0 & 0 \\ 0 & 1 & 0 & 0 \\ 0 & 0 & -1 & 0 \\ 0 & 0 & 0 & -1 \end{pmatrix}, \quad G_4 = \begin{pmatrix} 1 & 0 & 0 & 0 \\ 0 & -1 & 0 & 0 \\ 0 & 0 & -1 & 0 \\ 0 & 0 & 0 & -1 \end{pmatrix},$$

$$G_5 = \begin{pmatrix} -1 & 0 & 0 & 0 \\ 0 & -1 & 0 & 0 \\ 0 & 0 & -1 & 0 \\ 0 & 0 & 0 & -1 \end{pmatrix}.$$

The matrices G_1, G_2, and G_5 determine, respectively, Euclidean four-space, Minkowski space, and negative Euclidean space of dimension four. We refer to the geometry determined by G_4 as *negative Minkowski space* since the geometry is obtained from Minkowski space by replacing AB by $-AB$ for all $A, B \in V$ (see Exercise 166.14c). Finally, the geometry of G_3 is called *Artinian four-space* for reasons which will become clear in Section 35 on "Artinian Spaces." We will now show that these five geometries are not isometric without using Sylvester's theory. The difficulties encountered will be testimony to the power of Sylvester's theory which gives the same result in one line.

We first use the fact that isometric spaces represent the same set of scalars (Exercise 134.4). Euclidean four-space is not isometric to any of the other spaces because it represents only nonnegative real numbers while the other four geometries represent the number -1 and, therefore, it is not isometric to any of the other geometries.

Minkowski space cannot be isometric to Artinian four-space since they have discriminants -1 and 1, respectively. For the same reason, negative Minkowski space is not isometric to Artinian four-space.

There only remains to show that the two Minkowski spaces are not isometric. Neither of the arguments used thus far can be used since both of these spaces have discriminant -1 and they both represent all real numbers. Suppose these spaces were isometric and let us produce a contradiction. Since negative Minkowski space contains a negative Euclidean three-space (why?), Minkowski space would also have to contain a negative Euclidean three-space U. On the other hand, Minkowski space contains a Euclidean three-space E_3 (why?) and we know from linear algebra that

$$\dim(E_3 \cap U) \geq \dim E_3 + \dim U - 4 = 3 + 3 - 4 = 2.$$

Consequently, $E_3 \cap U$ is either a plane or a three-space, all of whose vectors A have a square A^2 which is simultaneously positive and negative. But this is impossible; hence the two Minkowski spaces are not isometric.

We conclude that Euclidean four-space, Minkowski space, Artinian four-space, negative Minkowski space, and negative Euclidean four-space are the only nonisometric, nonsingular real four-spaces. This concludes our discussion of Example 2.

The heuristic methods used above to settle which real spaces of dimensions two and four are isometric get out of hand for large dimensions; here is where Sylvester's theory comes to the rescue. If tools no more sophisticated than rectangular coordinate systems are used, the classification of metric vector spaces (equivalently, of quadratic forms) cannot be pushed appreciably beyond the two cases of this section.

Exercises

*1. Prove (without using Sylvester's theory) that the only nonisometric, nonsingular real three-spaces are Euclidean three-space, negative Euclidean three-space, the orthogonal sum of a Lorentz plane with a Euclidean line, and the orthogonal sum of a Lorentz plane with a negative Euclidean line.

*2. Let V be a nonsingular real space of dimension ≥ 2 which is neither Euclidean space nor negative Euclidean space.

a. Prove that V contains a Lorentz plane.

b. Prove that V represents all scalars. (Hint: Use Exercise 130.2e.)

3. Assume V is a real space (which may be singular).

a. Prove that V has a rectangular coordinate system such that the matrix of V is

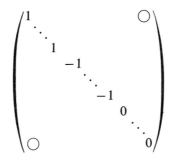

b. Prove that the number of zeros is the dimension of the radical of V and the number of ones and negative ones is the rank of V.

4. Let G be a symmetric $n \times n$ matrix with real numbers as entries. Prove that G is congruent to a matrix of the form given in Exercise 3 above where the number of ones and negative ones is the rank of G.

*5. Assume V is a real space which contains a Euclidean space E and a negative Euclidean space U. Prove that

a. $E \cap U = \{\mathbf{0}\}$.

b. $\dim E + \dim U \leq n$.

6. Let V be the Lorentz plane and A_1, A_2 be a coordinate system for V such that the matrix of V is $\begin{pmatrix} 1 & 0 \\ 0 & -1 \end{pmatrix}$. Picture this plane, as usual, with A_1 and A_2 appearing as horizontal and vertical vectors and the two null lines $U = \langle A_1 + A_2 \rangle$ and $W = \langle A_1 - A_2 \rangle$ as "45° lines." See Figure 163.1. Consider the four quadrants I, II, III, and IV determined by U and W.

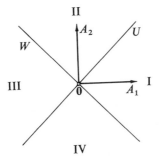

Figure 163.1

a. Prove that all lines (through **0**) which lie in quadrants I and III are Euclidean lines. An example is the coordinate axis $\langle A_1 \rangle$.
b. Prove that all lines which lie in quadrants II and IV are negative Euclidean lines. An example is the coordinate axis $\langle A_2 \rangle$.
*7. Let V be the real three-space whose metric is given by the matrix

$$\begin{pmatrix} 1 & 0 & 0 \\ 0 & 1 & 0 \\ 0 & 0 & -1 \end{pmatrix}$$

relative to the coordinate system A_1, A_2, A_3. We recall that the light cone has the equation $x_1{}^2 + x_2{}^2 - x_3{}^2 = 0$ and hence is a cone of revolution with $\langle A_3 \rangle$ as axis. See Figure 164.1. Prove that

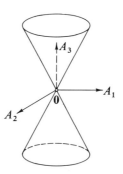

Figure 164.1

a. A plane which has two lines in common with the light cone is a Lorentz plane. (Hint: If P is a plane which has two lines in common with the light cone, choose two nonzero vectors A_1 and A_2 in P, one on each of the lines in question. Relative to the coordinate system A_1, A_2, the metric of P is given by the matrix

$$\begin{pmatrix} 0 & a \\ a & 0 \end{pmatrix}$$

where $a = A_1 A_2$. If $a = 0$, P would be a null plane and would, therefore, lie in its entirety on the light cone. But this is impossible.)
b. A plane which has only the point $\{0\}$ in common with the light cone is a Euclidean plane.
c. A plane which has precisely one line in common with the light cone (a tangent plane of the cone) is singular and has that line as radical. (Hint: If l is a line on the light cone, the only plane through

l which contains no other line of the cone is the tangent plane of the cone along l (see Figure 151.2). The orthogonal complement l^* of l is also a plane which contains l and, since l^* is singular, l^* contains no other line of the cone by Part a.)

  d. A line in V is Euclidean if and only if its orthogonal complement is a Lorentz plane.

  e. A line in V is a negative Euclidean line if and only if its orthogonal complement is a Euclidean plane.

  f. A line in V is a null line if and only if its orthogonal complement is a singular plane.

  g. All the lines outside the light cone are Euclidean lines. Examples are the coordinate axes $\langle A_1 \rangle$ and $\langle A_2 \rangle$.

  h. All lines inside the light cone are negative Euclidean lines. An example is the coordinate axis $\langle A_3 \rangle$.

8. Let V be a nonsingular real space of dimension ≥ 2 which is neither Euclidean space nor negative Euclidean space. Prove that

  a. V contains infinitely many Euclidean lines and infinitely many negative Euclidean lines.

  b. If $n = 2$, V contains two null lines; if $n > 2$, V contains infinitely many null lines.

  (Consequently, V contains all three kinds of real lines.)

9. In special relativity one uses the real four-space whose metric is given by the matrix

$$\begin{pmatrix} 1 & 0 & 0 & 0 \\ 0 & 1 & 0 & 0 \\ 0 & 0 & 1 & 0 \\ 0 & 0 & 0 & -c^2 \end{pmatrix}$$

where c is the speed of light. Prove that this space is Minkowski space.

10. Negative Euclidean space F_n was defined in Example 128.4 as an n-dimensional real space with a coordinate system relative to which the matrix is

$$\begin{pmatrix} -1 & & \bigcirc \\ & \ddots & \\ \bigcirc & & -1 \end{pmatrix}.$$

Prove that F_n may also be defined as an n-dimensional real metric vector space over $\mathbf{R}$ which is negative definite, meaning that $A^2 < 0$ for all $A \neq 0$. Observe that $\{0\}$ becomes a zero-dimensional negative Euclidean space. The fact that $\{0\}$ is both Euclidean (Exercise 159.6) and negative Euclidean does no harm.

*11. Assume that V is negative definite, that is, $A^2 < 0$ for all $A \neq 0$.

a. Prove that any subspace of V is also negative definite. Thus the property of being negative definite is inherited by all subspaces of V.

b. Prove that every m-dimensional subspace of negative Euclidean space is negative Euclidean m-space.

12. Assume $k = \mathbf{R}$. Prove that the only metric vector space which is both Euclidean and negative Euclidean is the zero-dimensional space $\{0\}$.

*13. Let $c \in k$ and assume that we change the metric of our n-dimensional space V to

$$A \circ B = c(AB) \qquad \text{for all} \quad A, B \in V.$$

a. Prove that $(A, B) \to A \circ B$ defines an inner product for V. We denote the new metric vector space corresponding to $A \circ B$ by V_o. Observe that V and V_o consist of the same vectors.

b. If $c \neq 0$, prove that $A \perp B$ in V if and only if $A \perp B$ in V_o. In other words, orthogonality is preserved.

c. If $c \neq 0$, prove that V and V_o have the same light cone.

d. If G is the matrix of V relative to some coordinate system, prove that cG is the matrix of V_o relative to the same coordinate system.

e. Prove that disc $V_o = c^n$ disc V.

f. If $c \neq 0$ and has a square root in k, prove that $V \simeq V_o$. We shall see in Exercise 14 below that V may be isometric to V_o without c having a square root in k, even if V is nonsingular.

g. Assume V is nonsingular, has positive dimension, and $V \simeq V_o$. Prove that $c \neq 0$ and c^n has a square root in k.

h. If V is nonsingular and n is odd, prove that $V \simeq V_o$ if and only if $c \neq 0$ and c has a square root in k.

*14. In Exercise 13 above, let $k = \mathbf{R}$ and $c = -1$.

a. Prove that V is Euclidean n-space if and only if V_o is negative Euclidean space.

b. We know Euclidean n-space and negative Euclidean n-space are nonisometric spaces. Nevertheless, prove that they have the same orthogonality relation and the same light cone.

c. Prove that V is Minkowski space if and only if V_o is negative Minkowski space.

d. Even though Minkowski space and negative Minkowski space are not isometric, prove they have the same orthogonality relation and the same light cone.

e. If V is the Lorentz plane, prove that $V \simeq V_o$. Observe that this is so even though -1 has no square root in $\mathbf{R}$.

f. If V is Artinian four-space, prove that $V \simeq V_o$. Observe again that -1 has no square root in $\mathbf{R}$.

We shall now rework Exercises 136.15, 137.16, and 137.17 in a much easier manner, using the fact that there are precisely three nonsingular planes over **R**.

*15. Let $n = 2$ and $k = \mathbf{R}$.

    a. If the metric of V is determined by the matrix $G = \left(\begin{smallmatrix} 3 & -1 \\ -1 & 2 \end{smallmatrix}\right)$, prove that V is the Euclidean plane.

    b. If the metric of V is determined by the matrix $G = \left(\begin{smallmatrix} 3 & -1 \\ -1 & -2 \end{smallmatrix}\right)$, prove that V is the Lorentz plane.

    c. Show that the matrices

$$G = \begin{pmatrix} 1 & \frac{3}{2} \\ \frac{3}{2} & 2 \end{pmatrix} \quad \text{and} \quad G' = \begin{pmatrix} -7 & \frac{3}{4} \\ \frac{3}{4} & -2 \end{pmatrix}$$

determine, respectively, the Lorentz plane and the negative Euclidean plane.

34. SYLVESTER'S THEORY

In Proposition 159.1, we showed that, if k is an ordered field whose positive elements have square roots, then a nonsingular space V over k has a rectangular coordinate system $A_1, \ldots, A_n$ where $A_1{}^2 = \cdots = A_r{}^2 = 1$ and $A_{r+1}^2 = \cdots = A_n{}^2 = -1$ for some r, $0 \le r \le n$. The main purpose of this section is to prove that the number r does not depend on the coordinate system, but only on the geometry of V. We begin with a remark on arbitrary fields and then concentrate on ordered fields.

It is clear that the radical of V is a linear subspace of the light cone, this being the same as saying that the vectors of rad V are isotropic. In general, the light cone contains many vectors which do not belong to the radical. For instance, all the nonzero isotropic vectors of the Lorentz plane or of Minkowski space are of this nature. In fact, if a nonsingular space has nonzero isotropic vectors, none of them lie in the radical. It is surprising, although very easy to prove, that, as soon as the light cone contains one vector which does not lie in the radical, V represents all scalars.

Proposition 167.1. If V contains an isotropic vector which does not belong to rad V, all scalars are represented by V.

Proof. Assume that the isotropic vector N does not belong to rad V. Then V contains a vector A such that $AN \ne 0$. The vectors A and N are linearly independent since, otherwise, $A = cN$ and hence $AN = cN^2 = 0$. It follows that A, N is a coordinate system for the plane $P = \langle A, N \rangle$; the geometry of P relative to this coordinate system is given by the matrix

$$\begin{pmatrix} a & b \\ b & 0 \end{pmatrix}$$

where $A^2 = a$ and $AN = b \neq 0$. The quadratic form of P is $ax_1{}^2 + 2bx_1x_2$ and it is entirely trivial that every scalar d is represented by this form; one simply puts

$$x_1 = 1 \quad \text{and} \quad x_2 = \frac{d - a}{2b}.$$

(Why is $2b \neq 0$?) We have shown that every scalar is already represented by a vector in P. Done.

The following corollary contains a surprising fact in the special case that V is nonsingular.

Corollary 168.1. If V is nonsingular and contains a nonzero isotropic vector, then V represents all scalars.

Proof. No nonzero isotropic vector belongs to rad V. Done.

Exercises

1. Let V be a nonsingular real space of dimension ≥ 2 which is neither Euclidean space nor negative Euclidean space. Prove that V contains a nonzero isotropic vector and hence V represents all scalars. (Note that this is a second proof of Exercise 163.2b.)
*2. One may ask whether there are nonsingular spaces which do not satisfy the conditions of Corollary 168.1 and, nevertheless, represent all scalars. In other words, are there anisotropic geometries which represent all scalars?
   a. Prove that the only anisotropic spaces over the real numbers are the Euclidean spaces and the negative Euclidean spaces. Thus there are no anisotropic geometries over the real numbers which represent all real numbers.
   b. Let $k = {}'\mathbf{Z}_3$, $n = 2$, and assume the matrix of V is $\begin{pmatrix} 1 & 0 \\ 0 & 1 \end{pmatrix}$ relative to some coordinate system. Prove that V is anisotropic and represents all scalars. (Actually, for every finite field, there is precisely one anisotropic plane and it represents all scalars. See Artin, [2], p. 143.)

*3. Let C denote the light cone of V.
   a. If V does not represent all scalars, prove that $C = \text{rad } V$.
   b. If V is nonsingular and does not represent all scalars, prove that V is anisotropic. The converse is false. Exercise 2b above provides an example of an anisotropic space which represents all scalars.

We assume for the remainder of this section up to Remark About Projective Geometry (p. 180) *that the field k is ordered.* Of course, the fields **Q** and **R** can serve as examples. The field **R** has the further property that every one of its positive elements has a square root, but it is easy to find positive elements of **Q** which have no square root in **Q**. Clearly, every ordered field induces an ordering on every one of its subfields. Hence not only **Q**, but every other subfield of **R** receives an ordering from **R**; an example of such a field is

$$\mathbf{Q}(\sqrt{2}) = \{a + b\sqrt{2} \,|\, a, b \in \mathbf{Q}\}.$$

Readers who are not acquainted with ordered fields should choose some subfield of **R** for k. Since we are not assuming that every element of k has a square root, the subfield should not be all of **R**. The field **Q** will do fine.

In the following definition, V may be singular.

Definition 169.1.   The metric vector space V is called

1. **Positive semidefinite** if $A^2 \geq 0$ for all $A \in V$.
2. **Positive definite** if $A^2 > 0$ for all $A \neq \mathbf{0}$ in V.
3. **Negative semidefinite** if $A^2 \leq 0$ for all $A \in V$.
4. **Negative definite** if $A^2 < 0$ for all $A \neq \mathbf{0}$ in V.

Observe that a "definite" space may be defined as a "semidefinite" space which is anisotropic. This shows that definite spaces are nonsingular.

Exercises

4. Prove that the only spaces which are both positive and negative semidefinite are the null spaces.
5. Prove that the only space which is both positive and negative definite is the zero-dimensional space $\{\mathbf{0}\}$.
6. Let U be a subspace of V. If V is positive semidefinite or positive definite, prove that U is the same kind. Prove the same for negative spaces.

7. Assume $V = U \oplus W$.
   a. Prove that V is positive semidefinite if and only if both U and V are positive semidefinite.
   b. Prove that V is positive definite if and only if both U and W are positive definite.
   c. Prove Parts a and b for negative spaces.

Proposition 170.1. If V is definite or semidefinite (of any kind), its radical and light cone coincide.

Proof. The proof is based on the fact that, if V does not represent all scalars, its radical and light cone coincide (Exercise 169.3). If V is positive definite or positive semidefinite, V represents only nonnegative scalars and hence its radical and light cone coincide. Similar reasoning holds for the negative case. Done.

Corollary 170.1. A semidefinite space is definite (of the same kind) if and only if the space is nonsingular.

Proof. We have already pointed out that definite spaces are nonsingular. Conversely, if a semidefinite space is nonsingular, its light cone is equal to its radical $\{0\}$. This shows that the space is anisotropic and hence definite. Done.

Exercise

*8. Let $A_1, \ldots, A_n$ be a rectangular coordinate system of V.
   a. Prove that V is positive definite if and only if $A_i^2 > 0$ for $i = 1, \ldots, n$.
   b. If V is positive semidefinite but not definite, prove that at least one coordinate vector A_i is isotropic while the nonisotropic vectors have positive squares. (Do not use Corollary 170.1, but use Part a above.)
   c. Formulate and prove Parts a and b for negative spaces.
   d. Use Parts a–c to give another proof of Corollary 170.1.

We now come to Sylvester's theory proper. Its subject is the study of the positive and negative definite subspaces of an arbitrary metric vector space V. Since $\{0\}$ is positive definite, V always contains positive definite subspaces. A positive definite subspace is called **maximal** if it is not contained in a strictly larger positive definite subspace. If V itself is positive definite, V is its only maximal positive definite subspace; in general, V contains many maximal positive definite subspaces. The finite dimensionality of V guarantees that every positive definite subspace is contained in a maximal positive definite subspace and, consequently, that maximal positive definite subspaces always exist. Maximal negative definite subspaces are defined similarly and exist for the same reasons. Suppose now that V is nonsingular. The main assertion of Sylvester's theory is that all the maximal positive definite subspaces have the same dimension r and that all the maximal negative definite subspaces have the same dimension $n - r$. Furthermore, if every positive element of k has a square root in k, this number r is the same as occurs in Proposition 159.1. This gives the geometric interpretation of the number r of Proposition 159.1 and shows that this number does not depend on the choice of coordinate system. All this is part of Sylvester's theory which we now develop.

Proposition 171.1.   If U is a positive definite subspace and W is a negative definite subspace of V (V may be singular), then

$$U \cap W = \{0\} \quad \text{and} \quad \dim U + \dim W \leq n.$$

The proof is precisely the result of Exercise 163.5 with $\mathbf{R}$ replaced by the arbitrary ordered field k and will be left for the reader to review.

As always, the orthogonal complement of a subspace U is denoted by U^*. If V is nonsingular and U is positive definite or negative definite, U is nonsingular and hence $V = U \oplus U^*$.

Proposition 171.2.   Let V be nonsingular. Assume that $V = U \oplus U^* = W \oplus W^*$ where the subspaces U and W are positive definite while U^* and W^* are negative definite. Then

$$\dim U = \dim W \quad \text{and} \quad \dim U^* = \dim W^*.$$

Proof. We put $\dim U = r$ and $\dim W = s$. Since $\dim U^* = n - r$ and $\dim W^* = n - s$, we conclude from Proposition 171.1 that $r + n - s \leq n$ and $s + n - r \leq n$. This shows that $r = s$. Done.

We are now ready for the main theorem of Sylvester's theory which states that U and W of the above proposition are maximal positive definite while U^* and W^* are maximal negative definite.

Theorem 172.1. Let V be nonsingular. A positive definite subspace U of V is maximal positive definite if and only if U^* is negative definite; in that case, U^* is maximal negative definite. A negative definite space U of V is maximal negative definite if and only if U^* is positive definite; in that case, U^* is maximal positive definite. All maximal positive definite subspaces have the same dimension r and all maximal negative definite subspaces have the same dimension $n - r$.

Proof. 1. Let U be maximal positive definite. We must show that U^* is negative definite. Let $A \in U^*$; it is not possible that $A^2 > 0$ since the line $\langle A \rangle$ would then be positive definite and hence $U \oplus \langle A \rangle$ would be a positive definite subspace of V which is strictly larger than U. Consequently, $A^2 \leq 0$ which, together with the fact that U^* is nonsingular, shows that U^* is negative definite (Corollary 170.1).

2. Assume U is positive definite and U^* is negative definite. Let us show that U is maximal positive definite. If U were not maximal positive definite, U would be contained in a strictly larger maximal positive definite subspace W. Then W^* would be negative definite (by 1) and hence dim U = dim W by Proposition 171.2. This is absurd.

Thus far, we have shown that a positive definite subspace U is maximal positive definite if and only if U^* is negative definite. One shows in the same way that a negative definite subspace U is maximal negative definite if and only if U^* is positive definite. Consequently, if $V = U \oplus U^*$ where U is positive definite and U^* is negative definite, both subspaces are maximal definite.

3. Let U and W be maximal positive definite subspaces of V. Then U^* and W^* are negative definite subspaces and hence dim U = dim W. This proves that all maximal positive definite subspaces have the same dimension r. One shows in the same way that all maximal negative definite subspaces have the same dimension s. Finally, if U is maximal positive definite, $V = U \oplus U^*$ and U^* is maximal negative definite. It follows that $r + s = n$. Done.

The common dimension of the maximal positive definite subspaces of a nonsingular space V is called the **signature** of V. (Some books define the signature as the pair $(r, n - r)$ where r is the dimension of the maximal positive definite subspaces.) A space is positive definite if and only if its signature is n and a space is negative definite if and only if its signature is 0.

The signature is determined by any rectangular coordinate system $A_1, \ldots, A_n$ of V. None of the vectors A_i can be isotropic (since V is non-singular) and hence every A_i^2 is either positive or negative.

Theorem 173.1. Let V be nonsingular and $A_1, \ldots, A_n$ a rectangular coordinate system of V. The number of vectors A_i whose square A_i^2 is positive is the signature of V.

Proof. We assume the vectors $A_1, \ldots, A_n$ have been ordered in such a way that $A_1^2, \ldots, A_r^2$ are positive and $A_{r+1}^2, \ldots, A_n^2$ are negative. Then

$$V = \langle A_1, \ldots, A_r \rangle \oplus \langle A_{r+1}, \ldots, A_n \rangle$$

where $\langle A_1, \ldots, A_r \rangle$ is positive definite and $\langle A_{r+1}, \ldots, A_n \rangle$ is negative definite. We conclude from Theorem 172.1 that $\langle A_1, \ldots, A_r \rangle$ is maximal positive definite and hence r is the signature of V. Done.

We see from the above theorem that the number of vectors with a positive square which occurs in a rectangular coordinate system of a nonsingular space does not depend on the coordinate system. This number, being the signature of V, is completely determined by the geometry of V. It is reasonable to conjecture from our experience with numbers that two nonsingular spaces with the same dimension and signature are isometric. As is the case with many reasonable things, this is false (see Exercise 10 below). The conjecture is correct, however, if every positive element of k has a square root in k (see Theorem 177.1).

Exercises

9. Prove that nonsingular, isometric spaces have the same signature.
*10. Let $k = \mathbf{Q}$ and $n = 2$. Assume that two metrics for the "rational" plane (the plane over $\mathbf{Q}$) are given by the matrices

$$G = \begin{pmatrix} 1 & 0 \\ 0 & 1 \end{pmatrix} \quad \text{and} \quad G' = \begin{pmatrix} 2 & 0 \\ 0 & 1 \end{pmatrix}.$$

(G represents a metric of V relative to one coordinate system and G' represents a metric relative to a possibly different coordinate system. When the coordinate systems are not important to the discussion, they will usually not be mentioned.)
a. Prove that both planes are nonsingular and have signature two.
b. Prove that the planes are not isometric. Here then is an example of two nonsingular spaces with the same dimension and signature which are not isometric.
11. This exercise discusses the part of Sylvester's theory which still holds in the singular case. Let V be singular. The signature of V is defined as the signature of the nonsingular space $V/\text{rad } V$.

    a. If a subspace U of V is positive definite or negative definite, prove that $U \cap (\mathrm{rad}\ V) = \{0\}$.

    b. Let $V = W \oplus \mathrm{rad}\ V$ where $W \simeq V/\mathrm{rad}\ V$ and let $p: V \to W$ denote the projection from V onto W which comes from the orthogonal splitting. Hence if $A = B + C$ where $A \in V$, $B \in W$, and $C \in \mathrm{rad}\ V$, then $pA = B$. If the subspace U of V is positive or negative definite, prove that the restriction of p to U is an isometry from U onto a subspace of W. (Hint: This only requires the fact that $U \cap (\mathrm{rad}\ V) = \{0\}$.)

    c. Prove that the maximal positive definite subspaces of V all have the same dimension, namely, the signature of V. (Hint: Use Part b to show that the dimension of a positive definite subspace of V is, at most, the signature of V.)

    d. Prove that the maximal negative definite subspaces of V all have the same dimension, namely, rank V − signature V.

    e. Let $A_1, \ldots, A_n$ be a rectangular coordinate system of V. Prove that the number of vectors A_i whose square $A_i{}^2$ is positive is equal to the signature of V.

    f. Prove that the number of vectors with positive square which occur in a rectangular coordinate system of V does not depend on the coordinate system.

We see from Parts c, d, and e that Theorem 173.1 and the last part of Theorem 172.1 still hold when V is singular. Parts d and e of the next exercise show that the rest of Theorem 172.1 is false when V is singular.

12. Let $n = 2$ and A_1, A_2 be a coordinate system of V. Suppose that the metric of V is given by the matrix $G = \left(\begin{smallmatrix} 1 & 0 \\ 0 & 0 \end{smallmatrix}\right)$ relative to the coordinate system A_1, A_2.

    a. Prove that $\langle A_1 \rangle \simeq V/\mathrm{rad}\ V$ and $\mathrm{rad}\ V = \langle A_2 \rangle$. Hence V is singular and has rank 1.

    b. Prove that the signature of V is 1.

    c. Prove that the maximal positive definite subspaces of V are the lines through $\mathbf{0}$ different from $\langle A_2 \rangle$.

    d. If U is a maximal positive definite subspace of V, prove that U^* is the null line $\langle A_2 \rangle$.

    e. Prove that $\{0\}$ is the only negative definite subspace and hence the only maximal negative definite subspace of V. Prove that $\{0\}^*$ is not positive definite.

13. Let G be a symmetric $n \times n$ matrix with entries in k.

a. Prove that there exists a nonsingular $n \times n$ matrix P with entries in k such that $a_1, \ldots, a_r > 0$, $a_{r+1}, \ldots, a_m < 0$, and

$$
P^{\mathrm{T}}GP = \begin{pmatrix}
a_1 & & & & & & & \bigcirc \\
& \ddots & & & & & & \\
& & a_r & & & & & \\
& & & a_{r+1} & & & & \\
& & & & \ddots & & & \\
& & & & & a_m & & \\
& & & & & & 0 & \\
\bigcirc & & & & & & & \ddots \\
& & & & & & & & 0
\end{pmatrix}.
$$

b. Prove that the integers r and m in Part a do not depend on the choice of P, but are completely determined by the matrix G. (The number r is called the **signature of the matrix** G and is, therefore, the signature of the geometry represented by G; of course, m is the rank of G. The fact that r does not depend on P is called "Sylvester's law of inertia.")

*14. A symmetric matrix is called **positive definite** if the geometry it represents is positive definite.

a. Prove that a symmetric $n \times n$ matrix G is positive definite if and only if

$$
G = P^{\mathrm{T}} \begin{pmatrix} a_1 & & \bigcirc \\ & \ddots & \\ \bigcirc & & a_n \end{pmatrix} P
$$

where P is a nonsingular $n \times n$ matrix and $a_1, \ldots, a_n$ are positive scalars.

b. Define positive semidefinite, negative definite, and negative semi-definite symmetric matrices. For each case, formulate and prove the analogue of Part a.

It often happens that a field can be ordered in more than one way. If k is such a field, the metric vector space V has a signature for every ordering of k and these signatures may be different from one another. Of course, two metric vector spaces over k can be isometric only if they have the same signature for each ordering. Sometimes, this fact can be used to show that two spaces are not isometric. We give an example of this situation.

Example 1

Let $k = \mathbf{Q}(\sqrt{2}) = \{a + b\sqrt{2} \,|\, a, b \in \mathbf{Q}\}$. Assume that metrics for the plane V are given by the matrices

$$G = \begin{pmatrix} 1 & 0 \\ 0 & 1 \end{pmatrix} \quad \text{and} \quad G' = \begin{pmatrix} 1 + \sqrt{2} & 0 \\ 0 & -1 + \sqrt{2} \end{pmatrix}.$$

Are these planes over k isometric? They certainly are nonsingular and both have the same discriminant 1; hence they could well be isometric.

Let us first give k its natural ordering $<$ as a subfield of $\mathbf{R}$. Since 1, $1 + \sqrt{2}$, and $-1 + \sqrt{2}$ are all positive numbers, both planes are positive definite and have signature 2. The pendulum swings toward isometry.

Now let us give k an unnatural ordering $<'$ by defining

$$a + b\sqrt{2} <' c + d\sqrt{2} \quad \text{if} \quad a - b\sqrt{2} < c - d\sqrt{2}$$

for all $a, b, c, d \in \mathbf{Q}$. The reason $<'$ defines an ordering for k is that the mapping $a + b\sqrt{2} \to a - b\sqrt{2}$ is an automorphism of k. In general, if $(k, <)$ is an ordered field and $\mu: k \to k'$ is an isomorphism from k onto some field k', one can give k' an ordering $<'$ by the definition: For $x, y \in k'$,

$$x <' y' \quad \text{if} \quad \mu^{-1}(x) < \mu^{-1}(y).$$

This is the ordering of k' which makes μ into an "order isomorphism" and one says that one has "transported" the ordering of k to k' under the isomorphism μ. This is precisely what we did, except that, using an expression of Artin, we were in the "psychologically difficult case where $k = k'$." In our case, when $k = k' = \mathbf{Q}(\sqrt{2})$, the ordering of k is the natural ordering of $\mathbf{Q}(\sqrt{2})$ and $\mu: k \to k'$ is the automorphism $a + b\sqrt{2} \to a - b\sqrt{2}$.

Observe that the rational numbers are ordered by $<'$ in the same way as by $<$ since μ leaves $\mathbf{Q}$ pointwise fixed. Consequently, $0 <' 1$ and hence the plane defined by G still has signature 2 in the ordering $<'$. However,

$$1 - \sqrt{2} < 0 \quad \text{implies that} \quad 1 + \sqrt{2} <' 0$$

and

$$-1 - \sqrt{2} < 0 \quad \text{implies that} \quad -1 + \sqrt{2} <' 0.$$

This shows that the plane defined by G' is negative definite and has signature 0 in the ordering $<'$. Therefore, the planes are not isometric!

Remark about Algebraic Number Fields. An algebraic number field is a finite algebraic extension of $\mathbf{Q}$; examples are $\mathbf{Q}$ itself and $\mathbf{Q}(\sqrt{2})$. Suppose that k is an algebraic number field and that two nonsingular spaces over k of the same dimension have the same signature for every ordering of k. Can we now claim that the spaces are necessarily isometric? The answer is still no. An example is furnished by the two planes over $\mathbf{Q}$ given by the matrices $\left(\begin{smallmatrix}1&0\\0&1\end{smallmatrix}\right)$ and $\left(\begin{smallmatrix}2&0\\0&1\end{smallmatrix}\right)$ (Exercise 173.10). In this case, the field $\mathbf{Q}$ has no orderings other than the natural one, but the planes are not isometric because they have different discriminants. However, algebraic number fields also have "valuations" and, if one takes them into account in addition to the orderings, one does arrive at a complete set of invariants for metric vector spaces over algebraic number fields (see O'Meara, [18]). This ends our remark about algebraic number fields.

We now turn our attention to ordered fields in which every positive element has a square root. Such fields can be ordered in only one way and hence give rise to only one signature. We do not prove this fact because it is not needed, but it clarifies Theorem 177.1. Of course, $\mathbf{R}$ is an example of such a field; other examples are the real closed fields mentioned on page 159.

Theorem 177.1. Assume that every positive element of k has a square root. Two metric vector spaces over k are isometric if and only if they have the same dimension, rank, and signature. In particular, two nonsingular spaces are isometric if and only if they have the same dimension and signature.

Proof. We only have to show that spaces with the same dimension, rank, and signature are isometric.
Case 1. V is nonsingular. Let r be the signature of V. If $n \geq 1$, V has a rectangular coordinate system relative to which the matrix of V is

$$\begin{pmatrix} 1 & & & & & & \\ & \ddots & & & & \bigcirc & \\ & & 1 & & & & \\ & & & -1 & & & \\ & & & & \ddots & & \\ \bigcirc & & & & & -1 \end{pmatrix}$$

with r ones and $n - r$ negative ones (Proposition 159.1 and Theorem 173.1). Two nonsingular spaces with the same positive dimension and signature have coordinate systems which give rise to identical matrices and hence the spaces are isometric. The isometry of zero-dimensional spaces is trivial.

Case 2. V is singular. Let m be the rank and r the signature of V. (This means r is the signature of $V/\mathrm{rad}\ V$.) We put $V = W \oplus \mathrm{rad}\ V$ where $W \simeq V/\mathrm{rad}\ V$ and apply Case 1 to the nonsingular space W. It follows immediately that V has a rectangular coordinate system relative to which its matrix is

$$
\begin{pmatrix}
1 & & & & & & & & \\
& \ddots & & & & & & \bigcirc & \\
& & 1 & & & & & & \\
& & & -1 & & & & & \\
& & & & \ddots & & & & \\
& & & & & -1 & & & \\
& & & & & & 0 & & \\
& & & & & & & \ddots & \\
& \bigcirc & & & & & & & 0
\end{pmatrix}
$$

with r ones, $m - r$ negative ones, and $n - m$ zeros. Consequently, two singular spaces with the same dimension, rank, and signature have coordinate systems which give rise to identical matrices and hence are isometric. Done.

Corollary 178.1.   If every positive element of k has a square root, there are precisely $n + 1$ nonisometric, nonsingular metric vector spaces of dimension n. A nonsingular space of positive dimension n and signature r has a rectangular coordinate system relative to which its matrix is a diagonal matrix with r ones and $n - r$ negative ones on the main diagonal.

Proof. The last statement was proved in Case 1 of the previous theorem. The first statement follows from the fact that there are precisely $n + 1$, $n \times n$ diagonal matrices with only ones and negative ones on the main diagonal, together with the fact that spaces with different signatures are not isometric. Done.

Table 179.1 gives the signatures of the nonsingular real spaces of dimension ≤ 4. Examples 161.1, 161.2, and Exercise 162.1 can be used to check the table.

The last two columns of Table 179.1 show at a glance that no two of the geometries listed are isometric. This makes the laborious methods used in Example 161.2 superfluous (although not less instructive).

TABLE 179.1

Nonsingular real space	Dimension	Signature
0-dimensional space $\{0\}$	0	0
Euclidean line	1	1
Negative Euclidean line	1	0
Euclidean plane	2	2
Lorentz plane	2	1
Negative Euclidean plane	2	0
Euclidean three-space	3	3
Orthogonal sum of the Lorentz plane and a Euclidean line	3	2
Orthogonal sum of the Lorentz plane and a negative Euclidean line	3	1
Negative Euclidean three-space	3	0
Euclidean four-space	4	4
Minkowski space	4	3
Artinian four-space	4	2
Negative Minkowski space	4	1
Negative Euclidean four-space	4	0

Exercises

In this exercise set, k is an ordered field whose positive elements have square roots.

15. Prove that there are $\frac{1}{2}(n + 2)(n + 1)$ nonisometric n-dimensional spaces (including the singular ones).

16. Let G be a symmetric $n \times n$ matrix with entries in k. Let m be the rank of G and r its signature. (We recall that r is the signature of the geometry determined by G.)

a. Prove that there is a nonsingular $n \times n$ matrix P with entries in k such that

$$P^{\mathrm{T}}GP = \begin{pmatrix} 1 & & & & & & & \\ & \ddots & & & & & \text{\Large O} & \\ & & 1 & & & & & \\ & & & -1 & & & & \\ & & & & \ddots & & & \\ & & & & & -1 & & \\ & & & & & & 0 & \\ & \text{\Large O} & & & & & & \ddots \\ & & & & & & & 0 \end{pmatrix}$$

where there are r ones, $m - r$ negative ones, and $n - m$ zeros. (Since this diagonal matrix is uniquely determined by G, it is called the canonical form of G under congruence transformations.)

b. Prove that G is positive definite (Exercise 175.14) if and only if there is a nonsingular $n \times n$ matrix P with entries in k such that $G = P^T P$.

c. Prove that G is negative definite if and only if there exists a nonsingular $n \times n$ matrix P with entries in k such that $-G = P^T P$.

Remark about Projective Geometry. This remark and Exercises 183.17–183.19 are meant only for readers who are familiar with projective geometry. This material is not needed for the sequel. Let $\mathscr{W}$ be an m-dimensional vector space over a field k where $m \geq 1$. $\mathscr{W}$ has not been given a metric; also, k is not necessarily ordered and may have characteristic two.

The points of the $(m - 1)$-dimensional projective space P over k are defined as the lines of $\mathscr{W}$, that is, one-dimensional subspaces of $\mathscr{W}$. Hence the points of P should be visualized as the lines through the origin and this is a very classical way of looking at projective geometry. We shall refer to P as "the space at infinity of $\mathscr{W}$" and, indeed, the space at infinity of an *affine* space should be *projective* space of one dimension lower. One barely needs a book on projective geometry to figure out how a coordinate system $A_1, \ldots, A_m$ of $\mathscr{W}$ induces a projective coordinate system for the points of P (is it not true that the coordinates of the nonzero vectors on a line of $\mathscr{W}$ form a class of proportional, nonzero m-tuples $(x_1, \ldots, x_n)$?); how a linear automorphism of $\mathscr{W}$ induces a projective transformation of P onto itself (does a linear automorphism of $\mathscr{W}$ not transform each line of $\mathscr{W}$ onto some other line?); and, in general, how an affine notion for $\mathscr{W}$ induces the corresponding projective notion for P.

Look at the force of the concept of vector space for geometry. It defines affine space (Chapter 1), metric affine space (Chapter 3) projective space (above), and now we say a few words about how it defines metric projective space.

We return to our metric vector space V over the field k. We drop the assumption that k is ordered and assume that $n \geq 1$. The space at infinity of V is the $(n - 1)$-dimensional projective space P whose points are the lines of V. The light cone of V is an $(n - 1)$-dimensional quadratic cone. Its generators (that is, the lines of the cone) are points P and these points form an $(n - 2)$-dimensional quadratic hypersurface (not a cone!) S of P. The metric of P induced by the metric of V is based on S as the "fundamental quadric" (quadric stands for quadratic hypersurface). For details, in particular, on how S makes it possible to measure distances between points of P, see R. Artzy, *Linear Geometry*, Ch. 3, [3].

We now look a little closer at the real projective lines and planes.

Let $k = \mathbf{R}$ and $n = 2$. The metrics of the three nonisometric, nonsingular real planes are given by the matrices

$$G_1 = \begin{pmatrix} 1 & 0 \\ 0 & 1 \end{pmatrix}, \qquad G_2 = \begin{pmatrix} 1 & 0 \\ 0 & -1 \end{pmatrix}, \qquad G_3 = \begin{pmatrix} -1 & 0 \\ 0 & -1 \end{pmatrix}.$$

The matrix G_1 determines the Euclidean plane V and its light cone has the equation $x_1{}^2 + x_2{}^2 = 0$. As long as we stick to the real numbers, the cone consists of $\mathbf{0}$ only; however, over the complex numbers, the cone consists of the two conjugate complex lines

$$l_1: x_1 - x_2 i = 0 \qquad \text{and} \qquad l_2: x_1 + x_2 i = 0 \qquad (i^2 = -1).$$

The metric projective line whose *points* are the lines of V and whose fundamental quadric consists of the two conjugate complex *points* l_1 and l_2 is called the *elliptic line*. If a two-dimensional Euclidean bug looks from $\mathbf{0}$ toward infinity in all possible directions, he sees a projective, elliptic line.

The matrix G_3 determines the negative Euclidean plane and its light cone has the equation $-(x_1{}^2 + x_2{}^2) = 0$. Consequently, this light cone is the same as the cone of the Euclidean plane and hence the line at infinity of the negative Euclidean plane is also the projective, elliptic line. No new projective metric is obtained here.

Finally, the matrix G_2 determines the Lorentz plane V. Its light cone has the equation $x_1{}^2 - x_2{}^2 = 0$ and hence consists of the two real lines

$$U: x_1 - x_2 = 0 \qquad \text{and} \qquad W: x_1 + x_2 = 0.$$

The metric projective line whose *points* are the lines of V and whose fundamental quadric consists of the two real points U and W is called the *hyperbolic line*. A two-dimensional, Lorentzian bug which looks from $\mathbf{0}$ toward infinity in all directions sees a projective, hyperbolic line.

We come to the conclusion that there are two nonisometric real projective lines, the elliptic line and the hyperbolic line. The terms "elliptic" and "hyperbolic" were introduced by Felix Klein and are explained in Footnote 1, page 170, of his book *Vorlesungen Uber Nicht-Euclidische Geometrie*, [14].

We now consider the real projective plane, that is, $k = \mathbf{R}$ and $n = 3$. The metrics of the four nonisometric nonsingular real three-spaces are given by the matrices

$$G_1 = \begin{pmatrix} 1 & 0 & 0 \\ 0 & 1 & 0 \\ 0 & 0 & 1 \end{pmatrix}, \qquad G_2 = \begin{pmatrix} 1 & 0 & 0 \\ 0 & 1 & 0 \\ 0 & 0 & -1 \end{pmatrix},$$

$$G_3 = \begin{pmatrix} 1 & 0 & 0 \\ 0 & -1 & 0 \\ 0 & 0 & -1 \end{pmatrix}, \qquad G_4 = \begin{pmatrix} -1 & 0 & 0 \\ 0 & -1 & 0 \\ 0 & 0 & -1 \end{pmatrix}.$$

The matrix G_1 determines Euclidean three-space V and its light cone has the equation $x_1{}^2 + x_2{}^2 + x_3{}^2 = 0$. Again, over the real numbers, the cone consists of $\mathbf{0}$ alone, but over the complex numbers the cone contains infinitely many complex lines. The metric projective plane whose points are the lines of V and whose fundamental conic has as points the above set of complex lines is called the *elliptic plane*. (Conic stands for one-dimensional quadric.) Hence the plane at infinity of Euclidean three-space is the elliptic plane. This agrees with our geometric intuition. If one looks from $\mathbf{0}$ toward infinity in all possible directions of Euclidean three-space, it is clear that the distance between two projective points (two lines through $\mathbf{0}$) is most simply defined as the measure of "the" Euclidean angle between those lines. Hence, using radians, it is always immediate what this distance is; moreover, this distance is unique only within sign and integral multiples of π. This very elementary metric geometry imposed on the lines through $\mathbf{0}$ by angular measurement is precisely plane elliptic geometry! That's all there is to it! Anyone who gazes at the celestial hemisphere of the stars at night sees a projective, elliptic plane; each star may be identified with a projective point. Try it! Ask yourself what the projective distance is between two bright stars which strike your romantic fancy. By the way, be sure to identify antipodal points on the horizon since they correspond to the same projective point.

We leave to the reader the task of showing that the matrix G_4 gives rise to the same light cone as G_1 and hence the plane at infinity of the negative Euclidean three-space is also the elliptic plane.

Consider now the real three-space V whose metric is given by the matrix G_2. The light cone C of this space has the equation $x_1{}^2 + x_2{}^2 - x_3{}^2 = 0$ and is the cone of revolution pictured in Figure 164.1. This cone contains infinitely many real lines (its generators) and they form a real conic in the plane at infinity of V. This metric projective plane whose points are the lines of V and whose fundamental conic has as points the generators of C is called the *hyperbolic plane*. In the classical literature, "hyperbolic geometry" means the geometry of either the hyperbolic projective plane or of the interior of the fundamental conic of the plane. One simply has to check which the writer means.

Finally, the light cone corresponding to G_3 has the equation

$$x_1{}^2 - x_2{}^2 - x_3{}^2 = 0.$$

By interchanging the first and third axis, the equation becomes

$$-(x_1{}^2 + x_2{}^2 - x_3{}^2) = 0.$$

Hence the plane at infinity of the three-space defined by G_3 is also the hyperbolic plane.

We conclude that there are two nonisometric real projective planes, the elliptic plane and the hyperbolic plane.

This remark about projective geometry brings into the open a basic difference between presentday geometry and the geometry as admirably exposed by Felix Klein, [14]. Klein establishes metric projective geometry first and then uses the theory of cross ratios as the foundation for all other metric geometries. He says on page 271: "In the first two parts of this book, we have developed non-Euclidean geometry on the foundation of projective geometry. This road is the most convenient procedure, ..." Nowadays, in the age of linear algebra, this position cannot be defended any longer. Instead, one should first establish the theory of metric vector spaces and then use the theory of inner products as the foundation of all other metric geometries, including projective geometry.

Exercises

*17. Prove that there are three nonisometric real projective three-spaces. In the classical terminology, the space at infinity of Euclidean four-space is called *elliptic* (projective) *three-space*; the space at infinity of Minkowski space is the *hyperbolic three-space*; the space at infinity of Artinian four-space has no special name.

18. Let $n \geq 1$.
    a. If n is odd, prove that there are $\frac{1}{2}(n + 1)$ nonisometric real projective spaces of dimension $n - 1$.
    b. If n is even, prove that there are $\frac{1}{2}(n + 2)$ such spaces.

19. Let P be the hyperbolic (projective) plane and ϕ its fundamental conic.
    a. Prove that a line of P which intersects ϕ in two distinct points is a hyperbolic line. (Hint: P is the space at infinity of the metric vector space V defined by the matrix

$$\begin{pmatrix} 1 & 0 & 0 \\ 0 & 1 & 0 \\ 0 & 0 & -1 \end{pmatrix}$$

(Exercise 164.7). Use the definition of ϕ and the fact that a line of P is a plane of V.
    b. Prove that a line of P which does not intersect ϕ in any point is an elliptic line.

35. ARTINIAN SPACES

We want to take the most instructive (not the shortest!) road toward two deep and beautiful theorems, the Witt theorem and the Cartan–Dieudonné theorem. Both theorems can be proved much easier for anisotropic spaces than for spaces with null lines. It is not true that all questions in geometry are made harder by the presence of null lines. On the contrary, certain questions concerning the groups attached to a metric vector space can actually be answered only when null lines are present. But it seems true that the more geometric a problem becomes, the more trouble and fun one can expect from isotropy. In any case, if V has null lines, its geometry takes on a highly non-Euclidean flavor and methods have to be developed which are foreign to Euclidean geometry. The purpose of the present section and the following one is to build the necessary techniques to deal effectively with nonzero isotropic vectors.

In this section, we return to the case that k is not necessarily ordered.

When $k = \mathbf{R}$, there are three nonsingular planes. (We should say "nonisometric nonsingular planes" but omit the word "nonisometric" when this omission cannot lead to confusion.) Of them, only the Lorentz plane has null lines and only the Lorentz plane has discriminant -1. The following proposition says that the analogous situation holds for plane geometries over any field.

Proposition 184.1.   Let $n = 2$ and assume that V is nonsingular. Then V has a nonzero isotropic vector if and only if disc $V = -1$.

Proof. Assume disc $V = -1$. Then V has a rectangular coordinate system A, B such that, if $A^2 = a$ and $B^2 = b$, then $ab = -1$. Thus $ab^2 + b = 0$ and the nonzero vector $bA + B$ is isotropic.

Assume V contains a nonzero isotropic vector N. We choose a rectangular coordinate system A, B for V and put $A^2 = a$, $B^2 = b$. Then $N = xA + yB$ and $x^2 a + y^2 b = N^2 = 0$. Since V is nonsingular, neither A nor B is isotropic and hence $a, b, x,$ and y are all nonzero. Because $x \neq 0$, we conclude that

$$ab = -\left(\frac{yb}{x}\right)^2.$$

Since $yb \neq 0$, it follows that disc $V = -1$. Done.

The above proposition says that a nonsingular plane is anisotropic if and only if its discriminant is not -1. For arbitrary fields, there are usually many nonisometric anisotropic planes. Over the real numbers, there are already two, the Euclidean and the negative Euclidean planes. Over the rational numbers, there are infinitely many such planes, as the following exercise shows.

Exercises

1. Let $n = 2$, $k = \mathbf{Q}$ and assume that the metric of the rational plane V is given by $\left(\begin{smallmatrix} q & 0 \\ 0 & 1 \end{smallmatrix}\right)$ where $q > 0$, $q \in \mathbf{Q}$.
   a. Prove that V is anisotropic. (This result also becomes obvious by writing the quadratic form of V.)
   b. By choosing q equal to prime numbers 2, 3, 5, 7, ... and so on, prove that one obtains infinitely many nonisometric anisotropic planes.
2. Give examples of nonisometric anisotropic planes over $\mathbf{Q}$ which have the same discriminant.

Everyone expects to find many nonisometric planes which are nonsingular and have null lines. Why does our intuition behave so miserably at times? We shall learn in Proposition 186.2 that, on the contrary, there is only one nonsingular plane with null lines and we work toward this proposition now. The investigation shows how important the null spaces of a nonsingular space are for the structure of that space.

Proposition 185.1. Let V be a nonsingular plane and N a nonzero isotopic vector of V. There exists one and only one isotropic vector M in V such that $MN = 1$.

Proof. Since N is an isotropic vector, $\langle N \rangle \subset \langle N \rangle^*$. Moreover, dim $\langle N \rangle^* = 1$ and hence $\langle N \rangle = \langle N \rangle^*$. If $A \notin \langle N \rangle$, $AN \neq 0$. The vectors A and N form a coordinate system of V and we now determine *all* vectors $M = xA + yN$ such that

1. $M^2 = x^2 A^2 + 2xyAN = 0.$
2. $MN = x A N = 1.$

Since $AN \neq 0$, Equation 2 determines x uniquely. If this value is c, Equation 1 becomes

$$c^2 A^2 + 2y = 0$$

which determines y uniquely (since char $k \neq 2$). Done.

In the following proposition, the metric vector space may be singular and can have any dimension.

Proposition 186.1.   Assume A and N are vectors of a metric vector space where N is isotropic and $AN \neq 0$. Then A and N are linearly independent.

Proof. Since $AN \neq 0$, $N \neq \mathbf{0}$ and $A \notin \langle N \rangle$; therefore, A and N are linearly independent. Done.

Proposition 186.2.   There is only one nonsingular plane, within isometry, which contains a null line.

Proof. Assume that the nonsingular plane V contains the nonzero isotropic vector N. Let M be such that $M^2 = 0$ and $MN = 1$. Then M and N form a coordinate system for V and the matrix of V relative to this coordinate system is

$$\begin{pmatrix} 0 & 1 \\ 1 & 0 \end{pmatrix}.$$

Hence all such planes are represented by the same matrix relative to appropriate coordinate systems; this shows that they are all isometric. Done.

The unique nonsingular plane which contains a null line can also be described as the plane with discriminant -1 (Proposition 184.1).

Definition 186.1.   The plane with discriminant -1 is called the **Artinian plane**.

When $k = \mathbf{R}$, we observe that the Artinian plane is the Lorentz plane.

Defense of Terminology. What we call the Artinian plane is called the hyperbolic plane by Artin, [2], Definition 3.8, p. 119. We have always felt that "hyperbolic plane" is a poor term for a metric affine plane since it causes confusion with the classical terminology where hyperbolic plane means a certain metric projective plane. (See Remark about Projective Geometry starting on page 180.) Only readers who have studied this remark should proceed with the discussion on terminology which follows. We do not know why Artin made his choice, but it may have been because the projective line at infinity of the Lorentz plane is the classical hyperbolic line. However, this argument already breaks down for dimension four. What Artin calls real four-dimensional hyperbolic space (and we have called real four-dimensional Artinian space) does not have the classical (projective) hyperbolic three-space as its space at infinity; this latter space is the space at infinity of Minkowski space (Exercise 183.17).

We are indebted to H. S. M. Coxeter for having pointed out the confusion caused by the hyperbolic terminology and for having suggested the name Artinian plane to us. As Coxeter said, Artin was the one who saw the importance of the Artinian plane for the classification of metric vector spaces.

One final word in defense of Artin. The only good term was not available to him.

We now study some properties of Artinian planes. They are all familiar to us from our study of the Lorentz plane.

Every Artinian plane P contains a nonzero isotropic vector N. There exists a vector M in P such that $M^2 = 0$ and $MN = 1$ (Proposition 185.1). Moreover, the vectors M, N are linearly independent (Proposition 186.1).

Definition 187.1. An ordered pair of vectors M, N of an Artinian plane with the property that

$$M^2 = N^2 = 0 \quad \text{and} \quad MN = 1$$

is called an **Artinian coordinate system**.

Our term "Artinian coordinate system" corresponds to Artin's "hyperbolic pair." (See Artin, [2], p. 119.)

Exercise

3. Let M and N be vectors of V such that $M^2 = N^2 = 0$ and $MN = 1$. Prove that the plane $\langle M, N \rangle$ is an Artinian plane. (Consequently, the ordered pair M, N is an Artinian coordinate system of $\langle M, N \rangle$.)

Proposition 187.1. An Artinian plane contains precisely two null lines.

Proof. Let P be an Artinian plane and M, N an Artinian coordinate system of P. Then $\langle M \rangle$ and $\langle N \rangle$ are two null lines of P. In order to show there are no others, consider an isotropic vector $X = xM + yN$ of P. Then $X^2 = 2xy = 0$ and hence $x = 0$ or $y = 0$. This shows that $X \in \langle N \rangle$ or $X \in \langle M \rangle$. Done.

The next proposition is an immediate consequence of Corollary 168.1.

Proposition 188.1.   An Artinian plane represents all scalars.

For the following result, V may be singular.

Proposition 188.2.   Let N be an isotropic vector of V which does not belong to the radical of V. Then N is contained in an Artinian plane.

Proof. Since $N \notin \operatorname{rad} V$, there is a vector $A \in V$ such that $AN \neq 0$. The vectors A and N are linearly independent (Proposition 186.1); hence they span a plane P which we show is Artinian. We put $A^2 = a$ and $AN = b$; the matrix of P relative to the coordinate system A, N is

$$\begin{pmatrix} a & b \\ b & 0 \end{pmatrix}.$$

Consequently, disc $P = -b^2$ which is -1 modulo squares since $b \neq 0$. This shows that P is Artinian. Done.

Exercises

4. Let P be an Artinian plane.
   a. Show that $\{0\}$ and the two null lines are the only null spaces of P.
   b. Prove that there exists a one-to-one correspondence between the nonzero elements of k and the set of Artinian coordinate systems of P.
5. Assume M, N is an Artinian coordinate system for the Artinian plane P. If one translates the line $\langle M \rangle$ by the vector N, one obtains the one-dimensional affine subspace (not a linear subspace)
$$N + \langle M \rangle = \{N + xM \,|\, x \in k\}$$
   of P.

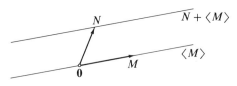

Figure 188.1

   a. Prove that the vectors of $N + \langle M \rangle$ represent all scalars.
   b. Let X be a vector of P. Prove that the vectors of the affine subspace $X + \langle M \rangle$ represent all scalars if and only if $X \notin \langle M \rangle$. (Of course, $b \Rightarrow a$.)

6. Prove that the vectors A and N of Proposition 186.1 span an Artinian plane.

*7. Let V be a nonsingular space and N a nonzero isotropic vector of V. Prove that N is contained in an Artinian plane.

Definition 189.1.   A metric vector space is called an **Artinian space** if it is the orthogonal sum

$$P_1 \oplus \cdots \oplus P_s$$

of Artinian planes $P_1, \ldots, P_s$.

The Artinian plane is the Artinian space of dimension two. If $k = \mathbf{R}$, we have already observed that Artinian four-space is the only nonsingular four-space which has signature 2. The interesting thing about the following proposition is that the three properties just mentioned characterize Artinian spaces. (See Corollary 195.2.)

Proposition 189.1.   An Artinian space V is nonsingular, has even dimension ≥ 2, and contains null spaces of dimension $\frac{1}{2}\dim V$.

Proof. Since by definition,

$$V = P_1 \oplus \cdots \oplus P_s$$

where $P_1, \ldots, P_s$ are Artinian planes, dim $V = 2s \geq 2$ and disc $V = (-1)^s \neq 0$. There only remains to be shown that V contains a null space of dimension s. For this, we choose a nonzero isotropic vector $N_i \in P_i$ and observe that $\langle N_1, \ldots, N_s \rangle$ is a null space of dimension s. Done.

The fundamental importance of Artinian spaces for the investigation of metric vector spaces will become apparent in the sequel.

Exercises

*8. Let M, N be an Artinian coordinate system for the Artinian plane P. Find scalars a, b, c, d such that the vectors $A_1 = aM + bN$ and $A_2 = cM + dN$ form a rectangular coordinate system relative to which the matrix of P is $\left(\begin{smallmatrix} 1 & 0 \\ 0 & -1 \end{smallmatrix}\right)$. This shows again that, if $k = \mathbf{R}$, the Artinian plane is the Lorentz plane.

*9. Let P be an Artinian plane with M, N, A_1, A_2 vectors, as given in Exercise 8 above. A "circle" with center $\mathbf{0}$ consists of the set of vectors $\{X \mid X^2 = c\}$ for some $c \in k$. Prove that the equation of this circle in the coordinate system A_1, A_2 is $x_1{}^2 - x_2{}^2 = c$. In the figure below, these circles were drawn in the case that $k = \mathbf{R}$ and $c = \pm 1$. ($M = A_1 + A_2$ and $N = \frac{1}{2}(A_1 - A_2)$; why?)

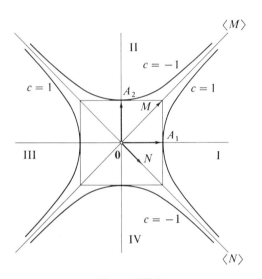

Figure 190.1

10. Let P be the Lorentz plane ($k = \mathbf{R}$) with M, N, A_1, A_2, again chosen as in Exercise 8 above. The lines $\langle M \rangle$ and $\langle N \rangle$ divide the plane into the four parts marked I, II, III, IV in Figure 190.1. We study the family of Lorentzian circles $\{X^2 \mid X^2 = c\}$ for $c \in \mathbf{R}$.
   a. If $c \neq 0$, prove that a Lorentzian circle is a Euclidean hyperbola.
   b. If $c > 0$, show that the circles lie in Parts I and III. Draw the circles for $c = \frac{1}{2}$ and $c = 2$.
   c. If $c < 0$, show that the circles lie in Parts II and IV. Draw the circles for $c = -\frac{1}{2}$ and $c = -2$.
   d. If $c = 0$, prove that the circle consists of the two null lines $\langle M \rangle$ and $\langle N \rangle$.

*11. Let P be a plane whose matrix is $\left(\begin{smallmatrix} 1 & 0 \\ 0 & 1 \end{smallmatrix}\right)$. Prove that P is Artinian if and only if -1 has a square root in k. (Hint: Use Proposition 184.1.)

12. Let P be the plane of the previous exercise. Prove that
    a. If $k = \mathbf{Z}_3$, P is anisotropic.
    b. If $k = \mathbf{Z}_5$, P is Artinian!
    c. If $k = \mathbf{Z}_7$, P is anisotropic.
    d. If $k = \mathbf{Q}$, P is anisotropic.
    e. If $k = \mathbf{R}$, P is anisotropic and is the Euclidean plane.
    f. If $k = \mathbf{C}$, P is Artinian!
    g. If k is any algebraically closed field, P is Artinian. (This is a generalization of Part f.)
13. Generalize Exercise 11 above by assuming that P is a plane whose matrix has determinant Δ. Prove that P is Artinian if and only if $\Delta \neq 0$ and $-\Delta$ has a square root in k.
14. Let P be a plane whose matrix is $\begin{pmatrix} -2 & 1 \\ 1 & -2 \end{pmatrix}$. Use the previous exercise to prove that
    a. If $k = \mathbf{Z}_3$, P is singular and has rank 1.
    b. If $k = \mathbf{Z}_5$, P is anisotropic.
    c. If $k = \mathbf{Z}_7$, P is Artinian.
    d. If $k = \mathbf{Q}$, P is anisotropic.
    e. If $k = \mathbf{R}$, P is anisotropic and is the negative Euclidean plane.
    f. If $k = \mathbf{C}$, P is Artinian.
    g. If k is algebraically closed, P is Artinian.
*15. Let P be a nonsingular plane. Prove that
    a. If k is algebraically closed, P is Artinian.
    b. More generally, if every element of k has a square root in k, P is Artinian.
*16. Assume that every element of k has a square root and that V is a nonsingular space of positive even dimension. Prove that V is Artinian. (Hint: We know from Proposition 157.1 that V has a coordinate system such that the matrix of V is the unit matrix. It follows that V is the orthogonal sum of s nonsingular planes where $s = \frac{1}{2}\dim V$. Now use the previous exercise.) Observe that all nonsingular spaces of positive even dimension over an algebraically closed field are Artinian.
17. Assume that every element of k has a square root in k and that V is a nonsingular space of odd dimension $2s + 1$, $s \geq 1$. Prove that V is the orthogonal sum of an Artinian space of dimension $2s$ and the nonsingular line over k. In particular, this is true when k is algebraically closed.
18. Prove that all Artinian spaces of the same dimension are isometric.
19. Assume V is Artinian space of dimension $2s$. Prove that V contains at least 2^s null spaces of dimension s. (If k is infinite and $s \geq 2$, V actually contains infinitely many null spaces of dimension s.)

*20. Let V be Artinian space of dimension $2s$ and U a null subspace of dimension s. Prove that $U^* = U$. This shows that the Euclidean picture of a subspace and its orthogonal complement intersecting only in the origin can be very wrong even in nonsingular spaces. As we know, the picture is only correct if U is also nonsingular.

21. Let $A_1, \ldots, A_h$ be Artinian spaces of dimensions $2s_1, \ldots, 2s_h$, respectively. Prove that the orthogonal sum $A_1 \oplus \cdots \oplus A_h$ is an Artinian space of dimension $2(s_1 + \cdots + s_h)$.

Assume that the metric of a space of dimension > 1 is given by the unit matrix

$$\begin{pmatrix} 1 & & \bigcirc \\ & \ddots & \\ \bigcirc & & 1 \end{pmatrix}.$$

It is clear from Exercises 16 and 17 above that one should not necessarily think of such a geometry as a generalization of Euclidean geometry. It depends on the field. If k is algebraically closed, the space is either Artinian or the orthogonal sum of an Artinian space and a line; such a geometry reminds us of the Lorentz plane and not at all of the Euclidean plane.

36. NONSINGULAR COMPLETIONS

We assume in this section that our n-dimensional metric vector space V is nonsingular and that U is a singular subspace of V.

Since $\{0\}$ is nonsingular, observe that $U \neq \{0\}$ and hence $n \geq 1$.

We shall show that every nonsingular subspace of V which contains U has dimension at least equal to dim U + dim(rad U). Furthermore, V always contains nonsingular subspaces which contain U and have this smallest dimension.

Definition 192.1. A nonsingular subspace of V which contains U and has dimension equal to dim U + dim(rad U) is called a **nonsingular completion of** U.

Usually, U has many nonsingular completions. The surprising fact is that they are all isometric and their common geometry has nothing to do with the geometry of V. The geometry of the nonsingular completions of U is entirely determined by the geometry of U alone! These nonsingular completions are a major tool for reducing proofs in which singular subspaces of V occur to the case when all the subspaces are nonsingular.

Theorem 193.1. Let W be a nonsingular subspace of U such that $U = W \oplus \text{rad } U$; assume $\dim(\text{rad } U) = s$ and that $N_1, \ldots, N_s$ is a coordinate system for rad U. Then there exist Artinian planes $P_1, \ldots, P_s$ in V with the property that

1. $N_i \in P_i$ for $i = 1, \ldots, s$.
2. The subspaces $W, P_1, \ldots, P_s$ of V are mutually orthogonal.

Proof. Case 1. $s = 1$. Then $U = W \oplus \langle N_1 \rangle$ and hence $N_1 \in W^*$. (Asterisks continue to denote orthogonal complements in V and not in U.) Since W^* is nonsingular and $N_1 \neq 0$, W^* contains an Artinian plane P_1 such that $N_1 \in P$ (Exercise 189.7). Because $P_1 \subset W^*$, we conclude that $W \perp P_1$.

Case 2. $s > 1$. We make the induction hypothesis that the theorem has been proved for $\dim(\text{rad } U) < s$. Let $U_0 = W \oplus \langle N_2, \ldots, N_s \rangle$ and observe that

1. $N_1 \in U_0^*$ since N_1 is perpendicular to W and all N_i, $i = 2, \ldots, s$.
2. $\text{rad } U_0 = \langle N_2, \ldots, N_s \rangle$ (Exercise 145.12a).
3. $N_1 \notin \text{rad } U_0^*$ (since $\text{rad } U_0^* = \text{rad } U_0$).

It follows from 1 and 3 that U_0^* contains an Artinian plane P_1 such that $N_1 \in P_1$ (Proposition 188.2) and it is clear that $W \perp P_1$. In order to choose planes $P_2, \ldots, P_s$, we conclude from $P_1 \subset U_0^*$ that $U_0 \subset P_1^*$. Since P_1^* is nonsingular and $\dim(\text{rad } U_0) = s - 1$, it follows from the induction hypothesis that P_1^* contains Artinian planes $P_2, \ldots, P_s$ satisfying

1. $N_i \in P_i$ for $i = 2, \ldots, s$.
2. The spaces $W, P_2, \ldots, P_s$ are mutually orthogonal.

It is trivial that $P_1 \perp P_i$ for $i = 2, \ldots, s$. Done.

We now prove the assertions made about nonsingular completions of U which preceded the above theorem. "Art_{2d}" will always denote an Artinian space of dimension $2d$. We recall that Artinian spaces of the same dimension are isometric.

Lemma 193.1.

1. $\dim U + \dim(\text{rad } U) \leq n$.
2. If $U \subset U' \subset V$ where the subspace U' is nonsingular, then

$$\dim U + \dim(\text{rad } U) \leq \dim U'.$$

3. Nonsingular completions of U exist.
4. All nonsingular completions of U are isometric with

$$(U/\text{rad } U) \oplus Art_{2s} \qquad \text{where} \quad s = \dim(\text{rad } U).$$

Proof. 1. Let $W, P_1, \ldots, P_s$ be chosen as in Theorem 193.1 and put $\overline{U} = W \oplus P_1 \oplus \cdots \oplus P_s$. Since $W \simeq U/\mathrm{rad}\ U$, $\dim W = (\dim U) - s$ and hence

$$\dim \overline{U} = (\dim U) - s + 2s = (\dim U) + s.$$

Because $\overline{U} \subset V$, we conclude that $(\dim U) + s \le n$ and, therefore,

$$\dim U + \dim(\mathrm{rad}\ U) \le n.$$

2. This follows immediately by using U' as the nonsingular space V.

3. It is trivial that $U \subset \overline{U} \subset V$ and, since the spaces $W, P_1, \ldots, P_s$ are all nonsingular, so is their orthogonal sum $\overline{U}$. We saw in 1 that

$$\dim \overline{U} = \dim U + \dim(\mathrm{rad}\ U)$$

and hence $\overline{U}$ is a nonsingular completion of U.

4. Let U' be a nonsingular completion of U. Then U' is nonsingular, by definition, and we may use U' as the V in Theorem 193.1. It follows that U' contains Artinian planes $P_1', \ldots, P_s'$ where

$$U \subset W \oplus P_1' \oplus \cdots \oplus P_s' \subset U'.$$

The dimensions of $W \oplus P_1' \oplus \cdots \oplus P_s'$ and U' are both equal to

$$\dim U + \dim(\mathrm{rad}\ U)$$

and hence

$$U' = W \oplus P_1' \oplus \cdots \oplus P_s'.$$

This shows that $U' \simeq (U/\mathrm{rad}\ U) \oplus Art_{2s}$ where $Art_{2s} = P_1' \oplus \cdots \oplus P_s'$. Done.

   Statement 4 of the above lemma shows that the geometry of a nonsingular completion of U has nothing to do with the geometry of V, but depends solely on the geometry of U. Namely, $U/\mathrm{rad}\ U$ is the nonsingular geometry which is intrinsically determined by U while the geometry of Art_{2s} is completely determined by the integer $s = \dim(\mathrm{rad}\ U)$. It follows that the geometry of the orthogonal sum $(U/\mathrm{rad}\ U) \oplus Art_{2s}$ depends on U only and all nonsingular completions are isometric to this orthogonal sum.

   The fact that U may have many nonsingular completions is brought out in Exercises 196.5 and 196.6.

Exercises

   *1. Let U be a nonzero null space of V. Prove that every nonsingular completion of U is an Artinian space of dimension 2(dim U).

   *2. Let $W, N_1, \ldots, N_s$ be as given in Theorem 193.1 and assume that $\overline{U}$ is a nonsingular completion of U. Prove that $\overline{U}$ contains Artinian planes $P_1, \ldots, P_s$ such that $N_i \in P_i$ for $i = 1, \ldots, s$ and $\overline{U} = W \oplus P_1 \oplus \cdots \oplus P_s$.

We now come to two beautiful corollaries of Lemma 193.1. The first one says that the dimension of null spaces which lie embedded in a nonsingular space cannot be too large. The second one says that the largest possible dimension for null spaces is attained only in Artinian spaces. We recall that V is nonsingular and dim $V = n$.

Corollary 195.1.   The dimension of a null space of V is at most $n/2$.

Proof. If the null space is $\{0\}$, the corollary is trivial. Otherwise, the null space is singular and we may assume that it is our space U. Since U is a null space, a nonsingular completion of U is an Artinian space whose dimension is $2(\dim U)$ (Exercise 194.1) and hence $2(\dim U) \leq n$. Done.

Exercise

*3. Recall that a **hyperplane** of an n-dimensional vector space is a subspace of dimension $n - 1$. Prove that V contains a hyperplane which is a null space if and only if V is an Artinian plane.

It is surprising that the occurrence of a null space in V of dimension $n/2$ causes V to be Artinian.

Corollary 195.2.   A metric vector space is Artinian if and only if the space is nonsingular, has even dimension $n \geq 2$, and contains a null space of dimension $n/2$.

Proof. We know from Proposition 189.1 that Artinian spaces have the three stated properties. Conversely, assume that our nonsingular space V has even dimension $n \geq 2$ and contains a null space U of dimension $n/2$. A nonsingular completion $\overline{U}$ of U is an Artinian subspace of V and dim $\overline{U} = n$. Consequently, V is the Artinian space $\overline{U}$. Done.

Exercises

*4. a. If dim $U + \dim(\text{rad } U) = n$, prove that V is the only nonsingular completion of U.
   b. If V is Artinian and U is a null space of V of dimension $n/2$, prove that V is the only nonsingular completion of U. (Of course, a $\Rightarrow$ b.) Both these cases are exceptional; a singular subspace usually has

many nonsingular completions as the following two exercises show.)

*5. We return to Exercise 164.7 where $k = \mathbf{R}$, $n = 3$, and the matrix of V is

$$\begin{pmatrix} 1 & 0 & 0 \\ 0 & 1 & 0 \\ 0 & 0 & -1 \end{pmatrix}.$$

We choose a line on the light cone C as our U.

a. Prove that every plane which contains U, except for the tangent plane of C along U, is a nonsingular completion of U. (Hint: Use Exercise 164.7a) This is an example of a singular subspace with infinitely many nonsingular completions.

b. A null line is always perpendicular to itself. Prove that two distinct lines of C cannot be perpendicular to one another. (Hint: Use Corollary 195.1.)

*6. Let V be Minkowski space; we again denote the light cone by C.

a. Prove that two distinct lines of C cannot be perpendicular. (Hint: Corollary 195.2 prevents the occurrence of a null plane in V.)

b. If U is a line on C, we know from Exercise 152.5 that U^* is a singular three-space of rank two with U as radical. Prove that U is the only null line in U^*.

c. Prove that U^* is the tangent space of C along U. (Hint: This is a geometric consequence of Part b.)

d. Prove that the nonsingular completions of U are Lorentz planes and that there are infinitely many of them.

7. In general, two distinct lines on the light cone of a nonsingular space may be perpendicular. Prove that the light cone contains a pair of distinct perpendicular lines if and only if the space contains a null plane. (We shall see in Exercise 208.6 that more is true; namely, if V contains a null plane and U is a line on the light cone, there always exists at least one other line $W \neq U$ on the cone such that $W \perp U$.)

*8. This exercise shows that the converse of Exercise 191.20 is true. Precisely, assume that $n \geq 1$ and V contains a null space whose orthogonal complement U^* is also a null space.

a. Prove that V is an Artinian space.

b. Prove that $U = U^*$ and that $\dim U = n/2$. (Hint: Use the properties of null subspaces and the fact that $\dim U + \dim U^* = n$.)

We now derive some properties of nonsingular completions which refer to "extensions of isometries." Assume $S \subset S'$ and $T \subset T'$ are all metric vector spaces; if $\sigma: S \to T$ and $\bar{\sigma}: S' \to T'$ are isometries, $\bar{\sigma}$ is called an **extension of** σ if $\bar{\sigma}A = \sigma A$ for all $A \in S$. (See Figure 197.1.) In other words, $\bar{\sigma}$

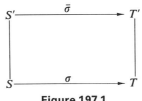

Figure 197.1

is an extension of σ if the restriction $\bar{\sigma}|S$ of $\bar{\sigma}$ to S is the isometry σ. (For the purists: We are committing "abuse of language" here by ignoring the ranges of σ and $\bar{\sigma}$.)

The following exercises discuss some light principles on isometries which will be used frequently in the sequel. In particular, we shall assume the reader has familiarized himself with the notion of the orthogonal sum $\sigma_1 \oplus \cdots \oplus \sigma_h$ of isometries introduced in Exercise 9 below.

Exercises

*9. Let S and T be metric vector spaces where S and T may be singular. Assume that
$$S = S_1 \oplus \cdots \oplus S_h \quad \text{and} \quad T = T_1 \oplus \cdots \oplus T_h$$
for subspaces $S_1, \ldots, S_h$ of S and $T_1, \ldots, T_h$ of T. Assume also that $\sigma_i: S_i \to T_i$ is an isometry for $i = 1, \ldots, h$. Define a linear mapping $\sigma: S \to T$ by $\sigma A = \sigma_i A$ if $A \in S_i$ and then extend σ "by linearity" to all of S. In other words, if $A = A_1 + \cdots + A_h$ where $A_i \in S_i$, then
$$\sigma A = \sigma_1 A_1 + \cdots + \sigma_h A_h.$$
The linear mapping σ is the usual "direct sum" of the linear mappings $\sigma_1, \ldots, \sigma_h$.
   a. Prove that σ is an isometry.
   b. Prove that σ is an extension of each σ_i for $i = 1, \ldots, h$.
   c. Prove that σ is the only isometry from S to T which is an extension of all the isometries $\sigma_1, \ldots, \sigma_h$.
   The isometry σ is called the **orthogonal sum of the isometries** $\sigma_1, \ldots, \sigma_h$ and is denoted by $\sigma_1 \oplus \cdots \oplus \sigma_h$.
10. Let P and P' be Artinian planes. Assume that M, N and M', N' are Artinian coordinate systems for P and P'. Define a linear mapping $\sigma: P \to P'$ by $\sigma M = M'$ and $\sigma N = N'$ and then extend σ by linearity to all of P. Prove that σ is an isometry from P to P'.
11. Let P and P' be Artinian planes. Assume that N is a nonzero isotropic vector of P and N' is a nonzero isotropic vector of P'. Prove that there exists one and only one isometry $\sigma: P \to P'$ such that $\sigma N = N'$.

The next lemma states that isometries between singular subspaces can always be extended to isometries between their nonsingular completion.

Lemma 198.1. Let $\overline{U}$ be a nonsingular completion of our singular space U. Assume that V' is also a nonsingular metric vector space, U' is a singular subspace of V', and $\overline{U}'$ is a nonsingular completion of U'. Then an isometry $\sigma: U \to U'$ can be extended to an isometry $\overline{\sigma}: \overline{U} \to \overline{U}'$.

Proof. In the diagram in Figure 198.1, the dots indicate that the isometry $\overline{\sigma}$ has to be constructed. We put $U = W \oplus \langle N_1, \ldots, N_s \rangle$ where $W \simeq U/\mathrm{rad}\, U$

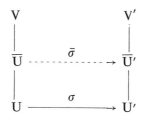

Figure 198.1

and $N_1, \ldots, N_s$ is a coordinate system for rad U. There exist Artinian planes $P_1, \ldots, P_s$ in $\overline{U}$ such that $N_i \in P_i$ for $i = 1, \ldots, s$ and $\overline{U} = W \oplus P_1 \oplus \cdots \oplus P_s$ (Exercise 194.2).

Since $\sigma: U \to U'$ is an isometry, $U' = \sigma W \oplus \langle \sigma N_1, \ldots, \sigma N_s \rangle$ where $\sigma W \simeq U'/\mathrm{rad}\, U'$ and $\sigma N_1, \ldots, \sigma N_s$ is a coordinate system of rad U' (Exercise 144.11). There exist Artinian planes $P_1', \ldots, P_s'$ in $\overline{U}'$ such that $\sigma N_i \in P_i'$ for $i = 1, \ldots, s$ and

$$\overline{U}' = \sigma W \oplus P_1' \oplus \cdots \oplus P_s'.$$

The isometry $\overline{\sigma}: \overline{U} \to \overline{U}'$ is defined by giving its action on W and on each P_i, namely,

$\overline{\sigma} A = \sigma A$ for all $A \in W$,
$\overline{\sigma}$ maps P_i isomorphically onto P_i' such that $\overline{\sigma} N_i = \sigma N_i$.

Clearly, $\overline{\sigma}$ maps W isometrically onto σW since $\overline{\sigma}|W = \sigma|W$; hence $\overline{\sigma}$ maps $\overline{U} = W \oplus P_1 \oplus \cdots \oplus P_s$ isometrically onto $\overline{U}' = \sigma W \oplus P_i' \oplus \cdots \oplus P_s'$ (Exercise 197.9). The fact that $\overline{\sigma}$ is an extension of σ follows from $\overline{\sigma}|W = \sigma|W$ and $\overline{\sigma} N_i = \sigma N_i$. Done.

Exercises

*12. Let $\overline{U}$ be a nonsingular completion of U and $\sigma: U \rightarrow U$ an isometry of U onto itself. Prove that σ can be extended to an isometry $\bar{\sigma}: \overline{U} \rightarrow \overline{U}$.

13. Let $\overline{U}$ be a nonsingular completion of U. Prove that rad U is the orthogonal complement of U in $\overline{U}$. (Hint: This exercise is not related to the theory developed in this section. Use only the definition of nonsingular completion and Proposition 148.1.)

37. THE WITT THEOREM

We assume for the remainder of this chapter that our n-dimensional metric vector space V is nonsingular.

The most fundamental question one can ask about the geometry of metric vector spaces is: If U and W are subspaces of V and $\sigma: U \rightarrow W$ is an isometry, can σ always be extended to an isometry $\bar{\sigma}: V \rightarrow V$, that is, such that $\bar{\sigma}A = \sigma A$ for all $A \in U$? One should think of an isometry of V onto itself as a rotation or a reflection. In Definition 217.1, a precise definition of rotation and reflection is given and we shall see that every isometry of V onto itself is either a rotation or a reflection. Using this classification of isometries now in our intuitive geometric thinking, the fundamental question can be formulated as follows. Can every isometry between subspaces of V be extended to either a rotation or a reflection, or, perhaps, even to both?

What does our knowledge of Euclidean three-space E_3 have to say about this? Let A_1, A_2, A_3 be an orthonormal coordinate system of E_3. We choose U and W both equal to the plane $\langle A_1, A_2 \rangle$ and let $\sigma: U \rightarrow U$ be the reflection

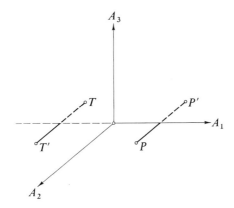

Figure 199.1

of U with respect to the line $\langle A_1 \rangle$. This means σ carries every point P of U onto its mirror image P' with respect to the line $\langle A_1 \rangle$ as shown in Figure 199.1. Under σ, the points of the coordinate axis $\langle A_1 \rangle$ are left fixed while the points "in front" of this axis are mapped on points "in back" of it and conversely.

There are two obvious ways in which σ can be extended to an isometry $\bar{\sigma}: E_3 \to E_3$. The most obvious one is the reflection of three-space with respect to the plane $\langle A_1, A_3 \rangle$. Thus $\bar{\sigma}$ maps each point of E_3 onto its mirror image with respect to the plane $\langle A_1, A_3 \rangle$; the points of the plane $\langle A_1, A_3 \rangle$ remain fixed while every point "in front" of it goes to a point "in back" of it and conversely. A second extension of σ is the rotation of E_3 over $180°$ around the line $\langle A_1 \rangle$ as axis of rotation. We see that, in this case, σ can be extended to both a reflection and a rotation of E_3.

Of course, one example does not tell the whole story. Mathematics is a very experimental science and, usually, much experimentation is necessary before a reasonable conjecture can be made. The reader will have to do many more experiments like the one above, both in Euclidean three-space and the Euclidean plane, before he can venture a conjecture about the fundamental question. The purpose of the following exercises is to help design such experiments.

In Exercises 1, 2, 5, and 6 of this section, the terms rotation and reflection are used in a nonrigorous, intuitive sense. Once the precise definition of these terms has been given, the reader will have an easy method for determining whether a given isometry is a rotation or a reflection.

Exercises

In each of the exercises below, check first that the given σ is an isometry.

*1. Let A_1, A_2, A_3 be an orthonormal coordinate system of Euclidean three-space E_3. For each of the following isometries σ from U onto W, describe both a rotation and a reflection of E_3 which are extensions of σ.

   a. $U = W = \langle A_1, A_2 \rangle$ and $\sigma: U \to U$ is the rotation of U onto itself over $180°$ around the point $\mathbf{0}$ as center of rotation. (Hint: For the reflection, try $\bar{\sigma}A = -A$ which is the reflection of E_3 in the point $\mathbf{0}$.)

   b. $U = W$ is a line of E_3 and $\sigma: U \to U$ is the isometry $\sigma A = -A$ where $A \in U$.

   c. $U = \langle A_1 \rangle$, $W = \langle A_3 \rangle$, and $\sigma: U \to W$ is the isometry $\sigma(t A_1) = t A_3$ for all $t \in k$.

2. Let A_1, A_2 be an orthonormal coordinate system for the Euclidean plane E_2. For each of the given isometries from U onto W, describe both a rotation and a reflection of E_2 which are extensions of σ.

a. $U = \langle A_2 \rangle$, $W = \langle A_1 \rangle$, and $\sigma: U \to W$ is the isometry $\sigma(tA_2) = tA_1$ for all $t \in k$.

b. $U = W$ is a line of E_3 and $\sigma: U \to U$ is the isometry $\sigma A = -A$ where $A \in U$.

Many more experiments like the ones above lead us to the following conjecture: If U and W are proper subspaces of Euclidean n-space E_n and $\sigma: U \to W$ is an isometry, σ can always be extended to both a rotation and a reflection of E_n. This conjecture is true not only for E_n, but for all aniso-tropic spaces. Moreover, the proof is relatively easy. Of course, when $U = V$ and hence σ is an isometry of V, the only extension of σ is σ itself. In this case, the extension exists but is unique and cannot be made at will into either a rotation or a reflection.

If V contains nonzero isotropic vectors, it is still true that every isometry $\sigma: U \to W$ between subspaces U and W of V can be extended to an isometry $\bar{\sigma}$ of V (see Theorem 202.1). The proof is harder than for anisotropic spaces, but can be reduced to that case by using nonsingular completions. However, it is no longer true, even if $U \neq V$, that σ can always be extended to both a reflection and a rotation of V. The precise amount of freedom one has in choosing the extension is given by Theorem 234.1.

The result which states that every isometry between subspaces of V can be extended to an isometry of V is due to the German mathematician E. Witt, [23]. One should be surprised that Witt's theorem was proved only as late as 1936. Artin called it "a scandal." If the most fundamental question which can be asked about the geometry of metric vector spaces has such a satis-factory answer and can be proved with such relative ease, why was the question not settled fifty years or so earlier? The reason is simple. This question is exposed only if we think about metric vector spaces geometrically. If we think about them as quadratic forms or, equivalently, as symmetric matrices and thereby blind ourselves to the underlying geometry, the question never appears in the first place. This is precisely what happened. As soon as the geometrization of algebra had begun, the fundamental question was asked and, of course, immediately answered.

Before we give the precise formulation of Witt's theorem, some exercises are given which take care of part of the proof of the theorem and are a good preparation for it.

Exercises

*3. Let U be a nonsingular subspace of V and $\sigma: U \to U$ an isometry. Prove that σ can be extended to an isometry $\bar{\sigma}$ of V. (Hint: Put $V = U \oplus U^*$ and use Exercise 197.9.)

*4. Let U and $\sigma: U \to U$ be as in the previous exercise. We found that σ can usually be extended in many ways to an isometry $\bar{\sigma}$ of V. Namely, each choice of an isometry $\rho: U^* \to U^*$ gives rise to the extension $\bar{\sigma} = \sigma \oplus \rho$. We now show there are no other ways to extend σ.
   a. If an isometry $\bar{\sigma}: V \to V$ is an extension of σ, prove that $\bar{\sigma} = \sigma \oplus \rho$ for some isometry $\rho: U^* \to U^*$.
   b. Prove there is a one-to-one correspondence between the set of all isometries of U^* onto itself and the set of those isometries of V which are extensions of σ.

5. Let us use the results of Exercise 4 above to generalize Exercise 200.1 where A_1, A_2, A_3 is an orthonormal coordinate system for Euclidean three-space E_3.
   a. Let $U = W = \langle A_1, A_2 \rangle$. Assume $\sigma: U \to U$ is the rotation of U over an arbitrary angle around the point $\mathbf{0}$ as center of rotation. Prove that σ can be extended to one and only one rotation of E_3 and to one and only one reflection of E_3. (Hint: U^* is the coordinate axis $\langle A_3 \rangle$.)
   b. Let $U = W$ be a line of E_3 and $\sigma: U \to U$ the isometry $\sigma A = -A$ for $A \in U$. Prove that σ can be extended to infinitely many rotations of E_3 and to infinitely many reflections of E_3. (Hint: U^* is now a Euclidean plane.)

*6. We now remove the restriction that U is nonsingular from Exercise 3 above by the technique of nonsingular completions. Let U be an arbitrary subspace of V and $\sigma: U \to U$ be an isometry. Prove that σ can be extended to an isometry $\bar{\sigma}$ of V. (Hint: The only case remaining is when U is singular. It is then impossible to form the orthogonal sum of U and U^* because $U \cap U^* = \text{rad } U \neq \{0\}$. Let $\bar{U}$ be a nonsingular completion of U. Then apply Exercise 199.12 and the fact that $V = \bar{U} \oplus \bar{U}^*$.)

We recall that V is a nonsingular space of dimension n. If S is any vector space, we denote the identity mapping of S by 1_S.

Theorem 202.1 (the Witt theorem). Every isometry $\sigma: U \to W$ between subspaces U and W of V can be extended to an isometry of V.

Proof. Case 1. dim $U =$ dim $W = 1$. If $U = W$, the theorem has been proved in Exercise 202.6. If $U \neq W$, choose a nonzero vector $A \in U$ and let $\sigma A = B$. Since $U = \langle A \rangle$ and $W = \langle B \rangle$ with $U \neq W$, the vectors A and B are linearly independent and hence span a plane $P = \langle A, B \rangle$. The linear transformation ρ of P defined by $\rho A = B$ and $\rho B = A$ is an isometry of P (why?). This isometry can be extended to an isometry γ of V (Exercise 202.6). Clearly, $\gamma A = B$ and hence γ is an extension of σ.

Case 2. The common dimension of U and W is arbitrary, but U and hence W are nonsingular. We shall proceed by induction on the dimension of V. The theorem is trivial for $n = 0$ and the induction hypothesis states that the theorem has been proved for dim $V < n$. Select a nonisotropic vector $A \in U$ and let $\sigma A = B$. Then

$$U = \langle A \rangle \oplus U' \quad \text{and} \quad W = \langle B \rangle \oplus \sigma U'$$

where U' is the orthogonal complement of $\langle A \rangle$ in U and hence $\sigma U'$ is the orthogonal complement of $\langle \sigma A \rangle = \langle B \rangle$ in W. By Case 1, there exists an isometry ρ of V such that $\rho A = B$ and it follows that $\rho U = \langle B \rangle \oplus \rho U'$. (Observe that ρ puts A just where we want it, but may act wildly on U'.) The two spaces $\sigma U'$ and $\rho U'$ both belong to the nonsingular space $\langle B \rangle^*$ of dimension $n - 1$ which now plays the role of V; hence the isometry $\sigma \rho^{-1} : \rho U' \to \sigma U'$ can be extended to an isometry $\gamma : \langle B \rangle^* \to \langle B \rangle^*$ by the induction hypothesis. Consider the isometry $\bar{\sigma} = (1_{\langle B \rangle} \oplus \gamma)\rho$ of V which is the product of the two isometries $1_{\langle B \rangle} \oplus \gamma : V \to V$ and $\rho : V \to V$. Clearly,

$$\bar{\sigma} A = (1_{\langle B \rangle} \oplus \gamma) B = B;$$

if $C \in U'$,

$$\bar{\sigma} C = \gamma \rho C = \sigma \rho^{-1} \rho C = \sigma C.$$

This shows that $\bar{\sigma}$ is an extension of σ to V.

Case 3. U and hence W are singular. Select nonsingular completions $\bar{U}$ of U and $\bar{W}$ of W. The isometry σ can be extended to an isometry $\sigma' : \bar{U} \to \bar{W}$ by Lemma 198.1. The isometry σ' can be extended to an isometry $\bar{\sigma} : V \to V$ by the previous case. Done.

When referring to Witt's theorem, we shall mean either Theorem 202.1 or the following equivalent theorem.

Theorem 203.1. Let V and V' be isometric vector spaces. Let U be a subspace of V, U' a subspace of V', and assume that $\sigma : U \to U'$ is an isometry. Then σ can be extended to an isometry from V to V'.

Proof. Observe that V' is necessarily nonsingular. If $V = V'$, one obtains Theorem 202.1 immediately and hence we need only show that, conversely, Theorem 202.1 implies the above statement. Since $V \simeq V'$, there exists an isometry $\rho: V \to V'$. The subspaces U and $\rho^{-1}U'$ are isometric subspaces of V under the isometry $\rho^{-1}\sigma: U \to \rho^{-1}U'$. By Theorem 202.1, $\rho^{-1}\sigma$ can be extended to an isometry $\gamma: V \to V$. Now consider the isometry $\rho\gamma: V \to V'$. If $A \in U$,

$$\rho\gamma A = \rho\rho^{-1}\sigma A = \sigma A$$

and hence $\rho\gamma$ is an extension of σ. Done.

The reader should try to visualize the various spaces which occur in Case 2 of the proof of Theorem 202.1. Let U and W be planes in Euclidean three-space E_3 and assume that σ is a rotation of U clockwise over $135°$ around the line of intersection m of U and W; see Figure 204.1 where we also drew the vectors A and $B = \sigma A$.

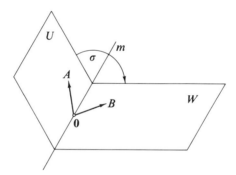

Figure 204.1

One extension of σ to E_3 is now obvious, namely, the rotation of E_3 over $135°$ with m as axis of rotation. However, this is not how the proof proceeds. Let l be the line through $\mathbf{0}$ which is perpendicular to both lines $\langle A \rangle$ and $\langle B \rangle$ and assume that the isometry ρ of Case 2 is the rotation of E_3 with l as axis of rotation and such that $\rho A = B$. In the figure, ρ rotates A clockwise approximately $45°$.

Exercises

7. Make a model of Figure 204.1 by opening a manilla folder $135°$ and drawing the point $\mathbf{0}$, the vectors A and B, and the lines U', $\sigma U'$, and m on the folder. Use the model to visualize the line l, the rotation ρ,

the line $\rho U'$, and the planes ρU and $\langle B\rangle^*$. Use the model to follow the various steps of Case 2. (We used E_3 only for the purpose of visualizing the general proof. There is a very short proof of the Witt theorem in case V is Euclidean n-space given in Exercise 10 below.)

8. Let S and T be metric vector spaces and $\rho: S \to T$ an isometry. Assume that J and K are subspaces of S such that $S = J \oplus K$. Prove that $T = \rho J \oplus \rho K$. Then indicate the places in Case 2 of the proof of Theorem 202.1 where this principle was used.

9. In case V is anisotropic, the proof of Theorem 202.1 can be greatly simplified. The whole theory of nonsingular completions can be disregarded; all that is needed is Exercise 202.3 along with Cases 1 and 2 of the proof of Witt's theorem. Give the simplest possible proof of this theorem in case V is anisotropic.

*10. For Euclidean n-space E_n, one can give a very short proof of Witt's theorem, the reason being that subspaces of E_n, having the same dimension, are isometric (Exercise 159.7b). In general, we shall say that V has "property P" if subspaces of V, having the same dimension, are isometric.

   a. If k is ordered and every positive element has a square root, prove that all positive definite and all negative definite spaces have property P.

   b. If V has property P, prove that V is anisotropic. (Hint: We recall that V is nonsingular.)

   c. Give examples of anisotropic spaces which do not have property P.

   d. Assuming that V has property P, give a short proof of the Witt theorem. (Hint: As a consequence of Part b above, $V = U \oplus U^* = W \oplus W^*$.)

   Remark. Readers who are interested in high school teaching should write the proofs of the Witt theorem for the Euclidean plane and for Euclidean three-space in the most elementary language possible, using the method of Part d. One will observe that the resulting proofs can very well be presented to an advanced geometry class in the eleventh or twelfth grades. Here then is excellent material for enriching our geometry offerings on that level.

The method of proof in Exercise 10d above is very interesting. It is based on the law that, if V has property P, isometric subspaces of V have isometric orthogonal complements. The next result which is an immediate corollary of the Witt theorem states that the same law holds whether or not V has property P.

Theorem 206.1. If U and W are isometric subspaces of V, their orthogonal complements U^* and W^* are also isometric.

Proof. Let $\sigma: U \to W$ be an isometry. If $\bar{\sigma}$ is an extension of σ, it is immediate that $\bar{\sigma}$ maps U^* onto W^*. Consequently, the restriction $\bar{\sigma}|U^*$ of $\bar{\sigma}$ to U^* is an isometry from U^* onto W^*. Done.

Using the proof method of Exercise 10d above, it should be very easy to prove the Witt theorem as a corollary of the above theorem (see Exercise 12 below). For this reason, we should consider these two theorems as equivalent.

Exercises

*11. We now formulate Theorem 206.1 for two different spaces V and V' just as Theorem 202.1 was given the equivalent formulation in Theorem 203.1. Let V and V' be isometric spaces (which implies that V' is nonsingular), U a subspace of V, and U' a subspace of V'. If U and U' are isometric, prove that their orthogonal complements U^* and U'^* are isometric.

*12. Prove Theorem 202.1 as a corollary of Theorem 206.1. (Hint: Use Theorem 206.1 to give a short proof of Case 2 of the proof of Theorem 202.1 and then repeat Case 3.)

*13. Let V be an Artinian space of dimension $2r$. If U is an Artinian subspace of dimension $2s$ and $s < r$, prove that U^* is an Artinian space of dimension $2r - 2s$. (Hint: Choose an arbitrary Artinian space W of dimension $2r - 2s$ and form the orthogonal sum $V' = U \oplus W$. Now show that $V \simeq V'$ and use Exercise 11 above.)

*14. Let V be an Artinian space of dimension $2r$ and U an Artinian space of dimension $2s$. Prove that U is contained in an Artinian subspace of V of dimension $2d$ for all $s \le d \le r$. (Hint: Put $V = U \oplus U^*$.)

If one translates the Witt theorem into matrix language, one obtains a rather dull and meaningless theorem on symmetric matrices. Nevertheless, we carry out this translation since the emptiness of the matrix theorem brings out dramatically the tremendous advantage, both in beauty and power, of the geometric approach over the matrix approach to metric vector spaces.

Let G and G' be $n \times n$ nonsingular symmetric matrices with entries in k. Assume that both matrices are decomposed into a diagonal $u \times u$ block S (S') and a diagonal $(n - u) \times (n - u)$ block T (T') giving

$$G = \left(\begin{array}{c|c} S & O \\ \hline O & T \end{array} \right) \quad \text{and} \quad G' = \left(\begin{array}{c|c} S' & O \\ \hline O & T' \end{array} \right).$$

Then S and S' are $u \times u$ nonsingular symmetric matrices while T and T' are $(n - u) \times (n - u)$ nonsingular symmetric matrices. Under these conditions, the matrix theorem states: If the matrices G, G' are congruent and the matrices S and S' are congruent, the matrices T, T' are also congruent. (See page 124 for the notion of congruent matrices.)

For the proof, choose a coordinate system $A_1, \ldots, A_n$ of V and define the metric of V by the matrix G. Choose some other n-dimensional vector space V' with coordinate system $A_1', \ldots, A_n'$ and define the metric of V' by G'. Since G and G' are nonsingular and congruent, the metric vector spaces V and V' are nonsingular and isometric. Furthermore,

$$\langle A_1, \ldots, A_u \rangle^* = \langle A_{u+1}, \ldots, A_n \rangle \quad \text{and} \quad \langle A_1', \ldots, A_u' \rangle^* = \langle A'_{u+1}, \ldots, A_n' \rangle;$$

the matrices S, T, S', and T' are the matrices of, respectively, the metric vector spaces $\langle A_1, \ldots, A_u \rangle$, $\langle A_{u+1}, \ldots, A_n \rangle$, $\langle A_1', \ldots, A_u' \rangle$, and $\langle A'_{u+1}, \ldots, A_n' \rangle$ (relative to the indicated coordinate systems). The congruence of S and S' implies that the spaces $\langle A_1, \ldots, A_u \rangle$ and $\langle A_1', \ldots, A_u' \rangle$ are isometric; we conclude from Exercise 11 above that $\langle A_1, \ldots, A_u \rangle^* \simeq \langle A_1', \ldots, A_u' \rangle^*$. This means that $\langle A_{u+1}, \ldots, A_n \rangle \simeq \langle A'_{u+1}, \ldots, A_n' \rangle$ and hence the matrices T and T' are congruent.

No wonder the fundamental question on extension of isometries was never asked in matrix language!

38. MAXIMAL NULL SPACES

The Witt theorem makes it possible to study the null spaces which are embedded in a metric vector space in greater detail. Every space contains a null space since the origin is a null space; we shall be interested in the "maximal" null spaces.

Definition 207.1. A null space N of V is called **maximal** if it is not properly contained in a larger null space of V.

Since $\{0\}$ is a null space and V has finite dimension, it is clear that V always contains maximal null spaces. Of course, $\{0\}$ is a maximal null space if and only if V is anisotropic.

The power of the Witt theorem will now become apparent.

Theorem 207.1. All maximal null spaces of V have the same dimension.

Proof. Let M and N be maximal null spaces of V and suppose that dim $M \neq$ dim N. Fix the notation so that dim $N <$ dim M. Choose a subspace U of M such that dim $U =$ dim N. Every nonsingular linear transformation $\sigma: U \to N$ from U onto N is an isometry and hence can be extended to an isometry $\bar{\sigma}$ of V (the Witt theorem). Then $\bar{\sigma} M$ is a null space which properly contains N, contrary to the hypothesis that N is maximal. Done.

Definition 208.1.   The common dimension of the maximal null spaces of V is called the **Witt index** or, simply, the **index** of V.

Clearly, a space has Witt index 0 if and only if the space is anisotropic.

Exercises

1. Prove that the Witt index of V is at most $n/2$.
2. Prove that V is Artinian if and only if n is a positive even integer and V has index $n/2$.
3. The table on page 179 lists the fifteen nonsingular nonisometric real spaces of dimension ≤ 4. Proceeding from top to bottom in the table, prove that the indices of these geometries are

$$0, 0, 0, 0, 1, 0, 0, 1, 1, 0, 0, 1, 2, 1, 0.$$

4. Prove that every null space of V is contained in a maximal null space. (Hint: Use only Definition 207.1 and the finite dimensionality of V.)
5. Let i be the index of V and N a null space of V of dimension d.
   a. Prove that $d \leq i$.
   b. If $d \leq d' \leq i$, prove that N is contained in a null space of V of dimension d'.
*6. Assume that V contains a null plane. Let U be a line on the light cone C of V. Prove that C contains a line $W \neq U$ such that $W \perp U$. (Hint: Since the index of V is at least two, show that U is contained in a null space M of dimension ≥ 2.)

39. MAXIMAL ARTINIAN SPACES

The real importance of the Witt index becomes evident if one studies the maximal Artinian subspaces of V instead of the maximal null spaces. In particular, this will lead us to the important conclusion that, in order to classify all metric vector spaces, one only has to classify the anisotropic ones.

We begin by observing that not all nonsingular spaces contain an Artinian space.

Proposition 209.1. V contains an Artinian space if and only if V is not anisotropic.

Proof. Every Artinian space contains a null space of positive dimension; hence, if V contains an Artinian space, V cannot be anisotropic. Conversely, if V is not anisotropic, V contains a null space N of positive dimension. The nonsingular completion $\bar{N}$ of N is an Artinian space contained in V (Exercise 194.1). Done.

Because of the above proposition, there is no need to talk about maximal Artinian subspaces of anisotropic spaces. We recall that an Artinian space is usually denoted by *Art* and Art_{2d} stands for an Artinian space of dimension $2d$.

Definition 209.1. Assume that V is not anisotropic. An Artinian subspace *Art* of V is called **maximal** if it is not contained in a properly larger Artinian subspace of V.

If V is not anisotropic, Proposition 209.1 guarantees the existence of Artinian subspaces and it follows immediately from the finite dimensionality of V that V then contains maximal Artinian subspaces. The following exercises will be used in the proof of Theorem 210.1.

Exercises

*1. Assume V is not anisotropic. Prove that every Artinian subspace of V is contained in a maximal Artinian subspace.
2. Prove that an Artinian space Art_{2d} contains an Artinian space Art_{2e} for all $1 \leq e \leq d$.

The next theorem gives the relation between the Witt index and the maximal Artinian subspaces. We recall that V is not anisotropic if and only if its Witt index is positive.

Theorem 210.1. Assume that V has Witt index $i > 0$. The maximal Artinian subspaces of V all have the same dimension $2i$.

Proof. We first prove that the maximal Artinian subspaces of V have the same dimension. Suppose that Art_{2d} and Art_{2e} are maximal Artinian subspaces of V and that $d \neq e$. We fix the notation so that $e < d$ and choose an Artinian space Art'_{2e} in Art_{2d}. By the Witt theorem, every isometry σ from Art'_{2e} onto Art_{2e} can be extended to an isometry $\bar{\sigma}$ of V. Clearly, $\bar{\sigma}$ maps Art_{2d} onto an Artinian space which properly contains Art_{2e}, contradicting the assumption that Art_{2e} is maximal. Thus $d = e$.

We now prove that the common dimension of the maximal Artinian subspaces of V is $2i$. V contains null spaces of dimension i and every non-singular completion of such a null space in V is an Artinian space of dimension $2i$. If $2t$ is the common dimension of the maximal Artinian subspaces of V, we conclude from Exercise 209.1 that $2i \leq 2t$ or $i \leq t$. On the other hand, every Artinian subspace of V of dimension $2d$ contains a null space of dimension d and $d \leq i$. In particular, $t \leq i$. This shows that $t = i$. Done.

Exercise

3. Let $i > 0$ be the index of V and Art_{2d} an Artinian subspace of V. If $d \leq e \leq i$, prove that Art_{2d} is contained in an Artinian subspace of dimension $2e$. (Hint: Use Exercise 206.14.)

We are now ready to see the importance of the maximal Artinian subspaces of V.

Theorem 210.2. An Artinian subspace Art of V is maximal if and only if its orthogonal complement Art^* is anisotropic. Different maximal Artinian subspaces of V have isometric orthogonal complements.

Proof. Let Art be maximal. Then $\dim(Art) = 2i$ where i is the index of V; moreover, Art contains a null space N of dimension i. If Art^* contained a null line U, then $N \oplus U$ would be a null space of dimension $i + 1$, contrary to the meaning of i. Hence Art^* is anisotropic.

Assume that Art^* is anisotropic. Put $\dim(Art) = 2d$ and suppose that Art is not maximal. Then $d < i$ and Art is a proper subspace of Art_{2i}. The orthogonal complement of Art in Art_{2i} is an Artinian space of dimension

$2i - 2d$ (Exercise 206.13) and hence contains nonzero isotropic vectors. Each such vector belongs to Art^*, contrary to the assumption that Art^* is anisotropic. Hence Art is maximal.

Finally, if Art and Art' are both maximal Artinian subspaces of V, they have the same dimension $2i$ and hence are isometric. Consequently, their orthogonal complements are isometric (Theorem 206.1). Done.

It follows from the above theorem that V can be written as

$$V = W \oplus Art_{2i}$$

where i is the index of V and W is anisotropic. The geometry of V determines i and hence the geometry of Art_{2i}. But V also determines the geometry of the anisotropic space W since a different choice of Art_{2i} changes W only by an isometry. The important thing is that V determines intrinsically an anisotropic geometry and this geometry, together with i, completely fixes the geometry of V. Consequently, in order to classify nonsingular spaces, one only has to classify the anisotropic ones! We saw on page 146 that a classification of all nonsingular spaces solves the classification problem of the singular ones as well; we now conclude that

In order to classify all metric vector spaces, it is sufficient to classify the anisotropic spaces.

The remainder of this section will not be used in the sequel and is meant only for readers who want to do research on the classification of quadratic forms.

A Research Idea of Artin. Artin has pointed out the similarity between the decomposition $W \oplus Art_{2i}$ and the Wedderburn theorem which states that every simple algebra is a total matrix ring over a division ring. He called the anisotropic space W "field like" since for him field meant division ring; he saw in Art_{2i} the analog of the total matrix ring whose structure is determined by one integer. Since anisotropic spaces then are like division rings, one should look at the classification of division rings to obtain insight as to how one might attempt to classify anisotropic spaces. Guided by these ideas, Artin felt that the first step toward a classification of anisotropic spaces should be a geometric characterization of their "splitting fields."

In order to understand the notion of splitting field, assume that k' is a field which contains our field k. The metric vector space V over k can be extended to a metric vector space V' over k' as follows. Simply choose a coordinate system $A_1, \ldots, A_n$ of V (over k) and define

1. The vector space V' consists of the vectors $a_1'A_1 + \cdots + a_n'A_n$ where $a_1', \ldots, a_n' \in k'$. Addition is defined "componentwise" and if $t' \in k'$, then

$$t'(a_1'A_1 + \cdots + a_n'A_n) = (t'a_1')A_1 + \cdots + (t'a_n')A_n.$$

In other words, $A_1, \ldots, A_n$ is also a coordinate system of V' over k'.

2. If G is the matrix of V relative to the coordinate system $A_1, \ldots, A_n$, the same matrix G is also the matrix of V' relative to its coordinate system $A_1, \ldots, A_n$.

It is possible to define the metric vector space V' over k' without the use of coordinates, but this is not necessary to convey Artin's ideas. It is immediate that the geometry of V' does not depend on the choice of coordinate system $A_1, \ldots, A_n$ of V. Observe, furthermore, that V' is still nonsingular since $\det G \neq 0$.

We assume now that V is anisotropic. A **splitting field** k' of V is a field which contains k and has the property that the vector space V' over k' is no longer anisotropic (and hence splits off an Artinian space). If $f \in k[x_1, \ldots, x_n]$ is a quadratic form of V relative to the coordinate system $A_1, \ldots, A_n$, f has no nonzero solutions in k; this is the same as saying that V is anisotropic. A splitting field k' of V is simply a field containing k and in which f has a nonzero solution.

If $n \geq 2$, V always has splitting fields. The algebraic closure of k is one of them (Exercises 191.15 and 191.16); furthermore, every extension field of a splitting field is again a splitting field. Clearly, then we want the smallest possible splitting fields of V. An essential part of Artin's idea is that these splitting fields should be characterized in terms of the *geometric* properties of V.

One might start this research by considering the following construction which produces a whole basketfull of splitting fields, constructed in a way that Artin would have accepted as geometric. We drop the assumption that V is anisotropic and consider an isometry σ of V. Since $\sigma: V \to V$ is a linear transformation, its characteristic roots and vectors are defined as usual. Namely, a vector $A \neq 0$ is called a **characteristic vector** or "eigenvector" of σ if $\sigma A = tA$ for some $t \in k$; the scalar t is then called a **characteristic root** of σ. Since σ is an isometry, $(\sigma A)^2 = A^2$, that is, $t^2 A^2 = A^2$. We conclude: *If an isometry has a characteristic root different from ± 1, the corresponding characteristic vector is isotropic.*

Returning to the case where V is anisotropic, we see that ± 1 are the only scalars of k which can occur as characteristic roots of isometries. Consequently, one can split V by splitting one of its isometries. Precisely, if one adjoins a characteristic root $\theta \neq \pm 1$ of an isometry σ of V to k, the resulting field $k(\theta)$ is a splitting field of V. The reason is that the characteristic vector A of σ which satisfies $\sigma A = \theta A$ must be isotropic. Of course, A belongs to the vector space V' over $k(\theta)$ and not to V. The first question one might investigate is when an anisotropic space has no splitting fields other than those which split isometries.

We advise the reader not to attempt this research unless he has made a thorough study of Artin's *Geometric Algebra* and of the theory of splitting fields of division rings. The following examples show that the research starts with dim $V = 3$.

Example 1

Dim $V = 1$. A line is nonsingular if and only if the line is anisotropic. Hence V is anisotropic and has no splitting fields. This is a shamefully complicated way of saying that, if $a \neq 0$, the quadratic form $ax_1{}^2$ has no nonzero solution in any field.

Example 2

Dim $V = 2$. Assume that V is anisotropic, equivalently, that disc $V \neq -1$ (Proposition 184.1). If disc $V = \Delta$, it is clear that an extension field k' of k splits V if and only if $\Delta = -t^2$ for some $t \in k'$. Since $t = \sqrt{-\Delta}$, this says that k' is a splitting field of V if and only if k' contains the field $k(\sqrt{-\Delta})$. Consequently, V has only one (smallest possible) splitting field, namely, $k(\sqrt{-\Delta})$. We shall see in Exercise 277.8 that this splitting field is obtained by adjoining to k a characteristic root of any rotation $\sigma \neq \pm 1_V$.

The fact that $k(\sqrt{-\Delta})$ is the smallest possible splitting field of V can also be explained in a completely elementary way. Let $ax_1{}^2 + bx_1x_2 + cx_2{}^2$ (where $a, b, c \in k$) be the quadratic form of V relative to some coordinate system. The fact that V is anisotropic means that this quadratic form has only the zero solution in k; we must show that an extension field k' of k contains a nonzero solution of this form if and only if $k(\sqrt{-\Delta}) \subset k'$. The two roots r_1 and r_2 of the quadratic equation

$$ax^2 + bx + c = 0$$

are given by the "quadratic formula," namely,

$$r_1, r_2 = \frac{-b \pm \sqrt{b^2 - 4ac}}{2a}$$

where $a \neq 0$ since V is anisotropic. It follows immediately that k' contains a nonzero solution of the quadratic form

$$ax_1{}^2 + bx_1x_2 + cx_2{}^2$$

if and only if k' contains the field $k(\sqrt{b^2 - 4ac})$. We finish the argument by showing that the field $k(\sqrt{b^2 - 4ac})$ is precisely the field $k(\sqrt{-\Delta})$.

The metric of V is given by the matrix

$$\begin{pmatrix} a & b/2 \\ b/2 & c \end{pmatrix}$$

and hence $\Delta = ac - (b^2/4)$. Consequently, $b^2 - 4ac = -4\Delta$ and

$$k(\sqrt{b^2 - 4ac}) = k(\sqrt{-4\Delta}) = k(\sqrt{-\Delta}).$$

40. THE ORTHOGONAL GROUP AND THE ROTATION GROUP

There are several groups built into a metric vector space. Without a thorough knowledge of them, one cannot penetrate very deeply into geometry. Does the study of groups belong in geometry or algebra? A general answer cannot be given because it depends on what kind of group properties one studies and how one goes about investigating them. However, there is no doubt in the authors' minds that *the group theoretic studies carried out in the remainder of this chapter are an integral part of geometry.*

Felix Klein's position that a geometry is characterized by "its groups" is no longer valid. Things have changed since 1872 when he stated his position in the Erlangen program during his famous inaugural address at the University of Erlangen. For instance, the geometries of metric vector spaces are best characterized by the inner products which define them and not by their groups. On the other hand, Klein's insistence that the study of the groups built into a geometry be an essential part of the study of that geometry is as valid today as when he formulated his Erlangen program about one hundred years ago. In the present section, we define the two most important groups associated with metric vector spaces, namely, the orthogonal group and the rotation group, and make some preliminary remarks about them.

Let W be an m-dimensional vector space over k without a metric. We know from linear algebra that the linear automorphisms (nonsingular linear transformations) of W form a group called the **general linear group** $GL(m, k)$. This group is one of the "classical groups" and the two groups we are going to define in this section are both subgroups of $GL(m, k)$.

Assume now that W has been given a metric, possibly singular. Since an isometry of W is a linear automorphism by definition, all isometries belong to $GL(m, k)$.

Exercise

*1. Assume W is an m-dimensional vector space (possibly singular) over k.
   a. Prove that the isometries of W form a subgroup of $GL(m, k)$.
   b. If W is the null space, prove that the group of isometries of W is the whole group $GL(m, k)$.

The group of isometries of W is called the **orthogonal group** of W and is denoted by $O(W)$. This subgroup of $GL(m, k)$ is highly dependent on the geometry of W, although it can happen that nonisometric geometries have the same orthogonal group (see Exercise 224.22). If W is the null space, we see by Exercise 1b above that $O(W) = GL(m, k)$. Orthogonal groups are also counted among the classical groups.

If $A_1, \ldots, A_m$ is a coordinate system of the vector space W, we recall from linear algebra that a linear transformation g of W can be expressed by an $m \times m$ matrix (a_{ij}), namely,

$$(gA_1, \ldots, gA_m) = (A_1, \ldots, A_m)(a_{ij}).$$

The determinant of (a_{ij}) depends only on g and not on the choice of the coordinate system $A_1, \ldots, A_m$ and is called the *determinant of g*; notation: det g. (This determinant can also be defined without the use of coordinate systems.) Whenever we speak of the determinant of a linear transformation, it is assumed that $n \geq 1$.

If the metric of W is singular, there is nothing special one can say about determinants of isometries. This follows from Exercise 3 below and from the fact that, if W is the null space, every linear automorphism of W is an isometry.

Exercises

*2. Assume that S_1 and S_2 are subspaces of W and $W = S_1 \oplus S_2$. (Since W may be singular, S_1 and S_2 are not necessarily orthogonal complements of one another. For instance, if W is the null space, $S_1^* = S_2^* = W$.) Assume that $\alpha \in O(S_1)$, $\beta \in O(S_2)$, and put $\sigma = \alpha \oplus \beta$. This means that if $X = A + B$ where $A \in S_1$ and $B \in S_2$, then $\sigma X = \alpha A + \beta B$. (See Exercise 197.9 where it was proved that σ is an isometry.) Prove that

$$\det \sigma = (\det \alpha)(\det \beta).$$

(Hint: This result has nothing to do with metrics, but only with the fact that $W = S_1 \oplus S_2$. Choose a coordinate system $A_1, \ldots, A_i$ for S_1 and $B_1, \ldots, B_j$ for S_2; then consider the matrix of σ relative to the coordinate system $A_1, \ldots, A_i, B_1, \ldots, B_j$ of W.)

*3. Assume that the metric of W is singular. Prove that, for every $a \in k^*$, there is an isometry σ of W such that det $\sigma = a$. (Hint: Put $W = T \oplus \mathrm{rad}\, W$; use the fact that $\dim(\mathrm{rad}\, W) \geq 1$ and every linear automorphism of rad W is an isometry of rad W.)

Determinants of isometries are severely restricted if the geometry is nonsingular. In order to study this, we return to our n-dimensional nonsingular space V over k.

Proposition 216.1.   If σ is an isometry of V, det $\sigma = \pm 1$.

Proof. Let $A_1, \ldots, A_n$ be a coordinate system of V and P the matrix of σ with respect to this coordinate system, that is,

$$(\sigma A_1, \ldots, \sigma A_n) = (A_1, \ldots, A_n)P.$$

Then det $\sigma = $ det P. Since σ is nonsingular, $\sigma A_1, \ldots, \sigma A_n$ is also a coordinate system of V. If the metric of V is given by the matrix G relative to the coordinate system $A_1, \ldots, A_n$, this metric is given by the matrix

$$G' = P^{\mathsf{T}}GP$$

relative to the coordinate system $\sigma A_1, \ldots, \sigma A_n$. Since σ is an isometry, it follows from Proposition 133.1 that $G' = G$ and hence

$$G = P^{\mathsf{T}}GP.$$

We conclude that

$$\det G = (\det G)(\det P)^2.$$

Using, for the first time, that V is nonsingular and, therefore, det $G \neq 0$, we conclude that $(\det P)^2 = 1$, equivalently, det $P = \pm 1$. Done.

Exercise

4. Let $A_1, \ldots, A_n$ be a coordinate system of V and $\sigma: V \to V$ be a linear transformation. Denote the matrix of σ relative to the coordinate system $A_1, \ldots, A_n$ by P and the matrix of the metric of V relative to the same coordinate system by G.

a. Prove that σ is an isometry if and only if $P^{\mathrm{T}}GP = G$. (The "only if" part is an immediate consequence of Proposition 133.1; therefore, just the "if" part remains to be proved.)

b. If the coordinate system $A_1, \ldots, A_n$ is orthonormal, prove that σ is an isometry if and only if $P^{\mathrm{T}}P = I$ where I is the $n \times n$ unit matrix. (A matrix P satisfying $P^{\mathrm{T}}P = I$ is called an "orthogonal matrix." Hence orthogonal matrices may be defined as matrices which represent isometries relative to orthonormal coordinate systems.)

According to Proposition 216.1, $O(V)$ is partitioned into two disjoint subsets, one set consisting of the isometries with determinant 1 and the other set those isometries with determinant -1. This is true only because of our assumption that char $k \neq 2$. In a field with characteristic two, $-1 = 1$.

Definition 217.1.   An isometry σ of V is called a **rotation** if det $\sigma = 1$ and a **reflection** if det $\sigma = -1$.

Whenever we speak of rotations or reflections, it is always understood that $n \geq 1$. It is clear that 1_V is a rotation since its matrix is the unit matrix. Rotations and reflections of singular spaces are defined in Exercise 230.10c; they cannot be defined in terms of the determinants of the given isometry because every nonzero scalar occurs as the determinant of an isometry if the space is singular (Exercise 216.3).

Exercise

5. Here we show that the determinant of a linear automorphism of V may be ± 1 even though σ is not an isometry. Let $n = 2$ and assume that V has an orthonormal coordinate system A_1, A_2. Consider the linear transformation $\sigma: V \to V$ defined by

$$\sigma A_1 = A_1 + A_2 \qquad \text{and} \qquad \sigma A_2 = A_1.$$

a. Prove that det $\sigma = -1$.

b. Prove that σ is not an isometry.

c. Give an example of a linear automorphism $\sigma: V \to V$ where det $\sigma = 1$ while σ is not an isometry.

We now study some special rotations and reflections. The linear automorphism $A \to -A$ of V is denoted by -1_V. Since $(-A)(-B) = AB$ for all $A, B \in V$, -1_V is an isometry.

Proposition 218.1. If n is even, -1_V is a rotation. If n is odd, -1_V is a reflection.

Proof. The matrix of -1_V is $-I$ where I is the $n \times n$ unit matrix, no matter what coordinate system is chosen for V. Since $\det(-I) = (-1)^n$, the proposition follows. Done.

Exercises

*6. a. Let $t \in k^*$ and consider the linear transformation $\sigma: V \to V$ defined by $\sigma A = tA$ for $A \in V$. If $n \geq 1$, prove that σ is an isometry if and only if $t = \pm 1$, equivalently, if and only if $\sigma = \pm 1_V$. (In the language of Chapter 1, this exercise says that $\pm 1_V$ are the only isometries among the magnification of V with $\mathbf{0}$ as center.)

b. Extend the result of Part a to singular spaces which are not null spaces. Precisely, let W be a singular metric vector space which is not a null space. For $t \in k^*$, let $\sigma: W \to W$ be the linear transformation defined by $\sigma A = tA$ for $A \in V$. Prove that σ is an isometry if and only if $\sigma = \pm 1_V$.

7. Assume that V is a line. Then $n = 1$ and every nonzero vector $A \in V$ is a coordinate system of V and a linear transformation $\sigma: V \to V$ is completely determined by its action on A; that is, $\sigma A = tA$ where $t \in k^*$.

a. Prove that $GL(1, k) \simeq k^*$.

b. Prove that $O(V)$ consists of only the rotation 1_V and the reflection -1_V. The reflection -1_V is shown in Figure 218.2.

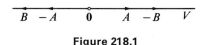

Figure 218.1

When $n = 2$, -1_V is a rotation (Proposition 218.1). This rotation is naturally called the **180° rotation**. See Figure 218.2.

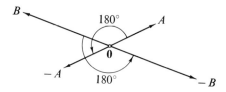

Figure 218.2

We do not associate a number of degrees with any other rotation in the plane. Measurement of rotations in degrees or other units can only be introduced if there exists a "winding function" which associates a rotation to every scalar; see Remark on Teaching Trigonometry, page 309, where the winding function for the Euclidean plane is discussed. Winding functions do not exist for arbitrary geometries.

If n is odd, there is both a rotation (1_V) and a reflection (-1_V) of V. If n is even, we have not yet observed a reflection. Even so, there is always a reflection of V.

Proposition 219.1. Let $A \in V$ where $A^2 \neq 0$ and put $V = \langle A \rangle \oplus \langle A \rangle^*$. Then $-1_{\langle A \rangle} \oplus 1_{\langle A \rangle^*}$ is a reflection which is denoted by τ_A.

Proof. We put $\rho = -1_{\langle A \rangle}$ and $\sigma = 1_{\langle A \rangle^*}$. Then $\tau_A = \rho \oplus \sigma$ is an isometry and, furthermore,

$$\det \tau_A = (\det \rho)(\det \sigma) = -1$$

by Exercise 215.2. Done.

The reflection τ_A is called the **symmetry** of V with respect to the non-singular hyperplane $\langle A \rangle^*$. (We recall that a *hyperplane* of V is an $(n-1)$-dimensional subspace of V.) The hyperplane $\langle A \rangle^*$ is nonsingular because the line $\langle A \rangle$ is nonsingular. The isometry τ_A is called a symmetry with respect to $\langle A \rangle^*$ because it sends every vector of V onto its "mirror image" with respect to the "mirror" $\langle A \rangle^*$. Vectors in $\langle A \rangle^*$ remain fixed; vectors in $\langle A \rangle$ are flipped over 0; in general, if a vector X looks into the mirror $\langle A \rangle^*$, it sees the vector $\tau_A X$. See Figures 219.1 and 220.1 where τ_A is pictured for $n = 2, 3$.

We caution the reader about the terms symmetry and reflection. Some authors reserve the term "reflection" for what we call a "symmetry." We are following the terminology which Artin uses in *Geometric Algebra*. There,

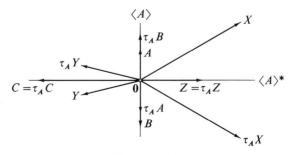

Figure 219.1

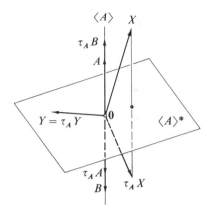

Figure 220.1

as here, a reflection is any isometry whose determinant is -1 while a symmetry is the very special kind of reflection which consists of "mirroring" V relative to a nonsingular hyperplane as mirror. We shall see that in the plane the notions of reflection and symmetry coincide (see Exercise 256.9a).

Exercises

8. Let $n = 1$ and $A \in V$ where $A \neq 0$. Prove that $\langle A \rangle^* = \{0\}$ and $\tau_A = -1_V$.

*9. Let $A_1, \ldots, A_n$ be a rectangular coordinate system of V. Choose an i with $1 \leq i \leq n$ and define the linear transformation $\tau: V \to V$ by $\tau A_j = A_j$ if $j \neq i$ and $\tau A_i = -A_i$. Prove that τ is the symmetry τ_{A_i} with respect to the nonsingular hyperplane $\langle A_1, \ldots, A_{i-1}, A_{i+1}, \ldots, A_n \rangle$.

10. Prove that V has at least n reflections. (Hint: Use the previous exercise.)

*11. Let A be a nonisotropic vector of V.
   a. Prove that a vector X of V is left fixed by τ_A if and only if $X \in \langle A \rangle^*$. (Hint: Since $V = \langle A \rangle \oplus \langle A \rangle^*$, write $X = Y + Z$ where $Y \in \langle A \rangle$ and $Z \in \langle A \rangle^*$.) Thus the fixed point space of τ_A is $\langle A \rangle^*$.
   b. Prove that a vector X of V is inverted by τ_A (that is, $\tau_A X = -X$) if and only if $X \in \langle A \rangle$.

*12. If A and B are nonisotropic vectors, prove that $\tau_A = \tau_B$ if and only if A and B are linearly dependent. (Hint: Use Exercise 11 above and the fact that A and B are linearly dependent if and only if $\langle A \rangle = \langle B \rangle$.)

13. Prove that there is a one-to-one correspondence between the set of symmetries of V and the set of lines of V which do not lie on the light cone.

*14. Rotations and reflections are elements of the multiplicative group $O(V) \subset GL(n, k)$. We recall that multiplication in $GL(n, k)$, and hence in $O(V)$, is the composition of functions.
   a. Prove that the product of rotations is a rotation.
   b. Prove that the product of a rotation and a reflection is a reflection.
   c. Prove that the product of m reflections, not necessarily distinct, is a reflection if and only if m is odd, equivalently, is a rotation if and only if m is even.

*15. Assume $V = U_1 \oplus \cdots \oplus U_m$ which implies that $U_1, \ldots, U_m$ are non-singular subspaces of V. Thus we can speak of rotations and reflections of U_i. Let σ_i be an isometry of U_i for $i = 1, \ldots, m$ and form the orthogonal sum $\sigma = \sigma_1 \oplus \cdots \oplus \sigma_m$. We know that σ is an isometry and, therefore, is either a rotation or a reflection of V.
   a. Prove that the orthogonal sum σ is a rotation if and only if an even number of the σ_i are reflections. (Hint: Use Exercise 215.2, generalized from two isometries to any finite number of isometries.)
   b. Conclude from Part a that
      i. The orthogonal sum of rotations is a rotation.
      ii. The orthogonal sum of an odd number of reflections is a reflection.
      iii. The orthogonal sum of an even number of reflections is a rotation.
      iv. The orthogonal sum of a reflection and a rotation is a reflection.

According to Exercise 220.12, V has a host of symmetries, one for each nonisotropic line. This gives us reflections galore and we shall see that, if $n \geq 3$, there are plenty of reflections which are not symmetries (see Exercises 223.19b and 224.20h). Furthermore, there are as many rotations as reflections (see Exercise 223.16) and hence there is no shortage of rotations either. We denote the set of rotations of V by $O^+(V)$ and derive the algebraic structure of $O^+(V)$.

Proposition 221.1. $O^+(V)$ is a normal subgroup of $O(V)$ of index two. The coset of $O^+(V)$, different from $O^+(V)$, is the set of reflections.

Proof. Consider the determinant mapping

$$\det: O(V) \to \{1, -1\}$$

which maps each isometry σ onto $\det \sigma$. We know from linear algebra that the determinant mapping from $GL(n, k)$ to k^* is a group homomorphism; hence det: $O(V) \to \{1, -1\}$ is a group homomorphism from $O(V)$ to the subgroup $\{1, -1\}$ of k^*. It follows that the kernel $O^+(V)$ of det is a normal subgroup of $O(V)$. Furthermore, since V has both a rotation and a reflection, det maps $O(V)$ *onto* the group $\{1, -1\}$ which shows that the index of $O^+(V)$ in $O(V)$ is two. The two cosets of $O^+(V)$ are hence $O^+(V)$ itself and the complement of $O^+(V)$ in $O(V)$; this latter set is the set of reflections. Done.

The rotation group $O^+(V)$ is also a "classical group" and hence we have the tower of classical groups

$$O^+(V) \subset O(V) \subset GL(n, k).$$

It is important to observe that det maps $O(V)$ onto the subgroup $\{1, -1\}$ of k^* and $O^+(V)$ is the kernel of det.

It is a surprising fact that the Greek geometers of antiquity, say from 500 B.C. until the year 0, failed to see the many groups which lie embedded in the geometry they themselves created. Thereby, they missed modern algebra and this should rank as one of the greatest of all Greek tragedies. Why didn't these men, who were mathematicians without peers, at least observe the orthogonal group and the rotation group in the case when V is the Euclidean plane? Perhaps the reason is that they did not formulate their geometric concepts explicitly in terms of sets and functions as we do today. For us, isometries are special *functions* on the set V into itself and it is obvious that functions on a set into itself can be "multiplied" by performing one after the other. Once the multiplication of isometries has been defined in this way, the construction of the orthogonal group $O(V)$ has already been accomplished!

Remark on Teaching High School Geometry. If the Greek mathematicians of antiquity failed to observe the great variety of groups in geometry because they did not take the attitude of sets and functions toward geometry, then it is important that high school geometry be taught as much as possible using the language of sets and functions. Do we not want to protect our children from tragedies? In a high school course in plane geometry, one should introduce rotations, reflections, and the two remaining types of Euclidean transformations (see Chapter 3) as special functions on the set of points of the plane into itself. These functions should then be multiplied by performing one after the other; in this way, the most important groups of Euclidean plane geometry should be constructed. This type of non-Greek mathematics constitutes a very simple, concrete, and important application of the notions of sets and functions to high school geometry.

It is curious that in many modern high school curricula, sets and functions are introduced in the algebra courses, but are allowed to die an ignominious death as soon as the geometry begins. It should be just the other way around!

The applications of sets and functions to high school geometry are impressive. They allow us to see the natural way to multiply Euclidean transformations and to construct the most important groups of plane geometry. On the contrary, the applications of sets and functions to high school algebra are practically nonexistent. An expression such as

   "the solution set of a quadratic equation consists of one or two roots"

is much clumsier than the classical expression

   "a quadratic equation has one or two roots."

"Solution sets" do not present us, in any sense, with an application of set theory to algebra. They merely trap us into using a clumsy language when a facile one is available. Sets and functions become important in algebra at the moment one starts the study of algebraic structures (groups, rings, fields, vector spaces, etc.), but not before. Even though groups belong in algebra, the way to expose high school students to them is not through algebra, but through geometry. The simple and elementary multiplication of Euclidean transformations is one of the best ways to integrate geometry and algebra.

Exercises

*16. Prove that V has as many rotations as reflections. (Hint: There always exists a one-to-one correspondence between two cosets of a subgroup.)

17. If σ is a reflection of V, prove that $O(V)$ is the disjoint union of $O^+(V)$ and $\sigma O^+(V)$, equivalently, that $\sigma O^+(V)$ is the set of reflections.

18. Since $O^+(V)$ is a group, the inverse of a rotation is always a rotation. Although reflections do not form a group, prove, nevertheless, that the inverse of a reflection is a reflection.

*19. Here we take a closer look at the isometry -1_V.
   a. Prove that -1_V leaves only the vector $\mathbf{0}$ fixed.
   b. If n is odd and at least three, prove that -1_V is a reflection which is not a symmetry. (Hint: Use Exercise 220.11a.)

*20. This exercise generalizes the notion of a symmetry. Let U be a nonsingular subspace of V and put $V = U \oplus U^*$. We study the isometry $\sigma = -1_U \oplus 1_{U^*}$.
   a. If $U = \{0\}$, prove that $\sigma = 1_V$.
   b. If $U = V$, prove that $\sigma = -1_V$.
   c. If $U = \langle A \rangle$, prove that $\sigma = \tau_A$.
   d. Prove that $\det \sigma = (-1)^{\dim U}$. Conclude that σ is a rotation if $\dim U$ is even and a reflection if $\dim U$ is odd.

e. Prove that U^* is the linear space of vectors left fixed by σ.

f. Prove that U is the linear space of vectors inverted by σ. In other words, $X \in U$ if and only if $\sigma X = -X$.

g. Prove that σ is a symmetry if and only if dim $U = 1$.

h. If dim U is odd and at least three, prove that σ is a reflection which is not a symmetry.

i. Prove that $\sigma^2 = 1_V$.

21. Assume V is a plane which has an orthonormal coordinate system A_1, A_2. Consider the linear transformation $\sigma: V \to V$ defined by $\sigma A_1 = -A_2$ and $\sigma A_2 = -A_1$.

a. Prove that σ is an isometry.

b. Prove that σ is a reflection.

c. Find a nonzero vector A such that $\sigma A = -A$.

d. Find a nonzero vector B such that $\sigma B = B$.

e. Prove that $\langle B \rangle = \langle A \rangle^*$ and that $V = \langle A \rangle \oplus \langle B \rangle$.

f. Prove that σ is the symmetry τ_A where A is the vector found in Part c. (We shall see in Exercise 256.9a that all reflections of a non-singular plane are symmetries whether or not the plane has an orthonormal coordinate system.)

g. Prove that $\sigma = -1_V \tau$ where τ is the reflection which interchanges A_1 and A_2.

*22. Let $c \in k^*$ and define a new inner product $A \circ B$ for V as follows: $A \circ B = c(AB)$ for $A, B \in V$. We denote the new metric vector space corresponding to the inner product $A \circ B$ by V_o. The vector spaces V and V_o were first compared in Exercises 166.13 and 166.14. We prove there that V and V_o are not necessarily isometric. Furthermore, since $c \neq 0$ and V is nonsingular, V_o is nonsingular.

a. Prove that $O(V_o) = O(V)$ and $O^+(V_o) = O^+(V)$. We are not being sloppy here, but mean equality of groups and not just isomorphism.

b. Let $k = \mathbf{R}$ and $c = -1$. Use Exercise 166.14 to give several examples of nonisometric spaces which have the same orthogonal group and the same rotation group.

41. COMPUTATION OF DETERMINANTS

In this section, we develop some linear algebra which is not very well known. In pure mathematics, it rarely happens that a determinant is computed directly. Instead, the underlying geometry usually tells us all we have to know about the determinant. The examples we shall see of this in the present section are necessary for the study of rotations and reflections.

We denote by $\mathscr{W}$ an m-dimensional vector space over an arbitrary field $\mathscr{k}$; $\mathscr{W}$ has not been given a metric. Here, the letter $\mathscr{k}$ is used for the field instead of k because we do not want to exclude the possibility that our field may have characteristic two. It is almost never necessary to restrict the characteristic of the field in investigations about vector spaces without a metric. Occasionally such a restriction is necessary, such as in Section 47 on involutions of the general linear group.

Let U be a linear subspace of $\mathscr{W}$. For each $X \in \mathscr{W}$, the coset $X + U$ of U is a *vector* of the quotient space $\mathscr{W}/U$ and this quotient space is again a vector space over $\mathscr{k}$. We saw on page 140 how these cosets should be visualized, namely, as affine spaces parallel to U. The only coset which is a linear subspace of $\mathscr{W}$ is the one which passes through the origin, that is, U itself.

Exercise

1. Assume that U and W are linear subspaces of $\mathscr{W}$ such that $\mathscr{W} = U \oplus W$. If $B_1, \ldots, B_w$ is a coordinate system of W, prove that

$$B_1 + U, \ldots, B_w + U$$

is a coordinate system of the vector space $\mathscr{W}/U$.

Let $\sigma: \mathscr{W} \to \mathscr{W}$ be a linear transformation of $\mathscr{W}$ which may be singular. A linear subspace U of $\mathscr{W}$ is called an **invariant subspace** of σ if σ transforms U into itself. This means only that $\sigma X \in U$ for each $X \in U$ and does not mean that σ leaves the vectors of U pointwise fixed. We denote the restriction of σ to U by $\sigma|U$ and hence $\sigma|U: U \to U$ is a linear transformation.

Exercises

In Exercises 2 and 3 below, U is an invariant subspace of the linear transformation $\sigma: \mathscr{W} \to \mathscr{W}$.

2. Prove that σ induces a linear transformation

$$\bar{\sigma}: \mathscr{W}/U \to \mathscr{W}/U$$

by the definition $\bar{\sigma}(X + U) = \sigma X + U$ for $X \in \mathscr{W}$. Do not forget to show that $\bar{\sigma}$ is well defined.

*3. Let W be a linear subspace of $\mathscr{W}$ such that $\mathscr{W} = U \oplus W$. Assume that $A_1, \ldots, A_u$ is a coordinate system for U and $B_1, \ldots, B_w$ is a coordinate system for W. We know that $m = u + w$ and $A_1, \ldots, A_u, B_1, \ldots, B_w$

is a coordinate system of $\mathscr{W}$. We denote the matrix of σ relative to this coordinate system by M.

a. Prove that M has the form

$$M = \left(\begin{array}{c|c} \alpha & \beta \\ \hline \bigcirc & \gamma \end{array} \right)$$

where α is a $u \times u$ matrix, β is a $u \times w$ matrix, $\bigcirc$ is a $w \times u$ matrix, and γ is a $w \times w$ matrix.

b. Prove that α is the matrix of the linear transformation $\sigma | U : U \to U$ relative to the coordinate system $A_1, \ldots, A_u$ of U.

c. Prove that γ is the matrix of the linear transformation

$$\bar{\sigma} : \mathscr{W}/U \to \mathscr{W}/U$$

relative to the coordinate system $B_1 + U, \ldots, B_w + U$ of $\mathscr{W}/U$. (Hint: The matrix (c_{ji}) of $\bar{\sigma}$ is defined by

$$\bar{\sigma}(B_i + U) = \sum_{j=1}^{u} (B_j + U)c_{ji}$$

for $i = 1, \ldots, w$. Denoting $\gamma = (\gamma_{ji})$, the problem is to show that $c_{ji} = \gamma_{ji}$ for $j, i = 1, \ldots, w$. To do this, compute $\bar{\sigma}(B_i + U) = \sigma B_i + U$ by expressing σB_i as a linear combination of $A_1, \ldots, A_u$, $B_1, \ldots, B_w$ using the matrix M and observing that $A_i + U = 0$ for $i = 1, \ldots, u$.)

We now see how the determinants of the three linear transformations σ, $\sigma | U$, and $\bar{\sigma}$ are related.

Proposition 226.1. Assume $\sigma : \mathscr{W} \to \mathscr{W}$ is a linear transformation of the vector space $\mathscr{W}$ and U is an invariant subspace of σ. The determinants of σ and the induced linear transformations $\sigma | U : U \to U$ and $\bar{\sigma} : \mathscr{W}/U \to \mathscr{W}/U$ are related by the formula

$$\det \sigma = (\det \sigma | U)(\det \bar{\sigma}).$$

Proof. By definition, $\det \sigma = \det M$ where M is the matrix of Exercise 3a above. It is well known that $\det M = (\det \alpha)(\det \gamma)$. Since α is the matrix of $\sigma | U$ (Exercise 3b above), $\det \alpha = \det \sigma | U$. Finally, $\det \gamma = \det \bar{\sigma}$ because γ is the matrix of $\bar{\sigma}$ (Exercise 3c above). Done.

This proposition often comes in handy when computing $\det \sigma$ since it is often difficult to compute this determinant directly, but easy to compute det $\sigma \,|\, U$ and det $\bar{\sigma}$. In particular, if $\bar{\sigma}$ is the identity mapping $1_{\mathscr{W}/U}$ of $\mathscr{W}/U$ and hence det $\bar{\sigma} = 1$, then det $\sigma = $ det $\sigma \,|\, U$. The problem of computing det σ has then been reduced from computing an $m \times m$ determinant to a $u \times u$ determinant. We now study when $\bar{\sigma} = 1_{\mathscr{W}/U}$.

For this, we consider the identity mapping $1_{\mathscr{W}}$ of $\mathscr{W}$. Since the difference between two linear transformations is a linear transformation, $\sigma - 1_{\mathscr{W}}$ is a linear transformation of $\mathscr{W}$. By definition,

$$(\sigma - 1_{\mathscr{W}})X = \sigma X - X$$

for all $X \in \mathscr{W}$. By denoting the image of a linear transformation by im, we see that

$$\text{im}(\sigma - 1_{\mathscr{W}}) = \{\sigma X - X \,|\, X \in \mathscr{W}\}.$$

The image of a linear transformation is a linear subspace and hence $\text{im}(\sigma - 1_{\mathscr{W}})$ is a linear subspace of $\mathscr{W}$. We can say when $\bar{\sigma} = 1_{\mathscr{W}/U}$ in such a way that it is unnecessary to check whether U is an invariant subspace of σ.

Proposition 227.1. Let σ be a linear transformation of $\mathscr{W}$ and U a linear subspace of $\mathscr{W}$. Then

$$\text{im}(\sigma - 1_{\mathscr{W}}) \subset U$$

if and only if U is an invariant subspace of σ and $\bar{\sigma} = 1_{\mathscr{W}/U}$. In that case, det $\sigma = $ det $\sigma \,|\, U$.

Proof. The last sentence needs no proof and we turn to the if and only if condition.

Assume that $\text{im}(\sigma - 1_{\mathscr{W}}) \subset U$. If $X \in \mathscr{W}$,

$$(\sigma - 1_{\mathscr{W}})X = \sigma X - X \in U$$

which is the same as saying that the two cosets $\sigma X + U$ and $X + U$ of U are the same. This uses only the group theoretic criterion for equality of cosets of a subgroup (U) of an additive group $(\mathscr{W})$. If $X \in U$, then

$$U = X + U = \sigma X + U$$

which is the same as $\sigma X \in U$. This shows that U is an invariant subspace of σ while

$$X + U = \sigma X + U = \bar{\sigma}(X + U)$$

shows that $\bar{\sigma} = 1_{\mathscr{W}/U}$.

Conversely, assume that U is an invariant subspace of σ and $\bar{\sigma} = 1_{\mathscr{W}/U}$; we must show that $\sigma X - X \in U$ for all $X \in \mathscr{W}$. It is given that $\bar{\sigma}(X + U) = $

$X + U$, equivalently, that $\sigma X + U = X + U$. But this last equality is the same as $\sigma X - X \in U$. Done.

Since the vectors of $\mathscr{W}/U$ are the affine subspaces $X + U$ of $\mathscr{W}$, the above proposition says that $\mathrm{im}(\sigma - 1_{\mathscr{W}}) \subset U$ if and only if σ transforms each affine subspace $X + U$ into itself. This is also clear geometrically since $X + U$ is parallel to U and $\sigma X - X \in U$ means that $\sigma X = X + A$ for some vector A in U (see Figure 228.1 where it is assumed that dim $U = 2$).

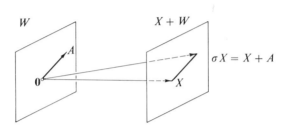

Figure 228.1

Proposition 227.1 gives us a particularly simple way to compute the determinant of σ when U is a line through $\mathbf{0}$ because det $\sigma|U$ is then trivial to compute. This is worked out in the exercise below.

Exercises

4. Assume dim $\mathscr{W} = 1$ and σ is a linear transformation of $\mathscr{W}$. If A is a nonzero vector of $\mathscr{W}$, prove that $\sigma A = tA$ for some $t \in \mathscr{k}$ and det $\sigma = t$. (This exercise has little to do with the propositions of this section, but only with the fact that A is a coordinate system of $\mathscr{W}$. It shows how trivial it is to compute the determinant of a linear transformation of a line.)

In Exercises 5–8, U is a line of $\mathscr{W}$ and A is a nonzero vector of U. Consequently, $U = \langle A \rangle$.

*5. Assume that σ is a linear transformation of $\mathscr{W}$ such that
$$\mathrm{im}(\sigma - 1_{\mathscr{W}}) \subset U.$$
a. If $X \in \mathscr{W}$, prove that $\sigma X = X + (gX)A$ where gX belongs to the field $\mathscr{k}$.
b. Prove that the function $g: \mathscr{W} \to \mathscr{k}$ defined in Part a is linear where $\mathscr{k}$ is considered a vector space over itself.

6. Let $g: \mathcal{W} \to \mathcal{k}$ be a linear function and define $\sigma: \mathcal{W} \to \mathcal{W}$ by $\sigma X = X + (gX)A$ for $X \in \mathcal{W}$. Prove that σ is a linear transformation and that $\text{im}(\sigma - 1_{\mathcal{W}}) \subset U$.

7. Prove that there exists a one-to-one correspondence between the linear functions $g: \mathcal{W} \to \mathcal{k}$ and those linear transformations σ of $\mathcal{W}$ for which $\text{im}(\sigma - 1_{\mathcal{W}}) \subset U$.

*8. Assume that σ is a linear transformation of $\mathcal{W}$ such that $\text{im}(\sigma - 1_{\mathcal{W}}) \subset U$. Let $g: \mathcal{W} \to \mathcal{k}$ be the linear function discussed in Exercise 5 above. Prove that $\det \sigma = 1 + gA$.

9. Let $A_1, \ldots, A_m$ be a coordinate system of $\mathcal{W}$. Put $A = A_1 + \cdots + A_m$ and $U = \langle A \rangle$. Define the function $\sigma: \mathcal{W} \to \mathcal{W}$ as follows. If $X = x_1 A_1 + \cdots + x_m A_m$, then

$$\sigma X = X + (x_1 + \cdots + x_m)A.$$

a. Prove that σ is a linear transformation of $\mathcal{W}$ and $\text{im}(\sigma - 1_{\mathcal{W}}) \subset U$.
b. Prove that $\det \sigma = 1 + m$. (Hint: Use Exercise 8 above.)
c. Prove that the matrix of σ relative to the coordinate system $A_1, \ldots, A_m$ is the $m \times m$ matrix M with 2 everywhere on the main diagonal and 1 everywhere else.

$$M = \begin{pmatrix} 2 & 1 & 1 & \cdots & 1 \\ 1 & 2 & 1 & \cdots & 1 \\ & & \vdots & & \\ 1 & 1 & \cdots & 1 & 2 \end{pmatrix}.$$

We know from Part b that $\det M = 1 + m$. If the reader does not feel that this section gives lovely methods for computing determinants, he should, for punishment, compute $\det M$ directly.

The smallest U which can be used in Proposition 227.1 is $\text{im}(\sigma - 1_{\mathcal{W}})$ itself. Hence, if $\dim(\text{im}(\sigma - 1_{\mathcal{W}})) = s$, this proposition enables us to reduce the computation of the $m \times m$ determinant $\det \sigma$ to that of the $s \times s$ determinant $\det \sigma | U$. The smaller s is, the better. If $m = s$, the proposition does not help.

We review the following terms from linear algebra. If $\alpha: \mathcal{W} \to \mathcal{W}$ is a linear transformation, a vector $A \neq 0$ is called a **characteristic vector** or **eigenvector** of α if $\alpha A = tA$ for some $t \in \mathcal{k}$; the scalar t is then called a **characteristic root** or **eigenvalue** of α.

The investigation of the linear transformation $\sigma - 1_{\mathcal{W}}$ is, therefore, tantamount to investigating the characteristic root (eigenvalue) 1 of σ. Namely, denoting the kernel of a linear transformation by ker,

$$\ker(\sigma - 1_{\mathcal{W}}) = \{X \mid X \in \mathcal{W}, \ \sigma X = X\}.$$

We recall from linear algebra that the kernel of a linear transformation is a linear subspace; hence we see that $\ker(\sigma - 1_{\mathscr{W}})$ is the linear subspace F of $\mathscr{W}$ consisting of the vectors left fixed by σ. In other words, $\ker(\sigma - 1_{\mathscr{W}})$ is the linear subspace of $\mathscr{W}$ whose vectors are those characteristic vectors (eigenvectors) of σ which correspond to the characteristic root 1 of σ. For every linear transformation f of $\mathscr{W}$,

$$\dim(\ker f) + \dim(\operatorname{im} f) = \dim \mathscr{W}.$$

If we assume for the linear transformation $\sigma - 1_{\mathscr{W}}$ that $\operatorname{im}(\sigma - 1_{\mathscr{W}}) = s$, we have

$$\dim F + s = m.$$

Consequently, $s = m$ if and only if σ has no fixed vectors and, in that case, Proposition 227.1 lets us down.

Exercises

*10. Here we make use of quotient spaces to define rotations and reflections of *singular* spaces. Let W be a singular metric vector space which is not a null space. We saw in Exercise 216.3 that, for every nonzero scalar a, there exists an isometry σ of W such that $\det \sigma = a$. Consequently, $\det \sigma$ cannot be used to define rotations and reflections of W. Nevertheless, whenever W is not a null space, a natural notion of reflection and rotation can be defined as follows. Let σ be an isometry of W.

   a. Prove that rad W is an invariant subspace of σ.

   b. It follows from Part a that σ induces a linear transformation $\bar{\sigma}: W/\operatorname{rad} W \to W/\operatorname{rad} W$. This quotient space is the nonsingular space associated with W. Prove that $\bar{\sigma}$ is an isometry of $W/\operatorname{rad} W$ and $\det \bar{\sigma} = \pm 1$. (Since W is not the null space, $\dim(W/\operatorname{rad} W) \geq 1$ and $\det \bar{\sigma}$ is defined.)

   c. According to Part b, $\bar{\sigma}$ is either a rotation or a reflection of $W/\operatorname{rad} W$. We call σ a **rotation of** W if $\bar{\sigma}$ is a rotation of $W/\operatorname{rad} W$ and a **reflection of** W if $\bar{\sigma}$ is a reflection of $W/\operatorname{rad} W$. Prove that σ is a rotation of W if $\det \bar{\sigma} = 1$ and a reflection if $\det \bar{\sigma} = -1$.

   d. Prove that the assertions of Exercise 221.14 are also true for W.

   e. Prove that 1_W is a rotation.

   f. Prove that the rotations of W form a group.

   g. Define the notion of symmetry for W.

   h. Prove that -1_W is a rotation if $\dim W - \dim(\operatorname{rad} W)$ is even and is a reflection if $\dim W - \dim(\operatorname{rad} W)$ is odd.

42. REFINEMENT OF THE WITT THEOREM

We return to our nonsingular metric vector space V of dimension n over the field k of characteristic $\neq 2$. The Witt theorem tells us that an isometry σ between two subspaces of V can always be extended to an isometry of V. We expect from Euclidean geometry that σ can very often be extended to both a rotation and a reflection. In this section, we derive the necessary and sufficient condition which assures that σ has both kinds of extensions (see Theorem 234.1).

We begin by preparing the way for the application of Proposition 227.1 to the computation of determinants of isometries. Two linear subspaces U and W of V may very well be orthogonal without being orthogonal complements. For instance, if $n = 3$ and A_1, A_2, A_3 is a rectangular coordinate system of V, the lines $\langle A_1 \rangle$ and $\langle A_2 \rangle$ are orthogonal, but

$$\langle A_1 \rangle^* = \langle A_2, A_3 \rangle \neq \langle A_2 \rangle^* \qquad \text{and} \qquad \langle A_2 \rangle^* = \langle A_1, A_3 \rangle \neq \langle A_1 \rangle.$$

The situation changes when $\dim U + \dim W = n$, as the following exercise shows.

Exercise

*1. Let U and W be subspaces of V where $\dim U + \dim W = n$. Prove that $U \perp W$ if and only if $U^* = W$, equivalently, $U = W^*$.

To every linear transformation g of V are associated the linear subspaces $\ker g$ and $\operatorname{im} g$. For applications of the following result, it is important that g is not assumed to be an isometry and may be singular.

Proposition 231.1. Let g be an arbitrary linear transformation of V. Then $\ker g$ and $\operatorname{im} g$ are orthogonal if and only if they are orthogonal complements.

Proof. We know from linear algebra that $\dim(\ker g) + \dim(\operatorname{im} g) = n$ and hence the desired result follows directly from Exercise 231.1. Done.

We now consider an isometry σ of V and investigate the linear transformations $\sigma + 1_V$ and $\sigma - 1_V$. As indicated at the end of the previous section, the study of the linear transformation $\sigma - 1_V$ is equivalent to the study of the characteristic root 1 of σ. Similarly, the study of the linear transformation $\sigma + 1_V$ is the study of the characteristic root -1 of σ. Namely,

$$\ker(\sigma + 1_V) = \{X \mid X \in V, \quad \sigma X = -X\}$$

and hence $\ker(\sigma + 1_V)$ is the subspace of V whose vectors are those characteristic vectors of σ which correspond to the characteristic root -1. The reason the characteristic roots ± 1 of an isometry are of particular importance is brought out in the exercise below.

Exercise

2. Let σ be an isometry of V and $a \in k$ a characteristic root of σ whose corresponding characteristic vector is not isotropic. Prove that $a = \pm 1$.

Proposition 232.1. Let σ be an isometry of V. Then $\ker(\sigma - 1_V)$ and $\mathrm{im}(\sigma - 1_V)$ are orthogonal complements and $\ker(\sigma + 1_V)$ and $\mathrm{im}(\sigma + 1_V)$ are orthogonal complements.

Proof. The transformations $\sigma + 1_V$ and $\sigma - 1_V$ are often singular (see Exercise 3 below), in which case they are certainly not isometries. Nevertheless, we use them as the linear transformation g of Proposition 231.1 and, therefore, only have to show that

$$\ker(\sigma - 1_V) \perp \mathrm{im}(\sigma - 1_V) \quad \text{and} \quad \ker(\sigma + 1_V) \perp \mathrm{im}(\sigma + 1_V).$$

In order to show that $\ker(\sigma - 1_V) \perp \mathrm{im}(\sigma - 1_V)$, we choose $X \in \ker(\sigma - 1_V)$ and $Y \in \mathrm{im}(\sigma - 1_V)$ and prove that $XY = 0$. We are given $\sigma X = X$ and $Y = \sigma Z - Z$ for some $Z \in V$. Consequently,

$$XY = X(\sigma Z) - XZ = (\sigma X)(\sigma Z) - XZ = 0$$

since σ is an isometry and, therefore, $(\sigma X)(\sigma Z) = XZ$. We leave it to the reader to prove that $\ker(\sigma + 1_V) \perp \mathrm{im}(\sigma + 1_V)$. Done.

Exercise

*3. a. We have seen that $\ker(\sigma - 1_V)$ is the fixed space of σ. Prove that $\ker(\sigma + 1_V)$ is the space of vectors inverted by σ.
b. Prove that $\sigma - 1_V$ is singular if and only if σ leaves a nonzero vector fixed.

c. Prove that $\sigma + 1_V$ is singular if and only if σ inverts a nonzero vector.

The "refinement of the Witt theorem" states when it is possible to extend an isometry between subspaces of V to both a rotation and a reflection. The proof is based on the following lemma which pertains to a subspace U of V with the property that dim $U +$ dim(rad U) $= n$.

Lemma 233.1.   Let U be a subspace of V where dim $U +$ dim(rad U) $= n$. If σ is an isometry of V which leaves every vector of U fixed, σ is a rotation.

Proof. We have to show that det $\sigma = 1$.
  Case 1. V is an Artinian space of dimension $2d$ and U is a null space of dimension d. Then $U =$ rad U and the condition dim $U +$ dim(rad U) $= n$ is satisfied. Since σ leaves every vector of U fixed, $U \subset \ker(\sigma - 1_V)$ which is the same as $\ker(\sigma - 1_V)^* \subset U^*$. But $\ker(\sigma - 1_V)^* = \text{im}(\sigma - 1_V)$ (Proposition 232.1) and $U^* = U$ (Exercise 191.20). Thus $\text{im}(\sigma - 1_V) \subset U$ and we conclude, from Proposition 227.1, that det $\sigma = $ det $\sigma | U$. However, $\sigma | U = 1_U$ and, consequently, det $\sigma | U = 1$. Is this not an elegant way to compute a determinant?
  Case 2. U is arbitrary. If U is nonsingular, then $U = V$ and $\sigma = 1_V$ and, therefore, σ is a rotation. If U is singular, put $U = W \oplus$ rad U where W is nonsingular. Since dim $U +$ dim(rad U) $= n$, V is the only nonsingular completion of U (Exercise 195.4a). It follows that $V = W \oplus Art_{2d}$ where rad U is a null subspace of dimension d of the Artinian space Art_{2d}. Since $\sigma | U = 1_U$ and $W \subset U$, $\sigma | W = 1_W$. The fact that σ transforms W onto itself implies that σ transforms $W^* = Art_{2d}$ onto itself, equivalently, the restriction σ' of σ to Art_{2d} is an isometry of Art_{2d}. This isometry σ' leaves rad U pointwise fixed and we conclude, from Case 1, that σ' is a rotation of Art_{2d}. Thus $\sigma = 1_W \oplus \sigma'$ is the orthogonal sum of two rotations which proves that σ is a rotation (Exercise 221.15b). Done.

We shall see in Section 43 on rotations of Artinian space around maximal null spaces that the rotation σ of the above lemma is not necessarily 1_V.
  Let U and W be subspaces of V and $\sigma\colon U \to W$ an isometry. The question whether σ can be extended to both a rotation and a reflection of V does not depend on σ but only on the way U is embedded in V.

Proposition 233.1.   Let U and W be subspaces of V and $\sigma\colon U \to W$ be an isometry. The isometry σ can be extended to both a rotation and a reflection of V if and only if there exists a reflection of V which leaves the vectors of U fixed.

Proof. If σ can be extended to both a rotation ρ and a reflection α of V, then $\alpha^{-1}\rho$ is a reflection of V which leaves U pointwise fixed. Conversely, let α be a reflection which leaves U pointwise fixed. By the Witt theorem, σ can be extended to an isometry $\bar{\sigma}$ of V and it is clear that $\bar{\sigma}\alpha$ is also an extension of σ. If $\bar{\sigma}$ is a rotation, $\bar{\sigma}\alpha$ is a reflection; if $\bar{\sigma}$ is a reflection, $\bar{\sigma}\alpha$ is a rotation. Done.

The previous proposition should not be regarded as answering the question when an isometry can be extended to both a rotation and a reflection. It is merely a reformulation of this question in terms of the embedding of U in V. The answer is now given.

Theorem 234.1 (refinement of the Witt theorem). An isometry $\sigma: U \to W$ between the subspaces U and W of V can be extended to both a rotation and a reflection of V if and only if dim $U + \dim(\text{rad } U) < n$.

Proof. We must show that dim $U + \dim(\text{rad } U) < n$ if and only if there exists a reflection of V which is the identity on U (Proposition 233.1). If such a reflection exists, then necessarily dim $U + \dim(\text{rad } U) < n$ by Lemma 233.1. Conversely, assume that dim $U + \dim(\text{rad } U) < n$. If U is nonsingular, $V = U \oplus U^*$ and, since dim $U < n$, dim $U^* \geq 1$. Consequently, U^* has a reflection α and the isometry $1_U \oplus \alpha$ is a reflection of V which is the identity on U. If U is singular, choose a nonsingular completion $\bar{U}$ of U. Then dim $\bar{U} = \dim U + \dim(\text{rad } U) < n$ and hence, as just shown, V has a reflection which is the identity on $\bar{U}$. This reflection is all the more the identity on U. Done.

This theorem shows in particular that, if U is nonsingular and $U \neq V$, every isometry $\sigma: U \to W$ can be extended to both a rotation and a reflection of V. Consequently, if V is anisotropic and $U \neq V$, every isometry $\sigma: U \to W$ can be so extended. It is precisely this fact to which we are accustomed from Euclidean geometry, the prototype of anisotropic geometry.

On the other hand, if V is an Artinian space Art_{2d} and U is a null space of dimension d, then dim $U + \dim(\text{rad } U) = 2d = n$. In this case, every nonsingular linear mapping $\sigma: U \to W$ is an isometry and the extensions of σ to isometries of Art_{2d} are either all rotations or all reflections. If we choose $W = U$ and $\sigma = 1_U$, the identity mapping of Art_{2d} is an extension of 1_U to a rotation of Art_{2d}. Hence all extensions of 1_U to Art_{2d} are rotations. This is precisely Case 1 of the proof of Lemma 233.1.

Even if $\sigma: U \to U$ is an arbitrary linear automorphism of U, we shall see that all extensions of σ to isometries of Art_{2d} are rotations. These are the "rotations around null spaces" discussed in Section 43.

Exercises

*4. Let V be the Euclidean plane and A_1, A_2 an orthonormal coordinate system of V. We put

$$A = \sqrt{3}A_1 + A_2, \qquad B = -A_1 + A_2,$$

and study the lines $U = \langle A \rangle$, $W = \langle B \rangle$.

  a. Prove that there is a unique isometry $\sigma: U \to W$ such that $\sigma A = \sqrt{2}B$.

  b. Prove that U and W are nonsingular lines. Thus

$$\dim U + \dim(\mathrm{rad}\ U) = 1 < 2$$

and it follows from Theorem 234.1 that σ can be extended to both a rotation and a reflection of V.

  c. Over what angle and in what direction should the plane be rotated around 0 so the resulting rotation is an extension of σ? (Hint: This kind of question is difficult to answer without a picture. With a picture, it is child's play; see Figure 235.1.)

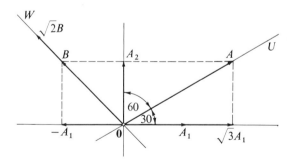

Figure 235.1

  d. We shall see in Exercise 256.9a that every reflection of a non-singular plane is a symmetry with respect to a nonsingular line. In what line through 0 should V be reflected so that the resulting symmetry is an extension of σ?

5. We generalize Exercise 4 above to the case when V is a plane. Assume that A and B are linearly independent vectors of V such that $A^2 = B^2 \neq 0$ and again let $U = \langle A \rangle$ and $W = \langle B \rangle$. Then U and W are non-singular lines and there is a unique isometry $\sigma : U \to W$ such that $\sigma A = B$. It follows from Theorem 234.1 that σ can be extended to both a rotation and a reflection of V.

   a. Find a nonisotropic vector C of V such that the symmetry τ_C is an extension of σ. (Hint: Draw an appropriate diagram.)

   b. Prove that $-1_V \tau_A$ is a reflection of V which is the identity on U.

   c. Prove that $\tau_C(-1_V)\tau_A$ is an extension of σ to a rotation of V.

6. Let $n = 3$ and let A_1, A_2, A_3 be an orthonormal coordinate system of V. Put $B_1 = A_2 + A_3$, $B_2 = A_1$, $C_1 = A_1 + A_3$ and $C_2 = -A_2$.

   a. Prove that the vectors B_1, B_2 are linearly independent and the vectors C_1, C_2 are linearly independent. Consequently, $U = \langle B_1, B_2 \rangle$ and $W = \langle C_1, C_2 \rangle$ are planes.

   b. Prove that there is a unique isometry $\sigma : U \to W$ such that $\sigma B_1 = C_1$ and $\sigma B_2 = C_2$.

   c. Prove that the planes U and W are nonsingular. Consequently, $\dim U + \dim(\mathrm{rad}\ U) = 2 < 3$ and σ can be extended to both a rotation and a reflection of V.

   d. Find a rotation of V which extends σ. (Hint: Draw a picture or make a model. Then write the matrix of such a "rotation" and check that its determinant is 1.)

   e. Find a reflection of V which is an extension of σ.

*7. Let V be an Artinian plane and M, N an Artinian coordinate system of V. We put $A = 2M + 3N$, $B = M + 6N$ and study the lines $U = \langle A \rangle$, $W = \langle B \rangle$.

   a. Prove that there is a unique isometry $\sigma : U \to W$ such that $\sigma A = B$. We assume in Parts b–f that the characteristic of $k \neq 3$.

   b. Prove that the lines U and W are nonsingular. Therefore, σ can be extended to both a rotation and a reflection of V.

   c. Find a rotation ρ of V which extends σ. (Hint: Under any isometry, nonzero isotropic vectors go into nonzero isotropic vectors. Hence there are nonzero scalars a, b such that $\rho M = aM$ or aN and $\rho N = bM$ or bN. The determinant of ρ is easy to compute relative to the coordinate system M, N and ρ is completely determined by its action on the coordinate vectors M and N.)

   d. Find a reflection of V which extends σ. (Explain how this extension acts on M and N.)

   e. Prove that σ can be extended to only one rotation and one reflection of V.

  f. Find a nonisotropic vector C such that the reflection of Part d is equal to the symmetry τ_C.

We assume in Parts g–i that the characteristic of $k = 3$.

  g. Prove that the lines U and W are singular.

  h. Find a rotation of V which is an extension of σ. (State how this rotation acts on M and N.)

  i. Prove that the rotation under Part h is the only extension of σ to an isometry of V.

*8. We generalize Exercise 7 above. Again, V is an Artinian plane and M, N an Artinian coordinate system of V. Assume that $a, b, c, d \in k$ where $ab = cd$; put $A = aM + bN$ and $B = cM + dN$. We assume that not both a and b are zero and not both c and d are zero so that $U = \langle A \rangle$ and $W = \langle B \rangle$ are lines.

  a. Prove that there is a unique isometry $\sigma: U \to W$ such that $\sigma A = B$.

Assume for Parts b–f that $ab \neq 0$.

  b. Prove that the lines U and W are nonsingular.

  c. Find a rotation of V which extends σ.

  d. Find a reflection of V which extends σ.

  e. Prove that σ can be extended to only one rotation and one reflection of V.

  f. Find a nonisotropic vector C in V such that the reflection of Part d is equal to the symmetry τ_C.

Assume for Parts g–k that $ab = 0$.

  g. Prove that the lines U and W are singular.

  h. If $ac \neq 0$ or $bd \neq 0$, find a rotation of V which extends σ.

  i. If $ad \neq 0$ or $bc \neq 0$, find a reflection of V which extends σ.

  j. Prove that the rotation of Part h or the reflection of Part i (depending on the scalars a, b, c, d) is the only extension of σ to an isometry of V.

  k. Find a nonisotropic vector C in V such that the reflection under Part i is the symmetry τ_C.

*9. Let U be a subspace of our nonsingular space V. Prove that

$$\dim U + \dim(\operatorname{rad} U) = n$$

if and only if U^* is a null space. This result shows that the condition on U in Lemma 233.1 can be formulated without mentioning dimensions.

10. The refinement of the Witt theorem is very easy to prove when V is anisotropic. In this case, the theorem says that, if $U \neq V$, every isometry $\sigma: U \to W$ can be extended to both a rotation and a reflection of V. Give an elementary proof for this special case. (Hint: Put $V = U \oplus U^*$ and use Proposition 233.1.)

Remark on High School Teaching. Readers interested in high school teaching should do Exercise 10 above in the special cases where V is the Euclidean line, plane, or three-space, using only the notions developed in high school courses in advanced geometry.

The geometry, centered around Witt's theorem and its refinement in the case of Euclidean spaces, offers excellent material for enriching high school courses in advanced geometry. The difficulty with enriching courses in geometry does not lie in finding suitable material, but only in choosing from the abundance of material.

43. ROTATIONS OF ARTINIAN SPACE AROUND MAXIMAL NULL SPACES

Rotations around null spaces give rise to such pretty geometry that we want to say a few words about them. These rotations do not occur in Artin's *Geometric Algebra*, but were discussed by him in a course at Princeton University in 1949.

We assume for this whole section that U is a null space of dimension d of an Artinian space Art_{2d}.

The linear automorphisms $\sigma: U \to U$ are the isometries of U. So far, all we know is that the extensions of such a σ to isometries of Art_{2d} are either all rotations or all reflections. Furthermore, if $\sigma = 1_U$, all extensions are rotations. We now obtain the additional information that, *independent of the choice of* σ, all extensions of σ to isometries of Art_{2d} are rotations. These are the rotations of Art_{2d} "around the null space U."

Let $M_1, \ldots, M_d$ be a coordinate system of U and $P_1, \ldots, P_d$ Artinian planes where $M_i \in P_i$ for $i = 1, \ldots, m$ and

$$Art_{2d} = P_1 \oplus \cdots \oplus P_d.$$

There exists a vector $N_i \in P_i$ such that M_i, N_i is an Artinian coordinate system of P_i for $i = 1, \ldots, d$ and the vectors

$$M_1, \ldots, M_d, N_1, \ldots, N_d$$

form a coordinate system of Art_{2d}.

It is clear that the matrix of Art_{2d} relative to this coordinate system is

$$G = \left(\begin{array}{c|c} \bigcirc & I \\ \hline I & \bigcirc \end{array} \right).$$

where I is the $d \times d$ unit matrix and $\bigcirc$ is the $d \times d$ zero matrix.

Assume that $\rho: Art_{2d} \to Art_{2d}$ is an arbitrary linear transformation (not necessarily an isometry) which has U as an invariant subspace. The matrix of ρ relative to the given coordinate system has the form

$$\theta = \left(\begin{array}{c|c} \alpha & \beta \\ \hline \bigcirc & \gamma \end{array}\right)$$

where α, β, and γ are $d \times d$ matrices (Exercise 225.3).

As before, the transpose of a matrix δ is denoted by δ^{T}; a matrix δ is called **skew symmetric** if $\delta^{\mathrm{T}} = -\delta$.

Proposition 239.1. The linear transformation $\rho: Art_{2d} \to Art_{2d}$ is an isometry if and only if $\alpha^{-1}\beta$ is skew symmetric and $\gamma = (\alpha^{-1})^{\mathrm{T}}$.

Proof. We know ρ is an isometry if and only if $\theta^{\mathrm{T}} G\theta = G$ (page 216), equivalently,

$$\left(\begin{array}{c|c} \alpha^{\mathrm{T}} & \bigcirc \\ \hline \beta & \gamma^{\mathrm{T}} \end{array}\right)\left(\begin{array}{c|c} \bigcirc & I \\ \hline I & \bigcirc \end{array}\right)\left(\begin{array}{c|c} \alpha & \beta \\ \hline \bigcirc & \gamma \end{array}\right) = \left(\begin{array}{c|c} \bigcirc & I \\ \hline I & \bigcirc \end{array}\right).$$

If the three matrices on the left are multiplied by "block multiplication," one obtains

$$\left(\begin{array}{c|c} \bigcirc & \alpha^{\mathrm{T}}\gamma \\ \hline \gamma^{\mathrm{T}}\alpha & \alpha^{\mathrm{T}}\beta + \beta\gamma \end{array}\right) = \left(\begin{array}{c|c} \bigcirc & I \\ \hline I & \bigcirc \end{array}\right).$$

Consequently, ρ is an isometry if and only if $\alpha^{\mathrm{T}}\gamma = I$, $\gamma^{\mathrm{T}}\alpha = I$, and $\gamma^{\mathrm{T}}\beta + \beta^{\mathrm{T}}\gamma = \bigcirc$. The equations

$$\alpha^{\mathrm{T}}\gamma = I \quad \text{and} \quad \gamma^{\mathrm{T}}\alpha = I$$

both say α is nonsingular and that $\gamma = (\alpha^{-1})^{\mathrm{T}}$. The equation $\gamma^{\mathrm{T}}\beta + \beta^{\mathrm{T}}\gamma = \bigcirc$ says that $\gamma^{\mathrm{T}}\beta$ is skew symmetric since

$$\gamma^{\mathrm{T}}\beta = -\beta^{\mathrm{T}}\gamma = -(\gamma^{\mathrm{T}}\beta)^{\mathrm{T}}.$$

We replace γ^{T} by α^{-1} and conclude that $\alpha^{-1}\beta$ is skew symmetric. Done.

Lemma 239.1. An isometry of Art_{2d} which transforms U into itself is a rotation.

Proof. We shall show that, under the conditions of Proposition 239.1, $\det \theta = 1$. Now

$$\det \theta = (\det \alpha)(\det \beta) = (\det \alpha)(\det(\alpha^{-1})^{\mathrm{T}}) = (\det \alpha)(\det \alpha^{-1}) = 1.$$

Done.

Proposition 239.1 gives good information about the set of rotations of Art_{2d} which are extensions of a given isometry $\sigma \colon U \to U$. Let ρ be such a rotation and

$$\theta = \left(\begin{array}{c|c} \alpha & \beta \\ \hline \bigcirc & \gamma \end{array} \right)$$

its matrix relative to the coordinate system of Art_{2d} given on page 238. The $d \times d$ matrix α is then the matrix of σ relative to the coordinate system $M_1, \ldots, M_d$ of U (Exercise 226.3b). Consequently, α is completely determined by σ and, since $\gamma = (\alpha^{-1})^{\mathrm{T}}$, γ is also determined by σ. However, the matrix β is not uniquely determined by σ since $\alpha^{-1}\beta$ can be any $d \times d$ skew symmetric matrix S; of course, $\alpha^{-1}\beta = S$ is the same as $\beta = \alpha^{-1}S$. We conclude that an isometry of Art_{2d} is an extension of σ if and only if its matrix has the form

$$\left(\begin{array}{c|c} \alpha & \alpha S \\ \hline \bigcirc & (\alpha^{-1})^{\mathrm{T}} \end{array} \right)$$

where S is an arbitrary $d \times d$ skew symmetric matrix.

Exercises

*1. Let (a_{ij}) be a $d \times d$ skew symmetric matrix. For this exercise, it is important that char $k \neq 2$, since, if char $k = 2$, the notions of skew symmetric matrix and symmetric matrix coincide. In working the exercises below, indicate where one uses char $k \neq 2$.
   a. Prove that $a_{ii} = 0$ for $i = 1, \ldots, d$.
   b. Prove that the only 1×1 skew symmetric matrix is $0 \in k$.
   c. Prove that the 2×2 skew symmetric matrices have the form $\left(\begin{smallmatrix} 0 & a \\ -a & 0 \end{smallmatrix} \right)$ where $a \in k$.
   d. Prove that the $d \times d$ skew symmetric matrices form a vector space of dimension $\frac{1}{2}d(d - 1)$ over k. (Hint: A skew symmetric matrix is completely determined by the elements above the main diagonal.)
   e. If d is odd, prove that every skew symmetric matrix is singular. (Hint: If S is a $d \times d$ skew symmetric matrix, compute the determinant of both sides of the equal sign in $S^{\mathrm{T}} = -S$.)
*2. Since U is a null space of Art_{2d}, the isometries of U are precisely the linear automorphisms of U. Prove that there exists a one-to-one correspondence between the set of isometries of Art_{2d} which extend a fixed isometry of U and the set of $d \times d$ skew symmetric matrices.
3. Consider the Artinian plane Art_2 $(d = 1)$ and hence U is one of the two null lines of Art_2. Choose a nonzero vector M in U and let N be the vector in Art_2 such that M, N is an Artinian coordinate system. Then

$U = \langle M \rangle$ and a linear automorphism $\sigma_a : U \to U$ is determined by the nonzero scalar a such that $\sigma_a A = aA$ for $A \in U$.

a. Prove that an isometry of Art_2 is an extension of σ_a if and only if its matrix relative to the coordinate system M, N is

$$\begin{pmatrix} a & 0 \\ 0 & a^{-1} \end{pmatrix}.$$

(Hint: Use the matrix of page 240 and Exercise 1b above.)

b. Prove that each σ_a can be extended to only one isometry of Art_2. Observe that this result agrees with Exercise 237.8j. In particular, 1_U can be extended to only the identity mapping of Art_2.

*4. We study the extensions of the identity mapping 1_U to rotations of Art_{2d}. Let $M_1, \ldots, M_d$ be a coordinate system of U and $M_1, \ldots, M_d$, $N_1, \ldots, N_d$ the coordinate system of Art_{2d} as chosen on page 238.

a. Prove that an isometry of Art_{2d} is an extension of 1_U if and only if its matrix relative to the above coordinate system has the form

$$\left(\begin{array}{c|c} I & S \\ \hline O & I \end{array} \right)$$

where I is the $d \times d$ unit matrix and S is an arbitrary $d \times d$ skew symmetric matrix.

b. Let $d = 2$. Prove that an isometry of Art_4 is an extension of 1_U if and only if its matrix is

$$\begin{pmatrix} 1 & 0 & 0 & a \\ 0 & 1 & -a & 0 \\ 0 & 0 & 1 & 0 \\ 0 & 0 & 0 & 1 \end{pmatrix}$$

where a can be any scalar. Observe that, if $a \neq 0$, the extension is not the identity mapping of Art_4.

c. Let $d = 2$. Prove that 1_U can be extended in many ways to rotations of Art_{2d} which are different from the identity mapping of Art_{2d}.

The geometry of Artinian spaces is very special, as evidenced, for example, by the following peculiar property of its maximal null spaces. We recall that all maximal null spaces of Art_{2d} have dimension d.

Let W and W' be maximal null spaces of Art_{2d}. If $\sigma : W \to W'$ is an isometry, the extensions of σ to isometries of Art_{2d} are either all rotations or all reflections. Nevertheless, we would expect these extensions to be all rotations for some one isometry from W to W' and all reflections for another one. This expectation is false. If one isometry from W to W' can be extended to a rotation of Art_{2d}, then all isometries from W to W' can only be extended to rotations; likewise, if the extension of one isometry $\sigma: W \to W'$ is a reflection, then all extensions of all isometries from W to W' are reflections. In other words, the type of extension depends only on the pair W, W' and not on the specific isometry from W to W'. This property, which is typical for the maximal null spaces of Artinian space, is formulated as a corollary of Lemma 239.1.

Corollary 242.1. Let W and W' be maximal null spaces of Art_{2d}. The isometries of Art_{2d} which carry W onto W' are either all rotations or all reflections.

Proof. Let ρ and σ be isometries of Art_{2d} such that $\rho W = W'$ and $\sigma W = W'$. The isometry $\rho \sigma^{-1}$ of Art_{2d} maps W onto itself and hence is a rotation (Lemma 239.1). Consequently, ρ and σ are either both rotations or both reflections. Done.

Exercises

5. In this exercise, we show that Corollary 242.1 holds only for maximal null spaces of Artinian spaces.
   a. Let W be a null space of V and assume V is not Artinian; W may or may not be maximal. Prove that dim $W + \dim(\text{rad } W) < n$. Conclude that every isometry from W onto another null space can be extended to both a rotation and a reflection.
   b. Assume that V is Artinian, but the null space W is not maximal. Prove that every isometry from W onto another null space can be extended to both a rotation and a reflection.
6. We shall call two maximal null spaces W and W' of Art_{2d} equivalent if the isometries of Art_{2d} which carry W onto W' are rotations. Notation: $W \sim W'$.
   a. Prove that $\sim$ is an equivalence relation on the set of maximal null spaces of Art_{2d}.
   b. The equivalence relation $\sim$ partitions the set of maximal null spaces of Art_{2d} into disjoint subsets. Prove there are precisely two such sets. (Hint: First, show that Art_{2d} contains two maximal null spaces W and W' which are not equivalent.)
7. Let W and W' be null planes of Art_4.
   a. If $W \sim W'$ and $W \neq W'$, prove that $W \cap W' = \{0\}$.
   b. If $W \nsim W'$, prove that $W \cap W'$ is a line.

Exercises 8, 9, 10 are meant only for readers who know some projective geometry and have studied the remark about projective geometry on page 180.

*8. a. Let S be a nondegenerate, quadratic surface of dimension $n - 2$ in $(n - 1)$-dimensional projective space P over k. If S contains a projective line, prove that every point of S lies on a projective line of S. (Hint: Consider P as the space at infinity of our vector space V. Assume that the metric of V is such that the lines on the light cone C are the points of S. A projective line of S is then a null plane of V and we conclude from Exercise 208.5b that every line of C is contained in a null plane.)

   b. As an example of Part a, let $V = Art_4$. S is now a quadratic surface in projective three-space. Since V contains null planes, S contains projective lines and every point of S lies on a projective line of S. In order to identify S, we choose a coordinate system for V such that the matrix of V is

$$\begin{pmatrix} 1 & 0 & 0 & 0 \\ 0 & 1 & 0 & 0 \\ 0 & 0 & -1 & 0 \\ 0 & 0 & 0 & -1 \end{pmatrix}.$$

The light cone of V then has the equation

$$x_1{}^2 + x_2{}^2 - x_3{}^2 - x_4{}^2 = 0$$

and this is also the equation of S in projective coordinates. Hence the part of S which lies in affine three-space has the equation $x_1{}^2 + x_2{}^2 = x_3{}^2 + 1$. If k is the field $\mathbf{R}$ of real numbers, prove that S is the hyperboloid of one sheet obtained by revolving the hyperbola $x_1{}^2 - x_2{}^2 = 1$ in the (x_1, x_2) plane around the x_3 axis; see Figure 243.1.

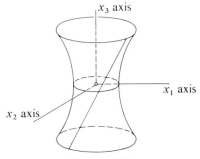

x_3 axis

x_1 axis

x_2 axis

Figure 243.1

*9. Let P be the projective space at infinity of Art_{2d}. Then P has dimension $2d - 1$ and we denote by S the quadratic surface of dimension $2d - 2$ of P whose points are the null lines of Art_{2d}.

    a. Prove that S contains projective spaces of dimension $d - 1$, but not of higher dimension.

    b. Prove that the projective spaces of dimension $d - 1$ of S can be partitioned in a natural way into two disjoint subsets, called "families" in classical geometry.

10. Assume $d = 2$ in the previous exercise; hence S is the hyperboloid of one sheet of Exercise 8 above (see Figure 243.1). We know from Exercise 9 above that S contains projective lines, but not planes, and that these lines are partitioned in a natural way into two disjoint families.

    a. Prove that two disjoint projective lines of the same family have no points in common. (Hint: This is the projective formulation of Exercise 8a above.)

    b. Prove that two projective lines which belong to different families have precisely one point in common.

44. ROTATIONS OF ARTINIAN SPACE WITH A MAXIMAL NULL SPACE AS AXIS

We assume in this section that U is a null space of dimension d of an Artinian space Art_{2d}.

In order to prepare for the proof of the Cartan–Dieudonné theorem, we study the isometries of Art_{2d} which leave U pointwise fixed. These isometries are all rotations and we call U their "axis." In other words, a rotation of Art_{2d} has U as axis if it is an extension of 1_U. Such a rotation may have fixed vectors, even isotropic ones, which do not lie on U; this happens, for instance, when the rotation is the identity mapping of Art_{2d}.

Exercise

*1. Prove that the rotations of Art_{2d} with U as axis form a subgroup of the rotation group $O^+(Art_{2d})$.

We denote the group of rotations of Art_{2d} with U as axis by $O^+(Art; U)$; we know from Exercise 1 above that this group is a subgroup of $O^+(Art_{2d})$. We expect a group which occurs in a natural way in geometry to be non-commutative, but, to our surprise, we shall find that $O^+(Art_{2d}; U)$ is commutative. In order to determine the structure of this group, we return to the coordinate system

$$M_1, \ldots, M_d, N_1, \ldots, N_d$$

of Art_{2d} on page 238. Hence $M_1, \ldots, M_d$ is a coordinate system of U, $M_i^2 = N_i^2 = 0$ and $M_i N_i = 1$ for $i = 1, \ldots, d$.

Relative to the above coordinate system, the matrix of a rotation $\rho \in O^+(Art_{2d}; U)$ has the form

$$\left(\begin{array}{c|c} I & S \\ \hline \bigcirc & I \end{array} \right)$$

where I is the $d \times d$ unit matrix and S is a $d \times d$ skew symmetric matrix (Exercise 241.4a). In order to indicate that S is uniquely determined by ρ, we write $S(\rho)$ and study the mapping $\rho \to S(\rho)$ from the group $O^+(Art_{2d}; U)$ to the vector space of $d \times d$ skew symmetric matrices; it was shown in Exercise 240.1d that this vector space has dimension $\frac{1}{2}d(d-1)$.

We know from linear algebra that the vectors of an s-dimensional vector space over k form a commutative group under addition and this group is isomorphic to the direct sum

$$nk^+ = k^+ \oplus \cdots \oplus k^+$$

of the additive group k^+ of k with itself s times. Consequently, the $d \times d$ skew symmetric matrices form a commutative group under addition which is isomorphic to the additive group $\frac{1}{2}d(d-1)k^+$.

Proposition 245.1.   The mapping $\rho \to S(\rho)$ is an isomorphism from the multiplicative group $O^+(Art_{2d}; U)$ onto the additive group of $d \times d$ skew symmetric matrices. Consequently, $O^+(Art_{2d}; U)$ is commutative and is isomorphic to the additive group $\frac{1}{2}d(d-1)k^+$.

Proof. We only have to prove the first statement. Let $\rho, \sigma \in O^+(Art_{2d}; U)$; we first prove that $S(\rho\sigma) = S(\rho) + S(\sigma)$. Relative to the given coordinate system of Art_{2d}, the matrices of ρ and σ are

$$\text{matrix of } \rho = \left(\begin{array}{c|c} I & S(\rho) \\ \hline \bigcirc & I \end{array} \right), \qquad \text{matrix of } \sigma = \left(\begin{array}{c|c} I & S(\sigma) \\ \hline \bigcirc & I \end{array} \right).$$

Using block multiplication of matrices, we see that

$$\text{matrix of } (\rho\sigma) = \left(\begin{array}{c|c} I & S(\rho) \\ \hline \bigcirc & I \end{array} \right)\left(\begin{array}{c|c} I & S(\sigma) \\ \hline \bigcirc & I \end{array} \right) = \left(\begin{array}{c|c} I & S(\rho) + S(\sigma) \\ \hline \bigcirc & I \end{array} \right)$$

which proves that $S(\rho\sigma) = S(\rho) + S(\sigma)$. This shows the mapping $\rho \to S(\rho)$ is a homomorphism. That this mapping is one-to-one and onto is a consequence of Exercise 240.2, considering the isometry 1_U of U. Done.

If $d = 1$, $O^+(Art_2; U)$ consists of only the identity mapping of Art_2. In that case, $\frac{1}{2}d(d - 1) = 0$ and, in the above proposition, $0k^+$ is to be understood as the additive group with just one element.

Let $\rho \in O^+(Art_{2d}; U)$. To say that ρ has U as axis is the same as to say that U is contained in the fixed space of ρ. We recall that the fixed space of ρ is always a linear subspace of Art_{2d} and now we ask when U is actually equal to this fixed space. Let us denote the identity mapping of Art_{2d} by 1 (leaving off the subscript Art_{2d} for obvious reasons) and recall that the fixed space of ρ is $\ker(\rho - 1)$. Hence we are given that $U \subset \ker(\rho - 1)$ and ask when $U = \ker(\rho - 1)$. The following exercise shows this is the same as asking when $\mathrm{im}(\rho - 1) = U$.

Exercise

*2. Let $\rho \in O^+(Art_{2d}; U)$.
    a. Prove that $\mathrm{im}(\rho - 1) \subset U \subset \ker(\rho - 1)$. (Hint: The inclusion $\mathrm{im}(\rho - 1) \subset U$ was proved in Case 1 of Lemma 233.1.)
    b. Prove that $U = \ker(\rho - 1)$ if and only if $\mathrm{im}(\rho - 1) = U$. (Hint: Start with one of the two equalities and take orthogonal complements.)

In order to settle when $U = \ker(\rho - 1)$, we return to the isomorphism $\rho \to S(\rho)$ of Proposition 245.1.

Proposition 246.1. Let $\rho \in O^+(Art_{2d}; U)$. Then U is the fixed space of ρ if and only if the $d \times d$ skew symmetric matrix $S(\rho)$ is nonsingular. This happens only when the dimension of Art_{2d} is divisible by four.

Proof. We saw in Exercise 240.1e that a $d \times d$ skew symmetric matrix is singular whenever d is odd. For $S(\rho)$ to be nonsingular, d must necessarily be even and hence the dimension of Art_{2d} is divisible by four. For the other part, we must prove that $S(\rho)$ is nonsingular if and only if $\mathrm{im}(\rho - 1) = U$ (Exercise 246.2b). We observe that the matrix of the linear transformation $\rho - 1$, relative to the given coordinate system $M_1, \ldots, M_d, N_1, \ldots, N_d$ of Art_{2d}, is

$$\left(\begin{array}{c|c} I & S(\rho) \\ \hline \bigcirc & I \end{array}\right) - \left(\begin{array}{c|c} I & \bigcirc \\ \hline \bigcirc & I \end{array}\right) = \left(\begin{array}{c|c} \bigcirc & S(\rho) \\ \hline \bigcirc & \bigcirc \end{array}\right).$$

Consequently, $\text{im}(\rho - 1)$ is spanned by the $2d$ vectors

$$(M_1, \ldots, M_d, N_1, \ldots, N_d)\left(\begin{array}{c|c} \bigcirc & S(\rho) \\ \hline \bigcirc & \bigcirc \end{array}\right),$$

equivalently, by the vectors

$$(M_1, \ldots, M_d)S(\rho).$$

This shows, again, that $\text{im}(\rho - 1) \subset U$ and, furthermore, that $\text{im}(\rho - 1) = U$ if and only if $S(\rho)$ is nonsingular. Done.

Exercise

3. Let $\rho \in O^+(Art_{2d}; U)$. Suppose that k is a finite field with q elements and $d = 2$.
   a. Prove that Art_4 has q rotations with U as axis.
   b. Prove that Art_4 has $q - 1$ rotations with U as fixed space.

It is important to reformulate the equivalent conditions

$$\text{im}(\rho - 1) = U \quad \text{and} \quad \text{ker}(\rho - 1) = U$$

of Exercise 246.2b in terms of the nonisotropic vectors of Art_{2d}.

Proposition 247.1.   Let $\rho \in O^+(Art_{2d}; U)$. The following three conditions are equivalent.

1. $\text{im}(\rho - 1) = U$.
2. $\text{ker}(\rho - 1) = U$.
3. If $X \in Art_{2d}$ and $X^2 \neq 0$, $(\rho - 1)X$ is a nonzero isotropic vector.

Moreover, these conditions occur only if $2d$ is divisible by four.

Proof. Only the equivalence of Conditions 2 and 3 above remains to be shown.

Assume Condition 2 holds. Then ρ has no fixed vectors outside of U. Consequently, if $X^2 \neq 0$ and hence $X \notin U$, then $\rho X \neq X$, equivalently, $(\rho - 1)X \neq \mathbf{0}$. Finally, Condition 1 implies that $(\rho - 1)X$ is isotropic and we conclude that Condition 3 holds.

Assume Condition 3 holds. Suppose $U \neq \text{ker}(\rho - 1)$. Since $U \subset \text{ker}(\rho - 1)$, $\text{ker}(\rho - 1)$ is properly larger than the maximal null space U. Therefore, $\text{ker}(\rho - 1)$ is not a null space and must contain a nonisotropic vector X. Then $X^2 \neq 0$ and $(\rho - 1)X = \mathbf{0}$ which contradicts Condition 3. It follows that $\text{ker}(\rho - 1) = U$. Done.

Observe that Condition 3 of the above proposition implies that ρ leaves no nonisotropic vector of Art_{2d} fixed. Furthermore, Condition 2 says that U is the fixed space of ρ.

Since U is a null space, U lies on the light cone C of Art_{2d}. The condition $\operatorname{im}(\rho - 1) \subset U$ hence implies that a line of Art_{2d} through $\mathbf{0}$ is transformed either onto a line of C or onto $\mathbf{0}$. It is clear that all lines of U are transformed onto $\mathbf{0}$ by $\rho - 1$. Condition 2 of Proposition 247.1 is equivalent to saying that, if a line does not belong to U, $\rho - 1$ transforms that line onto a line of C. Condition 3 is equivalent to saying that, if a line does not lie on C, $\rho - 1$ transforms that line onto a line of C. This gives a clear geometric picture of the implication $2 \Rightarrow 3$.

Exercise

4. If $\rho \in O^{+}(Art_{2d}; U)$, prove that the following three conditions are equivalent.
    1. $\operatorname{im}(\rho - 1) = U$.
    2. $\ker(\rho - 1) = U$.
    3'. ρ leaves no nonisotropic vector fixed.
    (Hint: All we did was remove the word "isotropic" from Condition 3 of Proposition 247.1. Hence only the implication $3' \Rightarrow 2$ has to be proved. Where did we use the fact that $(\rho - 1)X$ is isotropic in the proof that $3 \Rightarrow 2$?)

Condition 3 of Proposition 247.1 is very peculiar. It says that the isometry ρ is such that $\rho - 1$ maps every vector not on the light cone of Art_{2d} onto a nonzero vector on the light cone. Of course, this can never happen in anisotropic spaces where the light cone consists of $\mathbf{0}$ alone. Even if a space has isotropic vectors, it is very exceptional that an isometry exists which satisfies Condition 3. We shall see that this can happen only in Artinian spaces whose dimension is divisible by four and that the isometry must be a rotation which has a maximal null space as fixed space. The peculiarity of Condition 3 is further brought to light by the following exercises which will be used in the proof of Lemma 249.1.

Exercises

*5. Let W be a metric vector space over k; W may be singular. Let $X \in W$ and ρ an isometry of W. Prove that $(\rho - 1)X$ is isotropic if and only if $X \perp (\rho X - X)$. (This fact uses that char $k \neq 2$.)

Figure 249.1 displays the rhombus determined by the vectors X and ρX. We recall from Exercise 118.7c that the diagonals of a rhombus are orthogonal and we observe that the figure is misleading when X and $\rho X - X$ are orthogonal.

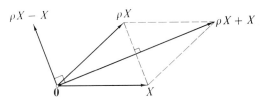

Figure 249.1

*6. Let W, X, and ρ be as in Exercise 5 above. Assume that $X^2 \neq 0$ and that $(\rho - 1_W)X$ is a nonzero isotropic vector.
  a. Prove that X and ρX are linearly independent.
  b. It follows from Part a that X and ρX span a plane P. Prove that P is singular. (Hint: Use Exercise 5 above when computing the discriminant of P relative to the coordinate system X, ρX.)
  c. Prove that the line $\langle \rho X - X \rangle$ is the radical of P.

We are now ready to show how very exceptional is an isometry which satisfies Condition 3 of Proposition 247.1. V is again our n-dimensional nonsingular space.

Lemma 249.1.   Let $n \geq 1$. An isometry ρ of V has the property that $(\rho - 1_V)X$ is a nonzero isotropic vector for every nonisotropic vector X of V if and only if the following conditions are satisfied:

1. V is an Artinian space Art_{4r} of dimension $4r$.
2. ρ is a rotation of Art_{4r} with a maximal null space U as fixed space.

Proof. The if part is contained in Proposition 247.1. Thus we assume that the isometry ρ of V has the property that $(\rho - 1_V)X$ is a nonzero isotropic vector for every nonisotropic vector X of V and show that the two stated conditions are satisfied.

Let us first show that $n \geq 3$. Since $n \geq 1$ and V is nonsingular, V contains a nonisotropic vector X. By assumption, $(\rho - 1_V)X$ is a nonzero isotropic vector; according to Exercise 249.6, the vectors X, ρX are linearly independent and span a singular plane. This plane cannot be equal to the nonsingular space V and hence $n \geq 3$.

Next, we prove that $\ker(\rho - 1_V)$ is a null space of V. If $X \in \ker(\rho - 1_V)$, $(\rho - 1_V)X = \mathbf{0}$. It follows that $X^2 = 0$ for, otherwise, $(\rho - 1_V)X \neq \mathbf{0}$ by hypothesis. Consequently, all the vectors of $\ker(\rho - 1_V)$ are isotropic which shows that $\ker(\rho - 1_V)$ is a null space.

We prove $\operatorname{im}(\rho - 1_V)$ is also a null space of V. For each $X \in V$, let us prove that $(\rho - 1_V)X$ is isotropic. This is trivial when $X = \mathbf{0}$ and is given by hypothesis when $X^2 \neq 0$. Hence we assume that X is a nonzero isotropic vector, equivalently, that $\langle X \rangle$ is a null line. The orthogonal complement $\langle X \rangle^*$ of $\langle X \rangle$ is a hyperplane of V which contains $\langle X \rangle$. Furthermore, $\langle X \rangle^*$ is not a null space since $n \geq 3$ (Exercise 195.3); hence $\langle X \rangle^*$ contains a nonisotropic vector Y. It follows from $X^2 = 0$ and $XY = 0$ that

$$(X + Y)^2 = X^2 + 2XY + Y^2 = Y^2 \neq 0$$

and

$$(X - Y)^2 = X^2 - 2XY + Y^2 = Y^2 \neq 0.$$

By assumption, the nonisotropic vectors Y, $X + Y$, and $X - Y$ give rise to isotropic vectors $(\rho - 1_V)Y$, $(\rho - 1_V)(X + Y)$, and $(\rho - 1_V)(X - Y)$. Consequently,

$$0 = [(\rho - 1_V)(X + Y)]^2 = [(\rho - 1_V)X + (\rho - 1_V)Y]^2$$
$$= [(\rho - 1_V)X]^2 + 2[(\rho - 1_V)X][(\rho - 1_V)Y]$$

and, similarly,

$$0 = [(\rho - 1_V)(X - Y)]^2 = [(\rho - 1_V)X]^2 - 2[(\rho - 1_V)X][(\rho - 1_V)Y].$$

We conclude by addition that $2[(\rho - 1_V)X]^2 = 0$ and hence $(\rho - 1_V)X$ is isotropic (since char $k \neq 2$).

We denote the null space $\operatorname{im}(\rho - 1_V)$ by U and its dimension by d.

Now we prove that V is an Artinian space Art_{2d}. By Proposition 232.1,

$$U^* = \operatorname{im}(\rho - 1_V)^* = \ker(\rho - 1_V)$$

which shows that $U^* = \ker(\rho - 1_V)$ is a null space. Since U is a null space such that U^* is also a null space, we know that V is an Artinian space whose dimension is $2d$ (Exercise 196.8).

Finally, we prove that ρ is a rotation of Art_{2d} with U as axis and $2d$ is divisible by four. The fixed space of ρ is $\ker(\rho - 1_V) = U^*$. But $U^* = U$ since U is a null space of dimension d; consequently, ρ is a rotation (Lemma 239.1). It follows from Proposition 246.1 that $2d$ is divisible by four. Done.

45. THE CARTAN–DIEUDONNÉ THEOREM

We return to our n-dimensional nonsingular space V and its orthogonal group $O(V)$. A large part of geometry and algebra centers around the investigation of this classical group. A nonempty subset of a group G is called a **set of generators of** G if no proper subgroup of G contains the set. Of course, G itself is a set of generators, but this is no help. Whenever a particularly simple set of generators for a group can be found, progress has been made in the investigation of that group. The Cartan–Dieudonné theorem gives such a set of generators for $O(V)$ and states, furthermore, that each isometry is the product of not more than n of these generators.

We begin with a few exercises on generators of groups.

Exercises

In these exercises, G denotes a multiplicative group with unit element 1.

1. Let K be a nonempty subset of G. Prove that K is a set of generators of G if and only if each $\sigma \in G$ can be written in at least one way as a finite product

$$\sigma = \sigma_1 \cdots \sigma_h$$

where either σ_i or σ_i^{-1} belongs to K for $i = 1, \ldots, h$; σ_i may be equal to σ_j for $i \neq j$ and h depends on σ.

*2. Let K be a nonempty subset of G which consists of elements of finite order, that is, if $\sigma \in K$, then $\sigma^i = 1$ for some $i \geq 1$. Prove that K is a set of generators of G if and only if each $\sigma \in G$ can be written in at least one way as a finite product

$$\sigma = \sigma_1 \cdots \sigma_h$$

where $\sigma_i \in K$ for $i = 1, \ldots, h$; σ_i may be equal to σ_j for $i \neq j$ and h depends on σ. This situation occurs, for instance, when G is finite.

3. Give an example of an infinite group with a finite number of generators.

An element of a multiplicative group is called an **involution** if its square is 1. Every nonisotropic vector A of V determines the symmetry τ_A (page 219); symmetries are involutions of the group $O(V)$ (Exercise 224.20i). There are always plenty of symmetries, namely, as many as there are lines not on the light cone. Symmetries are very simple isometries and the Cartan–Dieudonné theorem states that they form a set of generators of $O(V)$. In other words, since $\tau_A = \tau_A^{-1}$, every isometry is the product of a finite number of symmetries. But the Cartan–Dieudonné theorem says more; it says that it is possible to express each isometry as the product of not more than n symmetries. The following exercises will be used in the proof of this theorem.

Exercises

*4. Let U be a subspace of V. An isometry of V is said to have U as *axis* if it leaves the vectors of U fixed. Prove that the isometries of V with U as axis form a subgroup of $O(V)$. Notation: $O(V; U)$.

*5. Let U be a nonsingular subspace of V. It is now possible to compare the group $O(V; U)$ with the orthogonal group $O(U^*)$.

 a. Prove that $O(V; U) \simeq O(U^*)$. (Hint: If $\sigma \in O(V; U)$, then $\sigma = 1_U \oplus \sigma'$ where $\sigma' = \sigma | U$. The restriction mapping $\sigma \to \sigma'$ is the desired isomorphism.)

 b. The restriction mapping

$$\text{res}: O(V; U) \to O(U^*)$$

 is an isomorphism and hence has an inverse isomorphism

$$\text{ext}: O(U^*) \to O(V; U)$$

 (ext stands for "extension"). If $\sigma' \in O(U^*)$, prove that $\text{ext } \sigma' = 1_U \oplus \sigma'$.

 c. Let $A \in U^*$ where $A^2 \neq 0$. Then A gives rise to the symmetry τ_A' of U^* and the symmetry τ_A of V. Prove that $\text{res } \tau_A = \tau_A'$, equivalently, $\text{ext } \tau_A' = \tau_A$. (Hint: Consider the space of vectors inverted and the space of vectors left fixed by τ_A' and res τ_A.)

 d. Let $B_1, \ldots, B_m$ be nonisotropic vectors of U^*. Each vector B_i gives rise to a symmetry τ_{B_i}' of U^* and a symmetry τ_{B_i} of V. Prove that

$$\text{ext}(\tau_{B_1}' \cdots \tau_{B_m}') = \tau_{B_1} \cdots \tau_{B_m}.$$

 e. If $\sigma \in O(V; U)$, prove that σ is a rotation (reflection) of V if and only if res σ is a rotation (reflection) of U^*.

 f. We showed in Exercise 244.1 that the set $O^+(V; U)$ of rotations of V with U as axis is a subgroup of $O^+(V)$. Prove that $O^+(V; U) \simeq O^+(U^*)$.

One other preparation of a geometric nature is necessary for the proof of the Cartan–Dieudonné theorem. Let $A, B \in V$ where $A^2 = B^2$. The vectors $A + B$ and $A - B$ are then orthogonal and, if A and B are linearly independent, they determine a rhombus (see Figure 253.1). Since $A^2 = B^2$, it follows from the Witt theorem that there is an isometry ρ of V such that $A = \rho B$. Even so, we saw in Exercises 248.5 and 249.6 that all sorts of strange things happen

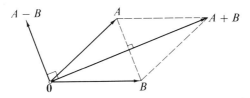

Figure 253.1

when the vector $A - B$ is isotropic. We now show that everything is "normal" when $(A - B)^2 \neq 0$. As always, "normal" means as in Euclidean geometry.

One expects from Euclidean geometry to be able to define the "bisector" of the vectors A and B; in Figure 253.1, this bisector is the line $\langle A + B \rangle$. Furthermore, one expects that it should be possible to choose a symmetry τ for the above isometry ρ; thus τ would have the properties that $A = \tau B$, equivalently (since $\tau = \tau^{-1}$), $B = \tau A$ and that τ would leave the bisector of A and B pointwise fixed. All of this is correct as long as the vector $A - B$ is not isotropic. In this case, one chooses $\tau = \tau_{A-B}$ and defines the **bisector** of A and B to be the orthogonal complement $\langle A - B \rangle^*$ of the nonsingular line $\langle A - B \rangle$. The proof that τ_{A-B} and $\langle A - B \rangle^*$ have the correct properties is asked for in the following exercise.

Exercise

6. Assume that $A, B \in V$ where $A^2 = B^2$ and $(A - B)^2 \neq 0$. Let τ be the symmetry τ_{A-B}. The bisector of A and B is $\langle A - B \rangle^*$.
   a. Prove that $\tau A = B$ and $\tau B = A$.
   b. Prove that the bisector of A and B is a nonsingular hyperplane which contains the vector $A + B$; also prove that $\langle A - B \rangle^*$ is left pointwise fixed by τ. See Figure 253.2 where it is assumed that $n = 3$ and hence the bisector is a plane.

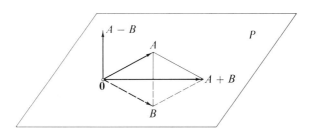

Figure 253.2

The following theorem by French mathematicians Elie Cartan and Jean Dieudonné is a cornerstone in the theory of metric vector spaces.

Theorem 254.1 (the Cartan–Dieudonné theorem). Every isometry of V is the product of not more than n symmetries.

Proof (by induction). Let σ be an isometry of V. If $n = 1$, $\sigma = \pm 1_V$ and -1_V is the only symmetry of V. Since $1_V = (-1_V)^0$, the theorem is trivial for $n = 1$. We make the induction hypothesis that the theorem has been proved for $\dim V < n$.

Case 1. σ leaves a nonisotropic vector X fixed. Then $V = \langle X \rangle \oplus \langle X \rangle^*$ and $\sigma = 1_{\langle X \rangle} \oplus \sigma'$ where σ' is the restriction of σ to $\langle X \rangle^*$. Since $\langle X \rangle^*$ is a nonsingular hyperplane, it follows from the induction hypothesis that there are nonisotropic vectors $B_1, \ldots, B_m$ in $\langle X \rangle^*$ such that $\sigma' = \tau'_{B_1} \cdots \tau'_{B_m}$ and $m \leq n - 1$. Here, τ'_{B_i} denotes the symmetry of $\langle X \rangle^*$ corresponding to B_i. By Exercise 252.5d, $\sigma = \tau_{B_1} \cdots \tau_{B_m}$ where τ_{B_i} denotes the symmetry of V corresponding to B_i. Consequently, in Case 1, σ is actually the product of at most $n - 1$ symmetries.

Case 2. There exists a nonisotropic vector B in V such that $\sigma B - B$ is also not isotropic. Since $(\sigma B - B)^2 \neq 0$, the vectors σB and B have the bisector $\langle \sigma B - B \rangle^*$. Furthermore, the symmetry $\tau = \tau_{\sigma B - B}$ is such that $\tau \sigma B = B$. Because $B^2 \neq 0$, it follows that the isometry $\tau \sigma$ falls under Case 1 and hence $\tau \sigma = \tau_1 \cdots \tau_m$ where each τ_i is a symmetry of V and $m \leq n - 1$. We conclude that $\sigma = \tau \tau_1 \cdots \tau_m$ (since $\tau = \tau^{-1}$) and Case 2 is done.

Case 3. $(\sigma - 1_V)X$ is a nonzero isotropic vector for every nonisotropic vector X of V. Observe that this is the same as to say that σ falls neither under Case 1 nor under Case 2 and, therefore, the present case will finish the proof. The isometry σ now has the property stated in Lemma 249.1; consequently, V is an Artinian space Art_{4r} of dimension $n = 4r$ and σ is a rotation. Choose any symmetry τ of V. Then $\tau \sigma$ is a reflection and hence $\tau \sigma$ does not fall under Case 3. Therefore, $\tau \sigma = \tau_1 \cdots \tau_m$ where $\tau_1, \ldots, \tau_m$ are symmetries of Art_{4r} and $m \leq 4r$. If m were even, $\tau_1 \cdots \tau_m$ would be a rotation; thus $m \leq 4r - 1 = n - 1$. Finally, we have $\sigma = \tau \tau_1 \cdots \tau_m$. Done.

Cases 1 and 2 of the above proof are not mutually exclusive. For example, if V is anisotropic, it is clear that an isometry σ falls under Case 2 if and only if $\sigma \neq 1_V$. Then all isometries $\sigma \neq 1_V$ which have a nonzero fixed vector fall under both Cases 1 and 2. Even if V contains nonzero isotropic vectors, it is easy to give examples of isometries which belong to both Cases 1 and 2 (see Exercise 255.8). Figure 255.1 diagrams the three subsets of $O(V)$ corresponding to the three cases.

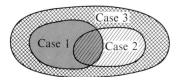

Figure 255.1

If an isometry is written as a product of symmetries, we often say that the isometry has been "factored" into the product of symmetries. One may ask how unique is the factorization of an isometry into a product of symmetries. Suppose that an isometry σ has been written as a product $\sigma = \tau_1 \cdots \tau_m$ of symmetries. If σ is a rotation, m is necessarily even and, if σ is a reflection, m is odd. Otherwise, there is little one can say about m. The Cartan–Dieudonné theorem states that m can be made less than or equal to n, but in many factorizations of σ, there will be more than n symmetries. Of course, the smallest possible m which can be used for a given σ is an invariant of σ, but it is not the kind of invariant which is much help in the analysis of the orthogonal group. Furthermore, the symmetries $\tau_1, \ldots, \tau_m$ themselves are by no means uniquely determined by σ, even when one keeps the m as small as possible (see Exercise 256.9).

Nevertheless, there is an important invariant connected with the factorization of isometries into symmetries. Let $B_1, \ldots, B_m$ be nonisotropic vectors of V such that $\sigma = \tau_{B_1} \cdots \tau_{B_m}$; we do not assume m is necessarily minimal for σ. It can be proved that the nonzero scalar $B_1{}^2 \cdots B_m{}^2$ is uniquely determined by σ modulo squares. In other words, the coset $B_1{}^2 \cdots B_m{}^2 k^{*2}$ of the group k^{*2} does not depend on how σ is written as a product of symmetries. Proofs of the above result involve the use of the "Clifford algebra" of V. Since we do not include a study of the Clifford algebra of V in this book, we refer the reader to Artin, [2], p. 186.

Exercises

*7. As always, $n = \dim V$.
    a. If n is odd, prove that a rotation of V can be written as a product of not more than $n - 1$ symmetries.
    b. If n is even, prove that a reflection of V can be written as the product of not more than $n - 1$ symmetries.

*8. Let U be a nonsingular subspace of V and put $\sigma = -1_U \oplus 1_{U^*}$. (This involution was studied in Exercise 223.20.) If $1 \leq \dim U < n$, prove that σ falls under both Cases 1 and 2 of the proof of the Cartan–Dieudonné theorem. Observe that this includes all symmetries.

*9. In order to show how highly nonunique is the factorization of iso-
metries into symmetries, we study the situation when V is a plane
$(n = 2)$.

a. Prove that every reflection is a symmetry. (Hint: Apply the Cartan–
Dieudonné theorem to a reflection.)

b. Prove that every rotation $\neq 1_V$ is the product of two and not less
than two symmetries. We recall that $1_V = \tau^0$ where τ is any
symmetry and hence 1_V is considered as the product of zero
symmetries.

c. Let ρ be a rotation and τ a symmetry of V. Since $O(V)$ is a group,
there exist unique isometries τ' and τ'' such that

$$\rho = \tau\tau' \quad \text{and} \quad \rho = \tau''\tau.$$

Prove that τ' and τ'' are symmetries. Observe that, consequently,
in the factorization $\rho = \tau_1\tau_2$ of ρ into a product of two symmetries,
one of the symmetries may be chosen at random. How nonunique
can a factorization be?

*10. Let V be the Euclidean plane. Assume that A_1, A_2 is an orthonormal
coordinate system of V and that the linear transformation $\sigma: V \to V$
has the matrix

$$\begin{pmatrix} \dfrac{1}{\sqrt{2}} & -\dfrac{1}{\sqrt{2}} \\ \dfrac{1}{\sqrt{2}} & \dfrac{1}{\sqrt{2}} \end{pmatrix}$$

relative to A_1, A_2.

a. Prove that σ is a rotation. (Warning: It is not sufficient to show
that $\det \sigma = 1$.)

b. Find two nonzero vectors

$$X = x_1 A_1 + x_2 A_2 \quad \text{and} \quad Y = y_1 A_1 + y_2 A_2$$

such that $\sigma = \tau_X \tau_Y$.

c. Use high school trigonometry to find the angle of the rotation σ.

11. Let V, A_1, and A_2 be as in the previous exercise. Assume that the
linear transformation $\sigma: V \to V$ has the matrix

$$\begin{pmatrix} \dfrac{1}{2} & \dfrac{\sqrt{3}}{2} \\ \dfrac{\sqrt{3}}{2} & -\dfrac{1}{2} \end{pmatrix}$$

relative to A_1, A_2.

    a. Prove that σ is a reflection.

    b. Find a nonzero vector $X = x_1 A_1 + x_2 A_2$ such that $\sigma = \tau_X$.

12. Assume that V is anisotropic. We discuss here how easy it is to prove the Cartan–Dieudonné theorem in this case. Case 1 should now be formulated as follows: "σ leaves a nonzero vector fixed." Observe that this includes 1_V. Case 2 should simply be "$\sigma \neq 1_V$." The difficult Case 3 does not occur. The proofs of Cases 1 and 2 cannot be simplified much; nevertheless, we urge the reader to write the simplest possible proof of the Cartan–Dieudonné theorem for anisotropic space.

Remark. Readers who are interested in high school teaching should go further and also write proofs of the Cartan–Dieudonné theorem for the Euclidean plane and Euclidean three-space, using only notions which can be developed in eleventh and twelfth grade courses in advanced geometry.

46. REFINEMENT OF THE CARTAN–DIEUDONNÉ THEOREM

Since an isometry of our nonsingular space V is the product of symmetries, it is natural to ask what is the minimum number of symmetries into which a given isometry can be factored. We know that this number is no larger than n, but what is it precisely? The question is completely answered by a very pretty theorem of Peter Scherk which can be regarded as a refinement of the Cartan–Dieudonné theorem. In this section, we prove the "nonsingular part" of Scherk's theorem, state the theorem completely and, along the way, discuss some further questions related to the Cartan–Dieudonné theorem.

The following exercises review some material from linear algebra which we shall need.

Exercises

In these exercises, $\mathscr{W}$ denotes an m-dimensional vector space without a metric over a field $\mathscr{k}$ which may have characteristic two. As always, a hyperplane of $\mathscr{W}$ is an $(m - 1)$-dimensional linear subspace of $\mathscr{W}$.

*1. If H is a hyperplane and U a linear subspace of $\mathscr{W}$, prove that either $U \subset H$ or $\dim(U \cap H) = (\dim U) - 1$.

2. If $H_1, \ldots, H_r$ are hyperplanes of $\mathscr{W}$, prove that

$$\dim(H_1 \cap \cdots \cap H_r) \geq m - r.$$

(Hint: Use Exercise 1 above and mathematical induction.)

In the remainder of this section, σ denotes an isometry of V with fixed space F. The dimension of the orthogonal complement F^* of F is denoted by s and hence $\dim F = n - s$.

Proposition 258.1.   The isometry σ cannot be factored into less than s symmetries.

Proof. Let $\sigma = \tau_1 \cdots \tau_r$, where τ_i is the symmetry of V with respect to the hyperplane H_i; we have to show that $r \geq s$. The vectors of H_i are left fixed by τ_i and hence $\tau_1 \cdots \tau_r$ leaves $H_1 \cap \cdots \cap H_r$ pointwise fixed. This is the same as saying that $H_1 \cap \cdots \cap H_r \subset F$ from which it follows that

$$\dim(H_1 \cap \cdots \cap H_r) \leq \dim F = n - s.$$

Since $\dim(H_1 \cap \cdots \cap H_r) \geq n - r$, we obtain the desired result $s \leq r$. Done.

The following corollary of the above proposition is useful.

Corollary 258.1.   If the isometry σ leaves no nonzero vector fixed, then σ is the product of n and not less than n symmetries.

Proof. According to the hypothesis, $F = \{0\}$ and hence $s = n$. By the above proposition, at least n symmetries are necessary to factor σ. The rest follows from the Cartan–Dieudonné theorem. Done.

The converse of Corollary 258.1 is false. There is one exceptional case in which an isometry has nonzero fixed vectors, but, nevertheless, cannot be factored into less than n symmetries (see Exercise 262.14). The fixed space of such an isometry must be a nonzero null space as Exercise 4 below shows.

Exercises

3. If σ has a nonisotropic fixed vector, prove that σ is the product of not more than $n - 1$ symmetries. (Hint: This is Case 1 of the proof of the Cartan–Dieudonné theorem.)

*4. If σ cannot be factored into less than n symmetries, prove that its fixed space is a null space.

*5. Assume V is anisotropic. Prove that the converse of Corollary 258.1 is now correct. In other words, show that σ leaves no nonzero vector fixed if and only if σ cannot be factored into less than n symmetries.

6. We study the factorization of -1_V into symmetries.

   a. Prove that -1_V cannot be factored into less than n symmetries.

   b. Let $A_1, \ldots, A_n$ be a rectangular coordinate system of V. Prove that

$$-1_V = \tau_{A_1} \cdots \tau_{A_n}.$$

(Hint: It is sufficient to show that $\tau_{A_1} \cdots \tau_{A_r} A_i = -A_i$ for $i = 1, \ldots, n$.) Figure 259.1 depicts the situation when $n = 2$. The

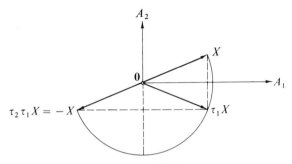

Figure 259.1

fact that every rectangular coordinate system of V gives rise to a factorization of -1_V into a minimal number of symmetries shows again how nonunique is the factorization of isometries.

*7. If $A, B \in V$, $A^2 \neq 0$, $B^2 \neq 0$, and $A \perp B$, prove that $\tau_A \tau_B = \tau_B \tau_A$. (Hint: Show that $P = \langle A, B \rangle$ is a nonsingular plane and study the factorization of -1_P.)

In Euclidean three-space E_3, we are accustomed to the fact that every rotation has an axis. This axis is the line around which E_3 is being rotated and the rotation is determined by its axis and the angle over which E_3 is rotated around the axis. This axis consists of the vectors left fixed by the rotation and hence every rotation of E_3 has nonzero fixed vectors. On the other hand, in the Euclidean plane, a rotation different from 1_V leaves only the origin (around which the plane is being rotated) fixed. These are special cases of the following rule.

Proposition 259.1. The isometry σ leaves a nonzero vector fixed in each of the following cases.

   1. σ is a rotation and n is odd.

   2. σ is a reflection and n is even.

Proof. In both cases, σ is the product of not more than $n - 1$ symmetries (Exercise 255.7). Hence, by Corollary 258.1, σ has a nonzero fixed vector. Done.

Proposition 258.1 states that it takes at least s symmetries to factor the isometry σ. Depending on σ, one can often do better and claim that σ is the product of exactly s symmetries. An example is the following result which is the "nonsingular part" of Scherk's theorem.

Proposition 260.1.   If F is nonsingular, σ is the product of s and not less than s symmetries.

Proof. There only remains to show that σ is the product of s symmetries. Since F is nonsingular,

$$V = F \oplus F^* \qquad \text{and} \qquad \sigma = 1_F \oplus \sigma'$$

where $\sigma' = \sigma | F^*$. We may apply the Cartan–Dieudonné theorem to the non-singular space F^* and conclude that there are nonisotropic vectors $B_1, \ldots, B_r$ in F^* such that $\sigma' = \tau'_{B_1} \cdots \tau'_{B_r}$. Here $r \leq s$ and τ'_{B_i} is the symmetry of F^* corresponding to the nonisotropic vector B_i. It follows that $\sigma = \tau_{B_1} \cdots \tau_{B_r}$ where τ_{B_i} is the symmetry of V corresponding to B_i (Exercise 252.5d). Since $s \leq r$, by Proposition 258.1, we conclude that $s = r$. Done.

Exercise

    8. Assume V is anisotropic.
        a. Prove that σ is the product of s and not less than s symmetries.
        b. Use Part a to give a second proof of the result in Exercise 258.5 which states that σ is the product of n and not less than n symmetries if and only if σ leaves no nonzero vector fixed.

Scherk's theorem states that Proposition 259.1 remains valid if F, and hence F^*, is singular as long as F^* is not a null space. However, if F^* is a null space and $\sigma \neq 1_V$, the minimum number of symmetries necessary to factor σ is $s + 2$. Isn't that beautiful?

Theorem 260.1 (Scherk's theorem).   If F^* is not a null space, σ is the product of s and not less than s symmetries. If F^* is a null space and $\sigma \neq 1_V$, σ is the product of $s + 2$ and not less than $s + 2$ symmetries.

The restriction $\sigma \neq 1_V$ is made because of our convention that 1_V is the product of zero symmetries. This convention was necessary in order to save the Cartan–Dieudonné theorem for $n = 1$.

For a proof of the above theorem, see P. Scherk, "On the Decomposition of Orthogonalities into Symmetries," *Proceedings of the American Mathematical Society*, Vol. 1, No. 4, August 1950, pp. 481–491, [20]. This paper is written in the language of matrices and the reader will benefit by translating the paper into the geometrical language of this book. Such a translation is by no means a mechanical exercise, but is a major effort which, however, pays off handsomely in mathematical insights, power, and beauty. The following exercises are designed to help the reader make the translation.

Exercises

*9. Let U be a subspace of V and assume U^* is a null space, equivalently, that $\dim U + \dim(\text{rad } U) = n$. If W is a subspace of V and $U \subset W \neq V$, prove that W is singular. (Hint: Take orthogonal complements.)

In the remaining exercises of this set, τ_i denotes the symmetry with respect to the nonsingular hyperplane H_i.

*10. If $\sigma = \tau_1 \cdots \tau_s$, prove that $F = H_1 \cap \cdots \cap H_s$.

*11. Assume F^* is a null space.
   a. Prove that σ is a rotation. (Hint: Use Exercise 237.9 and Lemma 233.1.)
   b. If $\sigma = \tau_1 \cdots \tau_r$, prove that r is even and that $s < r$. (Hint: If $s = r$, $F \subset H_i$ by Exercise 10 above; this contradicts Exercise 9 above.)

*12. If F^* is a null space, $F^* \subset F$. Assume now that F^* is a null space and $F^* = F$. Then V is an Artinian space Art_{2s} and σ is a rotation. These rotations were studied in Section 44. In particular, Proposition 246.1 states that s is even.
   a. If $\sigma = \tau_1 \cdots \tau_r$, prove that r is even and at least $s + 2$. (Hint: Use the previous exercise.)
   b. If $s = 2$, prove that σ is the product of four and not less than four symmetries.

Exercise 12 above shows that rotations of Art_4 which have a null plane as fixed space are the product of four and not less than four symmetries. Consequently, the converse of Corollary 258.1 is false. It is an easy consequence of Scherk's theorem that Art_4 is the only space in which Corollary 258.1 does not hold. Furthermore, the only isometries of Art_4 which leave nonzero vectors fixed, but cannot be factored into less than four symmetries, are precisely those rotations which have a null plane as fixed space. This is worked out in the following two exercises.

Exercises

*13. Let H be a hyperplane of V and α an isometry of V which leaves the vectors of H fixed.
   a. If H is nonsingular, prove that $\alpha = 1_V$ or α is a symmetry with respect to H. (Hint: $V = H \oplus H^*$.)
   b. If H is singular, prove that $\alpha = 1_V$. (Hint: rad H is a nonsingular line $\langle A \rangle$ and $H = W \oplus \langle A \rangle$ where W is nonsingular. Furthermore, $V = W \oplus P$ where P is an Artinian plane which contains $\langle A \rangle$ (Theorem 193.1). Now consider the restrictions $\alpha | W$ and $\alpha | P$.)

*14. Assume σ leaves a nonzero vector of V fixed, but cannot be factored into less than n symmetries. Use Scherk's theorem to prove that $V = Art_4$ and that σ is a rotation whose fixed space is a null plane. (Hint: V contains nonzero vectors and hence $n \geq 1$. Since 1_V is the product of zero symmetries, $\sigma \neq 1_V$. Furthermore, $F \neq \{0\}$, equivalently, $s < n$; therefore, σ can be factored into n but not s symmetries. It follows from Scherk's theorem that F^* is a null space and $s + 2 = n$. Because F^* is a null space, $s \leq n/2$. Conclude that $n \leq 4$. Finally, eliminate $n = 1, 2, 3$.)

47. INVOLUTIONS OF THE GENERAL LINEAR GROUP

We have two powerful theorems of metric geometry at our disposal, the Witt theorem and the Cartan–Dieudonné theorem. Before applying them to plane and three-dimensional geometry, it is convenient to have a geometric description of the involutions of the orthogonal group $O(V)$. We recall that an involution of a multiplicative group is an element of that group whose square is 1. Consequently, the involutions of $O(V)$ are the isometries σ with the property that $\sigma^2 = 1_V$. We prepare for their study in the present section by discussing involutions of $GL(n, k)$. The next section studies involutions of $O(V)$.

Assume $\mathscr{W}$ is an m-dimensional vector space over the field k where $\mathscr{W}$ is without metric. We are using k as our field since it is necessary that char $k \neq 2$. This is one of the few occasions when char $k \neq 2$ is needed in the theory of vector spaces without metric. It is needed here because we are going to investigate simultaneously the characteristic roots ± 1 of a linear transformation and for this it is necessary that $1 \neq -1$, equivalently, char $k \neq 2$.

Let $\sigma: \mathcal{W} \to \mathcal{W}$ be a linear transformation; σ may be singular. We shall study the space F of vectors left fixed by σ and the space I of vectors inverted by σ, that is,

$$F = \{X \mid X \in \mathcal{W}, \quad \sigma X = X\}$$
$$I = \{X \mid X \in \mathcal{W}, \quad \sigma X = -X\}.$$

The space F consists of those characteristic vectors of σ whose characteristic root is 1. The space I consists of the characteristic vectors of σ whose characteristic root is -1. The best way to study these roots is by investigating the linear subspaces F and I of $\mathcal{W}$. Finally, we recall that

$$F = \ker(\sigma - 1_{\mathcal{W}}) \quad \text{and} \quad I = \ker(\sigma + 1_{\mathcal{W}}).$$

Proposition 263.1. Let σ be a linear transformation of $\mathcal{W}$, F the fixed space of σ, and I the space of vectors inverted by σ. Then

1. $X = (\sigma - 1_{\mathcal{W}})(-\frac{1}{2}X) + (\sigma + 1_{\mathcal{W}})(\frac{1}{2}X)$ for all $X \in \mathcal{W}$.
2. $\mathcal{W} = \text{im}(\sigma - 1_{\mathcal{W}}) + \text{im}(\sigma + 1_{\mathcal{W}})$.
3. $F \subset \text{im}(\sigma + 1_{\mathcal{W}})$.
4. $I \subset \text{im}(\sigma - 1_{\mathcal{W}})$.
5. $F \cap I = \{\mathbf{0}\}$.

Proof. Statement 1 follows by clearing parentheses and Statement 2 is an immediate consequence of Statement 1. For Statement 3, we decompose a vector $X \in F$ according to Statement 1. Since

$$-\tfrac{1}{2}X \in F = \ker(\sigma - 1_{\mathcal{W}})$$

and hence

$$(\sigma - 1_{\mathcal{W}})(-\tfrac{1}{2}X) = \mathbf{0},$$

we conclude from Statement 1 that

$$X = (\sigma + 1_{\mathcal{W}})(\tfrac{1}{2}X) \in \text{im}(\sigma + 1_{\mathcal{W}}).$$

For Statement 4, we decompose a vector $X \in I$ according to Statement 1 and use the fact that

$$(\sigma + 1_{\mathcal{W}})(\tfrac{1}{2}X) = \mathbf{0}.$$

Finally, if $X \in F \cap I$, $\sigma X = X = -X$ and hence $X = \mathbf{0}$ since char $k \neq 2$. Done.

Observe that Statement 1 does not even make sense over a field of characteristic 2.

264 2. METRIC VECTOR SPACES

Exercises

1. Let $n = 2$ and A_1, A_2 be a coordinate system of the plane $\mathscr{W}$. Let the linear transformation $\sigma : \mathscr{W} \to \mathscr{W}$ be defined by

$$\sigma A_1 = A_1 + A_2 \qquad \text{and} \qquad \sigma A_2 = A_2 .$$

   Prove that
   a. The matrices of σ, $\sigma - 1_{\mathscr{W}}$, and $\sigma + 1_{\mathscr{W}}$ are, respectively,

$$\begin{pmatrix} 1 & 0 \\ 1 & 1 \end{pmatrix}, \quad \begin{pmatrix} 0 & 0 \\ 1 & 0 \end{pmatrix}, \quad \begin{pmatrix} 2 & 0 \\ 1 & 2 \end{pmatrix},$$

   relative to the coordinate system A_1, A_2.
   b. σ is nonsingular.
   c. $F = \langle A_2 \rangle$.
   d. $I = \{0\}$. (This uses that char $k \neq 2$.)
   e. $\mathrm{im}(\sigma - 1_{\mathscr{W}}) = \langle A_2 \rangle$.
   f. $\mathrm{im}(\sigma + 1_{\mathscr{W}}) = \mathscr{W}$. (This uses that char $k \neq 2$.)
   g. σ satisfies Conditions 2–5 of Proposition 263.1.
   h. $\mathscr{W}$ is not the direct sum of $\mathrm{im}(\sigma - 1_{\mathscr{W}})$ and $\mathrm{im}(\sigma + 1_{\mathscr{W}})$; also, prove that $F \neq \mathrm{im}(\sigma + 1_{\mathscr{W}})$ and $I \neq \mathrm{im}(\sigma - 1_{\mathscr{W}})$. Conclude that Conditions 2–4 of Proposition 263.1 cannot be "sharpened."
2. Assume W is an m-dimensional vector space over a field k of characteristic 2 and let $\sigma : W \to W$ be a linear transformation. Now Statement 1 of Proposition 263.1 does not make sense and we show that Statements 2–5 are false.
   a. Prove that $-1_W = 1_W$ and hence $\sigma - 1_W = \sigma + 1_W$.
   b. Choose $\sigma = 1_W$ and prove that for this σ, $F = I = W$ and

$$\mathrm{im}(\sigma - 1_W) = \mathrm{im}(\sigma + 1_W) = \{0\}.$$

   If $m \geq 1$, conclude that none of Statements 2–5 are true.

A linear transformation σ of $\mathscr{W}$ is called an involution if $\sigma^2 = 1_{\mathscr{W}}$. In this case, $(\det \sigma)^2 = 1$ and hence $\det \sigma = \pm 1$ which shows that σ is nonsingular and is an involution in the general linear group. Proposition 263.1 can be sharpened considerably when σ is an involution.

Proposition 264.1. Let σ be an involution of $\mathscr{W}$. Then

2′. $\mathscr{W} = F \oplus I$.
3′. $F = \mathrm{im}(\sigma + 1_{\mathscr{W}})$.
4′. $I = \mathrm{im}(\sigma - 1_{\mathscr{W}})$.

Proof. We first prove Statement 3'. Using Statement 3 of Proposition 263.1, there only remains to prove that $\text{im}(\sigma + 1_{\mathcal{W}}) \subset F$. If $Y \in \text{im}(\sigma + 1_{\mathcal{W}})$, $Y = \sigma X + X$ for some $X \in \mathcal{W}$ and hence

$$\sigma Y = \sigma^2 X + \sigma X = X + \sigma X = Y.$$

This shows that $Y \in F$.

Next, we prove Statement 4'. By applying Statement 4 of Proposition 263.1, there only remains to show that $\text{im}(\sigma - 1_{\mathcal{W}}) \subset I$. If $Y \in \text{im}(\sigma - 1_{\mathcal{W}})$, $Y = \sigma X - X$ for some $X \in \mathcal{W}$ and hence

$$\sigma Y = \sigma^2 X - \sigma X = X - \sigma X = - Y.$$

Thus $Y \in I$.

To prove Statement 2', we observe that Statement 2 of Proposition 263.1 combined with 3' and 4' give us that $\mathcal{W} = F + I$. But this sum is direct by Statement 5 of Proposition 263.1. Done.

We recall that a linear subspace U of $\mathcal{W}$ is called an invariant subspace of a linear transformation σ of $\mathcal{W}$ if $\sigma U \subset U$. One says "σ leaves U invariant" and this is, clearly, weaker than saying that every vector of U is fixed.

It is obvious that $1_{\mathcal{W}}$ and $-1_{\mathcal{W}}$ are involutions of $\mathcal{W}$ which leave all lines of $\mathcal{W}$ through $\mathbf{0}$ invariant. In order to show that they are the only involutions with this property, we study the linear transformations $r1_{\mathcal{W}}$ where $r \in k$. In the language of Chapter 1, if $r \neq 0$, $r1_{\mathcal{W}}$ is the magnification of $\mathcal{W}$ with center $\mathbf{0}$ and ratio r. Clearly, $r1_{\mathcal{W}}$ leaves all lines of $\mathcal{W}$ through $\mathbf{0}$ invariant.

Proposition 265.1. A linear transformation σ of $\mathcal{W}$ leaves all lines of $\mathcal{W}$ through $\mathbf{0}$ invariant if and only if $\sigma = r1_{\mathcal{W}}$ for some $r \in k$. This remains correct when the field has characteristic 2.

Proof. We have only to show that, if σ leaves all lines through $\mathbf{0}$ invariant, $\sigma = r1_{\mathcal{W}}$. If $m = 0$, there are no lines and $\sigma = 1_{\mathcal{W}}$; thus the proposition is "emptily" satisfied. Assume $m \geq 1$ and choose a nonzero vector $X \in \mathcal{W}$. Since σ leaves the line $\langle X \rangle$ invariant, $\sigma X = rX$ for some $r \in k$. We must prove that $\sigma Y = rY$ for all $Y \in \mathcal{W}$. Select $Y \in \mathcal{W}$.

Case 1. $Y \in \langle X \rangle$. Then $Y = aX$ for some $a \in k$ and

$$\sigma Y = \sigma(aX) = a(\sigma X) = a(rX) = r(aX) = rY.$$

Case 2. $Y \notin \langle X \rangle$. Since σ leaves the line $\langle Y \rangle$ invariant, $\sigma Y = bY$ for some $b \in k$; we must show that $b = r$. The line $\langle X + Y \rangle$ is also left invariant by σ, hence $\sigma(X + Y) = c(X + Y)$ for some $c \in k$. It follows that

$$c(X + Y) = \sigma X + \sigma Y = rX + bY,$$

equivalently,

$$(c - r)X + (c - b)Y = 0.$$

The vectors X and Y are linearly independent, therefore, $c = b = r$. Done.

Corollary 266.1.   An involution σ of $\mathscr{W}$ leaves all lines of $\mathscr{W}$ through $\mathbf{0}$ invariant if and only if $\sigma = \pm 1_{\mathscr{W}}$. This remains correct when the field has characteristic 2.

Proof. We only have to show that, if σ leaves all lines through $\mathbf{0}$ invariant, $\sigma = \pm 1_{\mathscr{W}}$. If $m = 0$, the corollary is emptily satisfied. If $m \geq 1$, choose a non-zero vector X in $\mathscr{W}$. By the previous proposition, $\sigma = r 1_{\mathscr{W}}$ and hence

$$\sigma^2 X = \sigma(\sigma X) = \sigma(rX) = r(\sigma X) = r^2 X.$$

Since σ is an involution, $\sigma^2 X = X$; we conclude from $r^2 X = X$ and $X \neq \mathbf{0}$ that $r^2 = 1$, equivalently, $r = \pm 1$. Done.

Exercise

3. Prove that a linear transformation σ of $\mathscr{W}$ leaves all lines of $\mathscr{W}$ through $\mathbf{0}$ invariant if and only if σ leaves all linear subspaces of $\mathscr{W}$ invariant. (Hint: Either use Proposition 265.1 or prove it directly.)

48. INVOLUTIONS OF THE ORTHOGONAL GROUP

Each nonsingular subspace U of V gives rise to the "orthogonal splitting" $V = U \oplus U^*$ and the isometry

$$\sigma = -1_U \oplus 1_{U^*}.$$

It was proved in Exercise 223.20 that σ is an involution, U^* is the space of vectors left fixed by σ, and U is the space of vectors inverted by σ. We now prove that every involution of the orthogonal group is obtained from an orthogonal splitting of V.

Lemma 266.1.   Assume σ is an isometry of V which is an involution. Let F denote the space of vectors left fixed by σ and I the space of vectors inverted by σ. Then

1. F and I are nonsingular and orthogonal complements of each other.
2. $\sigma = -1_I \oplus 1_F$.

Proof. Clearly, $1 \Rightarrow 2$ and, therefore, one only has to prove that $V = I \oplus F$. Proposition 264.1 states that $V = I \oplus F$ and $I = \text{im}(\sigma - 1_V)$. Since $F = \ker(\sigma - 1_V)$, it follows from Proposition 232.1 that $V = I \oplus F$. Done.

The fact that all involutions of $O(V)$ come from orthogonal splittings of V is a good example of what Artin calls "geometric algebra." The concept of an involution is strictly algebraic and holds for all groups. The pretty geometric characterization of involutory isometries, in terms of orthogonal splittings of V, is what makes this notion attractive and manageable for geometric investigations.

If an involution σ of the general linear group leaves all lines of V through **0** invariant, $\sigma = \pm 1_V$. The next proposition may be interpreted as saying that, if an isometry leaves all lines of V through **0** invariant, it is an involution. This is true even if the metric vector space is singular as long as it is not the null space.

Proposition 267.1. Let W be a metric vector space over k which is not a null space, but may be singular. An isometry σ of W leaves all lines of W through **0** invariant if and only if $\sigma = \pm 1_W$.

Proof. We only have to show that an isometry σ which leaves all lines through **0** invariant is equal to $\pm 1_W$. By Proposition 265.1, $\sigma = r1_W$ for some $r \in k$. If $X \in W$, $\sigma X = rX$ and $X^2 = (\sigma X)^2 = r^2 X^2$. If we choose a nonisotropic vector for X, it follows immediately that $r = \pm 1$. Done.

To each orthogonal splitting $V = U \oplus U^*$, there correspond the two involutions

$$\sigma = -1_U \oplus 1_{U^*} \quad \text{and} \quad \rho = 1_U \oplus -1_{U^*}.$$

Involutions are classified by the dimension of the space of inverted vectors.

Definition 267.1. If U is a nonsingular subspace of V, $\dim U$ is called the **type** of the involution $-1_U \oplus 1_{U^*}$.

There are $n + 1$ types of involutions. The only involution of type 0 is 1_V while the only involution of type n is -1_V. An involution $-1_U \oplus 1_{U^*}$ of type 1 is the symmetry with respect to the nonsingular hyperplane U^*.

Exercises

1. Let σ be an involution of type t. Prove that σ is a rotation if t is even and a reflection if t is odd.
2. Let σ be an involution of type 2. Prove that
   a. The space of vectors inverted by σ is a nonsingular plane.
   b. $\sigma | U$ is the 180° rotation of U.

By Exercise 2 above, an involution of type 2 is nothing but the 180° rotation of a nonsingular plane U extended to V by the identity on U^*. In other words, such an involution rotates V 180° around the $(n - 2)$-dimensional axis U^* (see Figure 268.1 where $n = 3$).

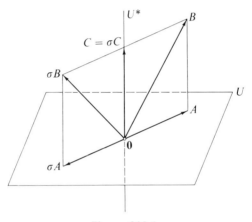

Figure 268.1

The following definition is now entirely natural.

Definition 268.1.   An involution of type 2 is called a **180° rotation.**

This definition agrees with our previous definition of 180° rotation of the plane. Namely, when $n = 2$, -1_V is the only involution of type 2 and hence the only 180° rotation.

Exercise

*3. Give an example of each of the following kinds of isometries.
   a. An involution which is not a symmetry but is a reflection. Prove that n must be at least 3.

b. An involution which is not a reflection. Prove that n may be 1.
c. An involution which is not a rotation. Prove that n may be 1.
d. A rotation which is not an involution. (Hint: The Euclidean plane is full of them.)
e. A reflection which is not an involution. Prove that n must be at least 3. (Hint: Consider an orthogonal splitting of Euclidean three-space $E_3 = U \oplus U^*$ where U is a line and U^* is a plane. Now use Part d above.)

49. ROTATIONS AND REFLECTIONS IN THE PLANE

How much do our fancy theorems tell us about plane geometry? Which theorems of Euclidean plane geometry hold for all planes? Sections 49–51 discuss the general theory of nonsingular planes while Sections 52–54 deal with some special questions of plane geometry.

We assume in this whole section that $n = 2$.

Thus V is a nonsingular plane over k and the hyperplanes of V are the lines through **0**. There is only one involution of type 0, namely, 1_V, and -1_V is the only involution of type 2; both of these are rotations. All other involutions have type 1 and are hence symmetries with respect to nonsingular lines. The following exercises will be used frequently.

Exercises

*1. Let σ be an isometry of V and A a nonzero vector such that $\sigma A = A$.
   a. If $A^2 \neq 0$, prove that $\sigma = 1_V$ or σ is the symmetry with respect to the nonsingular line $\langle A \rangle$. (Hint: Use Exercise 262.13.)
   b. If $A^2 = 0$, prove that $\sigma = 1_V$.
*2. Let σ be an isometry of V and A a nonzero vector such that $\sigma A = -A$.
   a. If $A^2 \neq 0$, prove that $\sigma = -1_V$ or σ is the symmetry with respect to the nonsingular line $\langle A \rangle^*$. (Hint: Consider the isometry $-\sigma = -1_V\sigma$.)
   b. If $A^2 = 0$, prove that $\sigma = -1_V$.
*3. Let ρ and σ be isometries of V and A a nonzero vector such that $\rho A = \sigma A$.
   a. If ρ and σ are both rotations or both reflections, prove that $\rho = \sigma$.
   b. If one of the isometries ρ and σ is a rotation and the other is a reflection, prove that $A^2 \neq 0$ and $\rho^{-1}\sigma$ is the symmetry with respect to $\langle A \rangle$. Also prove that, in this case, $\rho^{-1}\sigma = \sigma^{-1}\rho$.

It is possible to give special descriptions of rotations and reflections of the plane. These descriptions agree with our knowledge of rotations and reflections of the Euclidean plane.

Proposition 270.1 (plane reflections). If σ is an isometry of V, the following three statements are equivalent.

1. σ is a reflection.
2. σ is a symmetry.
3. $\sigma \neq 1_V$ and σ leaves a nonzero vector fixed.

Furthermore, a reflection is completely determined by its action on one nonzero vector.

Proof. Exercise 256.9a states that $1 \Rightarrow 2$; since a symmetry is a reflection, Statements 1 and 2 are equivalent. It is obvious that $2 \Rightarrow 3$ and it follows from Exercise 269.1 that $3 \Rightarrow 2$. Finally, if $\rho A = \sigma A$ for a nonzero vector A, $\rho = \sigma$ by Exercise 3a above. Done.

Proposition 270.2 (plane rotations). If σ is an isometry of V, the following three statements are equivalent.

1. σ is a rotation.
2. σ is the product of two symmetries.
3. Either $\sigma = 1_V$ or σ leaves no nonzero vector fixed.

Furthermore, a rotation is completely determined by its action on one nonzero vector.

Proof. Exercise 256.9b states that $1 \Rightarrow 2$ and hence Statements 1 and 2 are equivalent. The equivalence of Statements 1 and 3 follows from the fact that Statement 1 is the negation of the first statement of Proposition 270.1 while Statement 3 is the negation of the third statement of that proposition. Finally, if ρ and σ are rotations such that $\rho A = \sigma A$ for a nonzero vector A, then $\rho = \sigma$ (Exercise 269.3). Done.

If ρ is a rotation and τ a symmetry of V, there are symmetries τ' and τ'' such that $\rho = \tau\tau' = \tau''\tau$ (Exercise 256.9c). Thus, in the factorization of a rotation into the product of two symmetries, one symmetry may be chosen at random.

The reader should be careful to interpret the last sentences of Propositions 270.1 and 270.2 correctly. An isometry σ of V is not completely determined by the image σA of one nonzero vector A. In order to retrieve the isometry from the knowledge of A and σA, one also has to know whether σ is a rotation or a reflection. This fact is illustrated in the following exercise.

Exercise

*4. Let A and B be two vectors of V such that $A^2 = B^2 \neq 0$. Prove that there exists a unique rotation ρ and a unique symmetry τ of V such that $\rho A = \tau A = B$. (Hint: Use the refinement of the Witt theorem.)

50. THE PLANE ROTATION GROUP

We assume throughout this section that $n = 2$. This section is a study of the rotation group $O^+(V)$ of the nonsingular plane V. If V is the Euclidean plane, high school geometry alone tells us the group $O^+(V)$ is commutative. Namely, one associates to every real number θ the rotation $w(\theta)$ of the plane "over θ degrees"; then one shows that, if θ_1 and θ_2 are real numbers,

$$w(\theta_1 + \theta_2) = w(\theta_1)w(\theta_2).$$

Consequently, one has constructed a group homomorphism $\theta \to w(\theta)$ from the additive group $\mathbf{R}^+$ of real numbers onto the multiplicative rotation group of the Euclidean plane. Under a homomorphism, the image of a commutative group is always a commutative group; hence the rotation group of the Euclidean plane is commutative. (The function $\theta \to w(\theta)$ is basically the winding function of trigonometry. See Remark on Teaching Trigonometry, page 309.)

For an arbitrary plane, there is no group homomorphism from the additive group of k onto the rotation group $O^+(V)$ (see Exercise 273.2). This is the main reason why there is no satisfactory way to measure rotations of an arbitrary plane by means of degrees or similar units. Nevertheless, we shall prove that $O^+(V)$ is always commutative; the following exercise is used in the proof of this result.

Exercise

*1. Let G be a multiplicative group.
  a. Prove that the mapping $\sigma \to \sigma^{-1}$ where $\sigma \in G$ is an antiautomorphism of G. (A mapping $f: G \to G$ is called an **antiautomorphism** if f is one-to-one, onto, and $f(\rho\sigma) = f(\rho)f(\sigma)$.)

   b. Prove that G is commutative if and only if the antiautomorphism
      $\sigma \to \sigma^{-1}$ is an automorphism.
   c. Prove that an element $\alpha \in G$ is an involution if and only if $\alpha = \alpha^{-1}$.
   d. If $\alpha_1, \ldots, \alpha_r$ are involutions of G, prove that

$$(\alpha_1 \cdots \alpha_r)^{-1} = \alpha_r \cdots \alpha_1.$$

Theorem 272.1.   The plane rotation group $O^+(V)$ is commutative.

Proof. Choose a reflection τ of V. Since $O^+(V)$ is a normal subgroup of $O(V)$, the mapping

$$\rho \to \tau \rho \tau^{-1}$$

where $\rho \in O^+(V)$ is an automorphism of $O^+(V)$. Thus far, everything holds for arbitrary dimension; we now use the fact that $n = 2$ to show that $\tau \rho \tau^{-1} = \rho^{-1}$.

   In the plane, the reflection τ is a symmetry and hence there is a symmetry τ' such that $\rho = \tau \tau'$. We conclude from Exercise 1 above that

$$\tau \rho \tau^{-1} = \tau \rho \tau = \tau(\tau \tau')\tau = \tau' \tau = \rho^{-1}.$$

Consequently, the antiautomorphism $\rho \to \rho^{-1}$ of $O^+(V)$ is the automorphism $\rho \to \tau \rho \tau^{-1}$; it follows from Exercise 1b above that $O^+(V)$ is commutative. Done.

   It is very important for geometry that the plane rotation group is commutative. For example, there would have been no trigonometry if $O^+(V)$ were not commutative. When studying Section 53, Plane Trigonometry, the reader should take notice whenever the commutativity of $O^+(V)$ is used. For instance, it will be important in the proofs of Propositions 291.1 and 298.1.

   Another good result about the plane rotation group states that all its finite subgroups are cyclic. In order to obtain this result, we first derive the structure of the rotation group of the Artinian plane.

Lemma 272.1.   The rotation group of the Artinian plane is isomorphic to the multiplicative group $k*$.

Proof. Let M, N be an Artinian coordinate system of the Artinian plane Art_2. Also, choose an isometry σ of Art_2.

   Since σM is a nonzero isotropic vector, either $\sigma M = aM$ or $\sigma M = aN$ for some $a \in k*$. For the same reasons, $\sigma N = bM$ or $\sigma N = bN$ for some $b \in k*$.

Case 1. $\sigma M = aN$. Since σM and σN are linearly independent, it must be that $\sigma N = bM$. Furthermore,

$$1 = MN = (\sigma M)(\sigma N) = (aN)(bM) = abMN = ab,$$

equivalently, $a = b^{-1}$. Therefore, the matrix of σ relative to the coordinate system M, N is

$$\begin{pmatrix} 0 & b \\ b^{-1} & 0 \end{pmatrix}$$

and σ is a reflection.

Case 2. $\sigma M = aM$. Again, because σM and σN are linearly independent, $\sigma N = bN$. In this case,

$$1 = MN = (\sigma M)(\sigma N) = (aM)(bN) = abMN = ab,$$

equivalently, $b = a^{-1}$. Thus the matrix of σ relative to the coordinate system M, N is

$$\begin{pmatrix} a & 0 \\ 0 & a^{-1} \end{pmatrix}$$

and σ is a rotation.

We see from Cases 1 and 2 that the matrices

$$\begin{pmatrix} a & 0 \\ 0 & a^{-1} \end{pmatrix}$$

where $a \in k^*$ represent all the rotations of Art_2. The isomorphism

$$a \rightarrow \begin{pmatrix} a & 0 \\ 0 & a^{-1} \end{pmatrix}$$

from the multiplicative group k^* onto the multiplicative group of these matrices may, therefore, be interpreted as an isomorphism from k^* onto $O^+(Art_2)$. Done.

Exercise

*2. Let k be a finite field with q elements.
   a. Prove that $O^+(Art_2)$ has $q - 1$ elements.
   b. Prove that there exists no homomorphism from the additive group k^+ onto $O^+(Art_2)$. (Hint: $q - 1$ does not divide q.)

Let k' be a field which contains k. The proof of the next theorem is based on the fact that the metric vector space V over k can be extended to a metric vector space V' over k'. This was explained on page 211 in the research idea of Artin; we repeat the construction of V' here and go just as far as needed for the proof of Theorem 275.1.

Let A_1, A_2 be a coordinate system of V and denote by V' the vector space over k' which has A_1, A_2 as coordinate system. In other words, V' consists of the vectors $a_1'A_1 + a_2'A_2$ where a_1', $a_2' \in k'$. Observe that V and V' have the same dimension, two, but, if $k \neq k'$, V' contains more vectors than V.

The metric of V over k is given by a 2×2 symmetric matrix G relative to the coordinate system A_1, A_2 of V. The metric of V' over k' is *defined* by the same matrix G relative to the coordinate system A_1, A_2 of V'. Since $\det G \neq 0$, the metric of V' is nonsingular.

If B_1, B_2 is another coordinate system of V over k, the metric of V is given, relative to B_1, B_2, by the matrix $P^{\mathsf{T}}GP$ where P is a nonsingular 2×2 matrix with entries in k. Consequently, if the coordinate system B_1, B_2 had been used to construct the metric vector space V' over k', the metric of V' would have been defined by the matrix $P^{\mathsf{T}}GP$ relative to the coordinate system B_1, B_2 of V'. Since P has entries in k and hence in k', it follows that the metric of V' does not depend on the choice of the coordinate system of V.

Let σ be an isometry of V whose matrix relative to A_1, A_2 is D. We define the isometry σ' of V' as the linear automorphism of V' whose matrix relative to the coordinate system A_1, A_2 of V' is the same matrix D. This says simply that σ' acts the same way on the coordinate vectors A_1, A_2 of V' as σ, that is,

$$\sigma'A_1 = \sigma A_1 \quad \text{and} \quad \sigma'A_2 = \sigma A_2 .$$

Because σ is an isometry of V, we conclude that

$$(\sigma'A_1)^2 = A_1{}^2, \quad (\sigma'A_2)^2 = A_2{}^2, \quad (\sigma'A_1)(\sigma'A_2) = A_1A_2 .$$

These last equalities show that σ' is an isometry of V'.

The construction of the metric vector space V' and the isometry σ' can be carried out without the use of coordinate systems by using "tensor products" instead. This coordinate free construction of V' and σ' becomes an absolute necessity in more advanced usages of the "extended" metric vector space V'.

Exercise

3. Prove that
   a. The mapping $\sigma \to \sigma'$ is a homomorphism from $O(V)$ into $O(V')$ which is one-to-one.
   b. σ is a reflection (rotation) if and only if σ' is a reflection (rotation).
   c. The mapping $\sigma \to \sigma'$ when restricted to $O^+(V)$ is a homomorphism from $O^+(V)$ into $O^+(V')$ which is one-to-one.

The above exercise shows that $O(V)$ may be regarded as a subgroup of $O(V')$ and $O^+(V)$ may be regarded as a subgroup of $O^+(V')$. By choosing the field k' appropriately, one can sometimes use this observation to draw conclusions about the groups $O(V)$ and $O^+(V)$ from the "simpler" groups $O(V')$ and $O^+(V')$. We do precisely this in the proof of the next theorem.

Exercise

4. Prove that the whole "extension" theory of V' and σ' (page 273 to the present exercise) holds for arbitrary dimension n.

Theorem 275.1.   Every finite subgroup of the plane rotation group $O^+(V)$ is cyclic. Furthermore, $O^+(V)$ contains at most one cyclic subgroup of a given finite order.

Proof. Let k' be the algebraic closure of k. (As always, readers who are not familiar with algebraically closed fields should assume that $k = \mathbf{R}$ and hence $k' = \mathbf{C}$.) Then the metric vector space V' over k' is the Artinian plane Art_2 (Exercise 191.15). The rotation group $O^+(Art_2)$ is isomorphic to the multiplicative group k'^* and, therefore, we may regard the rotation group $O^+(V)$ as a subgroup of k'^*. It follows that every finite subgroup of $O^+(V)$ is a finite subgroup of k'^*. Finally, the multiplicative group of a field contains no finite subgroups except cyclic groups and, at most, one cyclic group of a given finite order (see Artin, [1], p. 49). Done.

We have not penetrated very deeply into the structure of the orthogonal group and the rotation group of the plane. For example, we have not shown that the rotation group $O^+(E_2)$ of the Euclidean plane E_2 contains a cyclic subgroup of order s for every positive integer s. Even so, we know from high school geometry that this is true. The rotation ρ of E_2 over $360/s$ degrees generates the cyclic subgroup

$$\{1_{E_2}, \rho, \rho^2, \dots, \rho^{s-1}\}$$

of order s of $O^+(E_2)$; see Figure 276.1 where $s = 6$ and ρ is the counterclockwise rotation of $60°$ with $\mathbf{0}$ as center of rotation. An elementary proof of this property of $O^+(E_2)$ is the content of Exercise 308.30e. Exercise 6 below shows that, in general, $O^+(V)$ does not contain a cyclic subgroup of order s for every positive integer s. The question as to which plane rotation groups do have this property can only be tackled with the help of Clifford algebras, a

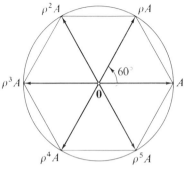

Figure 276.1

subject beyond the scope of this book. A good source for the study of Clifford algebras is Artin, [2].

Exercises

5. In this exercise, we expect the reader to use the fact that finite subgroups of the multiplicative group of a field are cyclic.
    a. Prove that $\{1\}$ and $\{1, -1\}$ are the only finite subgroups of $\mathbf{Q}^*$ or $\mathbf{R}^*$.
    b. Prove that $\mathbf{C}^*$ contains a cyclic subgroup of order s for every positive integer s.
    c. If k is a finite field with q elements, prove that k^* is a cyclic group with $q - 1$ elements. Consequently, we know from modern algebra that k^* contains a cyclic subgroup of order s if and only if s divides $q - 1$.
6. a. If V is the Artinian plane over $\mathbf{Q}$, prove that $O^+(V)$ contains only the two finite subgroups $\{1_V\}$ and $\{1_V, -1_V\}$. Conclude that, in particular, when V is the Lorentz plane $O^+(V)$ contains only the two finite subgroups $\{1_V\}$ and $\{1_V, -1_V\}$.
    b. If V is the Artinian plane over $\mathbf{C}$, prove that $O^+(V)$ contains a cyclic subgroup of order s for every positive integer s.
    c. If V is the Artinian plane over a finite field with q elements, prove that $O^+(V)$ is a cyclic group with $q - 1$ elements and hence contains a cyclic subgroup of order s if and only if s divides $q - 1$.
*7. Let ρ be a rotation of V where $\rho \neq \pm 1_V$.
    a. Prove that ρ leaves no nonzero vector fixed and inverts no nonzero vector. Conclude that ρ leaves no nonsingular line invariant.
    b. If V is anisotropic, prove that ρ has no characteristic root in k, that is, there is no $t \in k$ such that $\rho X = tX$ for a nonzero vector X. Consequently, the two characteristic roots of ρ belong to the algebraic closure of k, but not to k.

8. This exercise is meant for readers who have studied the research idea of Artin on pages 211–213. Let V be anisotropic and ρ a rotation of V where $\rho \neq \pm 1_V$. The two characteristic roots of ρ belong to the algebraic closure k' of k but not to k. If $\theta \in k'$ is a characteristic root of ρ, prove that the extension field $k(\theta)$ of k is the (smallest possible) splitting field of V (Example 213.2). (Hint: The fact that $k(\theta)$ is a splitting field of V follows from $\theta \neq \pm 1$ (page 212).)

9. Use the method of extending the field as done in Theorem 275.1 to give a second proof that the plane rotation group $O^+(V)$ is commutative.

10. Assume that V has an orthonormal coordinate system A_1, A_2. Let σ be an isometry of V whose matrix relative to A_1, A_2 is $\begin{pmatrix} a & b \\ c & d \end{pmatrix}$.

   a. If σ is a rotation, prove that $d = a, b = -c$, and $a^2 + c^2 = 1$. (Hint: $\sigma A_1 = aA_1 + cA_2$ and $\sigma A_2 = bA_1 + dA_2$. There are, basically, two ways to proceed. One is to express the given data in terms of the four scalars a, b, c, d and to solve the resulting four quadratic equations. The other is to observe that $A_1(\sigma A_1) = a$ and $A_2(\sigma A_2) = d$. In order to show that $d = a$, use the facts that there is a rotation ρ such that $A_2 = \rho A_1$ and the plane rotation group is commutative. Use the relation $(\sigma A_1)(\sigma A_2) = 0$ to show that $b = -c$.)

   b. If σ is a reflection, prove that $d = -a, b = c$, and $a^2 + c^2 = 1$. (Hint: Either use again the appropriate four quadratic equations or use the fact that the isometry defined by $A_1 \to A_2$ and $A_2 \to A_1$ is a reflection and hence the matrix

$$\begin{pmatrix} a & b \\ c & d \end{pmatrix}\begin{pmatrix} 0 & 1 \\ 1 & 0 \end{pmatrix} = \begin{pmatrix} b & a \\ d & c \end{pmatrix}$$

   is the matrix of a rotation.)

   c. If σ is a reflection, find scalars u and v such that the vector $A = uA_1 + vA_2$ is not isotropic and σ is the symmetry τ_A.

*11. Assume that V has a rectangular coordinate system A_1, A_2 where $A_1{}^2 = A_2{}^2 \neq 0$. Let σ again be an isometry of V whose matrix relative to A_1, A_2 is $\begin{pmatrix} a & b \\ c & d \end{pmatrix}$. Prove that the results of Parts a and b of the previous exercise remain correct.

51. THE PLANE ORTHOGONAL GROUP

We assume in this whole section that $n = 2$.

The fact that the rotation group of a nonsingular plane is always commutative presents a somewhat exceptional situation. It is much more common for an interesting group in geometry to be noncommutative. Our experience with high school mathematics leads us to believe that commutativity is more natural than noncommutativity because high school algebra is limited to strictly commutative parts of algebra such as elementary arithmetic and the theory of polynomials. It is a great pity that high school algebra shies away from the many noncommutative groups which abound in elementary geometry, thereby barring the student from much simple and beautiful geometry and giving him the false impression that commutativity is more natural than noncommutativity.

We now turn to the orthogonal group $O(V)$ of the nonsingular plane V. The following reasoning leads one to conjecture that $O(V)$ is never commutative; however, this conjecture is false, but only for one particular geometry.

Assume that A and B are linearly independent vectors of V where $A^2 = B^2 \neq 0$. By the Witt theorem, there is an isometry σ of V such that $\sigma A = B$. Let τ denote the symmetry with respect to the nonsingular line $\langle A \rangle$. Then $\tau A = A$ and, since $B \notin \langle A \rangle$, $\tau B \neq B$. We conclude that

$$\sigma \tau A = \sigma A = B$$
$$\tau \sigma A = \tau B \neq B.$$

Since $\sigma \tau$ and $\tau \sigma$ act differently on A, $\sigma \tau \neq \tau \sigma$ and the group $O(V)$ is not commutative. See Figure 278.1 where V is the Euclidean plane, A and B form an orthonormal coordinate system, σ is the counterclockwise rotation over $90°$ and τ is the symmetry with respect to the "x axis" $\langle A \rangle$.

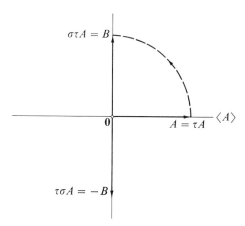

Figure 278.1

The above reasoning shows that, whenever V contains two linearly independent vectors with the same nonzero square, the group $O(V)$ is not commutative. It is tempting to conjecture that every nonsingular plane contains a pair of vectors of this kind. However, there is one plane for which this conjecture is false, namely, the Artinian plane over $\mathbf{Z}_3$. We refer to the Artinian plane over $\mathbf{Z}_3$ as the **exceptional plane** and investigate it in the following exercises.

Exercises

*1. Let k be a finite field with q elements.
   a. Prove that V has q^2 vectors.
   b. Prove that V has $q + 1$ lines through $\mathbf{0}$.
   c. Prove that each line through $\mathbf{0}$ has q points (vectors).
   d. If V is the Artinian plane, prove that V has $q - 1$ nonsingular lines through $\mathbf{0}$.
   e. If V is the Artinian plane, prove that V has $q - 1$ symmetries and $q - 1$ rotations.
   f. If V is the Artinian plane, prove that the order of $O^+(V)$ is $q - 1$ and the order of $O(V)$ is $2(q - 1)$.
   g. If V is not the Artinian plane, prove that the order of $O^+(V)$ is $q + 1$ and the order of $O(V)$ is $2(q + 1)$.

2. Let $k = \mathbf{Z}_3$ and denote the elements by $-1, 0, 1$. This gives away the addition and multiplication table of k, except that we must remember that $1 + 1 = -1$ and $(-1) + (-1) = 1$. If A_1, A_2 is a rectangular coordinate system of V, the matrix of V can only be one of the following four:

$$G_1 = \begin{pmatrix} 1 & 0 \\ 0 & 1 \end{pmatrix}, \qquad G_2 = \begin{pmatrix} -1 & 0 \\ 0 & -1 \end{pmatrix},$$

$$G_3 = \begin{pmatrix} 1 & 0 \\ 0 & -1 \end{pmatrix}, \qquad G_4 = \begin{pmatrix} -1 & 0 \\ 0 & 1 \end{pmatrix}.$$

Prove that
   a. The planes defined by G_3 and G_4 are isometric.
   b. The planes defined by G_1 and G_4 are isometric. (Hint: Consider the new rectangular coordinate system $A_1 + A_2$, $A_1 - A_2$.)
   c. Within isometry, there are precisely two nonisometric planes over $\mathbf{Z}_3$, one anisotropic plane, and the exceptional plane. It can be proved that there are only two planes over any finite field, one anisotropic plane and the Artinian plane; see Artin, [2], page 143.

3. Assume $k = \mathbf{Z}_3$. We know from Exercise 1 above that V has nine vectors, four lines through $\mathbf{0}$, and each line contains three points. If we again denote the elements of $\mathbf{Z}_3$ by $-1, 0, 1$, a convenient picture of V is given in Figure 280.1.

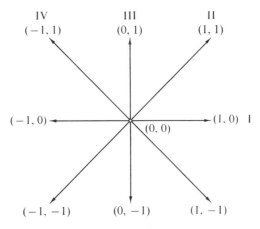

Figure 280.1

a. Prove that the nine points of V are distributed over the four lines I, II, III, IV as indicated in Figure 280.1. This must be proved and cannot be concluded from the fact that it "looks right" from the Euclidean picture. Part b will show the danger in drawing conclusions about finite planes from Euclidean pictures.

b. Suppose that the elements of $\mathbf{Z}_3$ had been denoted by 0, 1, 2. This would have been perfectly natural since, in $\mathbf{Z}_3$, $1 + 1 = 2\,(= -1)$. The most natural way to picture V would then have been as in Figure 280.2. Check that lines I, II, and III of Figures 280.1 and 280.2 are the same. The moral of the picture lies in the fact that line IV now consists of the vertices of the shaded triangle.

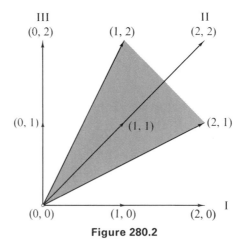

Figure 280.2

*4. Let V be the exceptional plane. We know from Exercise 1 above that V has two null lines, two nonsingular lines, two symmetries, and two rotations. Prove that

a. $O^+(V) = \{1_V, -1_V\}$ and hence $O^+(V)$ consists of two involutions.

b. $O(V)$ contains four elements, all of which are involutions.

c. $O(V)$ is commutative and, in fact, is the Klein four group. We remind the reader that there are only two groups with four elements, the cyclic group with four elements and the Klein four group. Both groups are commutative and, of these, only the Klein four group consists of involutions only.

*5. Let V be the exceptional plane and A_1, A_2 a rectangular coordinate system of V relative to which the matrix of V is $\left(\begin{smallmatrix} 1 & 0 \\ 0 & -1 \end{smallmatrix}\right)$.

a. Prove that the vectors

$$M = A_1 + A_2 \qquad \text{and} \qquad N = -A_1 + A_2$$

form an Artinian coordinate system of V. Check that Figure 280.1 becomes Figure 281.1.

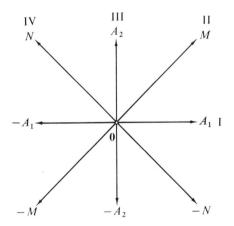

Figure 281.1

b. Since $O(V)$ is commutative, V cannot contain two linearly independent vectors with the same nonzero square. The square of a nonisotropic vector can only be ± 1. How many vectors of square 1 are there? On which of the four lines I–IV of Figure 281.1 do they lie? Answer the same questions about the vectors with square -1.

c. $O(V)$ consists of the four isometries 1_V, -1_V, τ_{A_1}, and τ_{A_2}. Give the image of each of the nine vectors

$$\pm A_1, \quad \pm A_2, \quad \pm M, \quad \pm N, \quad \mathbf{0}$$

of V under each of the four isometries. Visualize these isometries by using Figure 281.1.

We now show that $O(V)$ is noncommutative for all cases other than when V is the exceptional plane.

Theorem 282.1.   The following three statements are equivalent.

1. $O(V)$ is commutative.
2. V is the exceptional plane.
3. $O^+(V)$ consists of involutions only.

Proof. We shall prove that $1 \Rightarrow 3 \Rightarrow 2 \Rightarrow 1$.

Assume Statement 1 holds and let $\rho \in O^+(V)$. If τ is a symmetry, then $\tau\rho\tau^{-1} = \rho^{-1}$ (see the proof of Theorem 272.1). Since $O(V)$ is commutative, it follows that $\rho = \rho^{-1}$, equivalently, $\rho^2 = 1_V$. Thus ρ is an involution.

Assume Statement 3 holds. In Section 49, we saw that the only involutions which are rotations are $\pm 1_V$; hence $O^+(V) = \{1_V, -1_V\}$. Consequently, there are precisely two symmetries and hence two nonsingular lines. If these were all the lines of V, k would consist of one element (Exercise 279.1b) which is impossible. Thus V also has two null lines and must be an Artinian plane. Since the total number of lines through $\mathbf{0}$ is four, k has three elements, and, therefore, V is the exceptional plane.

Finally, the fact that $2 \Rightarrow 1$ was shown in Exercise 4c above. Done.

Exercise

6. Prove that V has two linearly independent vectors with the same non-zero square if and only if V is not the exceptional plane. (Hint: The "only if" part was proved in Exercise 5b above. For the "if" part, choose a nonisotropic vector $A \in V$. Since V is not the exceptional plane, V has a rotation $\rho \neq \pm 1_V$. Study the vectors A and ρA.)

52. RATIONAL POINTS ON CONICS

The present section and Sections 53, 54 discuss some special questions of plane geometry. These sections are not needed for the remainder of the book. *We assume throughout this section that $n = 2$.*

Whenever it is more natural to say "point" instead of "vector," we shall do so. A point should always be visualized as a dot, just the endpoint of the corresponding arrow—no shaft and no feathers. In many situations, people say "rational point" instead of point when they want to stress the fact that the field k reminds them of the rational numbers. Whatever the reason may be for all these terms, vector, point, and rational point all mean the same thing.

If $r \in k$, we refer to the set of points

$$\{X \mid X^2 = r\}$$

as the **circle with radius** r; notation: C_r. This circle is not empty if and only if r is represented by V. For instance, if V is the Euclidean plane and r is a negative real number, C_r is empty. For the same plane, if r is a positive real number, the customary radius of the Euclidean circle C_r is $\sqrt{r}$, so, perhaps, we should have called r the "square radius" of C_r. It is better to drop the clumsy square and remember the slight ambiguity which occurs in the Euclidean plane.

C_r should be visualized as an ordinary circle with radius r and center $\mathbf{0}$. Even so, one should realize that, from the point of view of nonmetric affine geometry, C_r is a *conic* which is defined over k. Here, a conic defined over k is simply a plane curve whose equation in a coordinate system of V is a quadratic polynomial with coefficients in k. Indeed, if $ax_1^2 + bx_1x_2 + cx_2^2$ is the quadratic form of V relative to a coordinate system, the equation of C_r in the same coordinate system is

$$ax_1^2 + bx_1x_2 + cx_2^2 = r.$$

Readers who are experienced with conics can easily prove that, by letting the geometry of V vary and keeping $r \neq 0$, one obtains, in this way, all non-singular conics which have a center.

The following proposition shows that all nonempty circles with nonzero radius have the same number of points. (Two sets are said to have the **same cardinality** if there exists a function from one set to the other which is one-to-one and onto. For finite sets, this means they have the same number of elements. For infinite sets, "the same number of elements" is often useful abuse of language for "the same cardinality.")

Proposition 283.1.   If r is a nonzero scalar which is represented by V, the circle C_r and the rotation group $O^+(V)$ have the same cardinality.

Proof. Choose a vector A on C_r; this can be done because r is represented by V. We now define a mapping from $O^+(V)$ to C_r. If $\rho \in O^+(V)$, the vector ρA also lies on C_r since $(\rho A)^2 = A^2 = r$. The resulting mapping $\rho \to \rho A$ from

$O^+(V)$ to C_r is one-to-one since different rotations act differently on every nonzero vector (Proposition 270.2). Finally, since $r \neq 0$, if $B \in C_r$, then $A^2 = B^2 = r$ and it follows from the refinement of the Witt theorem that there is a rotation of V which maps A onto B (Exercise 271.4). Thus the mapping from $O^+(V)$ to C_r is onto. Done.

Exercises

1. Prove that the following sets all have the same cardinality.
   a. The rotation group $O^+(V)$.
   b. The set of symmetries of V.
   c. The set of nonsingular lines of V.
   d. The circle C_r where $r \neq 0$ and r is represented by V.

*2. Let r be a nonzero scalar which is represented by V and choose a point $A \in C_r$. If Y is a nonisotropic vector, the symmetry τ_Y does not change the square of vectors and hence $\tau_Y A \in C_r$.
   a. Prove that every point of C_r is of the form $\tau_Y A$ for some vector Y. (Hint: Use Exercise 271.4.)
   b. If X, Y are both nonisotropic vectors of V, prove that $\tau_X A = \tau_Y A$ if and only if X and Y are linearly dependent.
   c. Consider the mapping $Y \to \tau_Y A$ from the set of nonisotropic vectors of V to C_r. Prove that this mapping is onto, but not one-to-one.

3. Let k be a finite field with q elements.
   a. Assume that V is anisotropic. If r is a nonzero scalar which is represented by V, prove that C_r has $q + 1$ points. Also prove that C_0 consists of one point.
   b. Assume V is the Artinian plane. We know, by Proposition 188.1, that all scalars are represented. If $r \neq 0$, prove that C_r has $q - 1$ points. Prove that C_0 has $2q - 1$ points.
   c. In Figure 281.1, determine which points make up, respectively, the circles C_0, C_1, and C_{-1}.

4. Let k be infinite and r a nonzero scalar which is represented by V. Prove that the circle C_r has infinitely many points.

We continue our study of the circle C_r where r is a nonzero scalar represented by V. It is very interesting to know how many points lie on C_r since this is related to deep questions in mathematics; we shall now explain.

Let A_1, A_2 be a coordinate system of V and suppose that the quadratic form of V is

$$ax_1{}^2 + bx_1 x_2 + cx_2{}^2$$

relative to this coordinate system. The equation of C_r is then

$$ax_1{}^2 + bx_1x_2 + cx_2{}^2 = r;$$

this means that a point $X = pA_1 + qA_2$ lies on C_r if and only if its coordinates (p, q) satisfy

$$ap^2 + bpq + cq^2 = r.$$

Consequently, we know the number of solutions (p, q) of the quadratic equation $ax_1{}^2 + bx_1x_2 + cx_2{}^2 = r$. Since $p, q \in k$, these solutions are called "rational points" on the quadratic curve C_r and hence we know the number of rational points on C_r. It is of particular interest that this number is infinite if k is infinite. Some of the deepest unsolved problems in mathematics are concerned with the number of rational points on plane algebraic curves of degree larger than two. A plane algebraic curve of degree m is simply a subset of V which, relative to a coordinate system of V, is given by an equation $f = 0$ where f is a polynomial of degree m of $k[x_1, x_2]$. In particular, C_r is given by the equation

$$ax_1{}^2 + bx_1x_2 + cx_2{}^2 - r = 0$$

and, therefore, is a plane algebraic curve of degree two (that is, a conic). The number of rational points on plane algebraic curves of degree larger than two often is finite even if k is infinite. For example, if $k = \mathbf{Q}$ and $m > 2$, the famous "last theorem" of Fermat states that the curve

$$x_1{}^m + x_2{}^m = 1$$

has no rational points besides $(1, 0)$ and $(0, 1)$. Fermat's proof of his theorem has been lost and today no one has been able to either prove or disprove it. The methods of metric vector spaces deal only with algebraic curves of degree two. For curves of higher degree, the methods of algebraic geometry are indispensable.

Because of the connection with the theorem of Fermat, we ask how the coordinates of the points of C_r can be represented parametrically. For this purpose, we construct a rectangular coordinate system of V which has one vector on the circle C_r.

Since r is represented by V, we can choose a vector A on C_r. Then $A^2 = r \neq 0$ and hence

$$V = \langle A \rangle \oplus \langle A \rangle^*.$$

If B is a nonzero vector on the line $\langle A \rangle^*$, then A, B is a rectangular coordinate

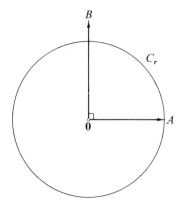

Figure 286.1

system of V (see Figure 286.1). We put $B^2 = b$ and disc $V = \Delta$. Then the matrix of V is

$$\begin{pmatrix} r & 0 \\ 0 & b \end{pmatrix}$$

and $\Delta = rb$ with $b \neq 0$.

Every point of C_r is of the form $\tau_Y A$ for some nonisotropic vector Y (Exercise 284.2a). Thus we are asking how the coordinates of the points $\tau_Y A$ can be represented parametrically. The answer is given in the following exercise.

Exercise

*5. Let $X = uA + vB$ be a nonisotropic vector of V.
  a. Prove that $ru^2 + bv^2 \neq 0$, equivalently, $r^2u^2 + \Delta v^2 \neq 0$.
  b. If $Y = bvA - ruB$, prove that $\langle Y \rangle = \langle X \rangle^*$. (Do not forget to say why $Y \neq \mathbf{0}$.)
  c. Show that

$$A = \frac{rv}{r^2u^2 + \Delta v^2} Y + \frac{r^2u}{r^2u^2 + \Delta v^2} X$$

and

$$\tau_Y A = \frac{-rv}{r^2u^2 + \Delta v^2} Y + \frac{r^2u}{r^2u^2 + \Delta v^2} X.$$

  d. Prove that the coordinates of $\tau_Y A = pA + qB$ are

$$p = \frac{r^2u^2 - \Delta v^2}{r^2u^2 + \Delta v^2}, \qquad q = \frac{2r^2uv}{r^2u^2 + \Delta v^2}.$$

We refer to these two formulas as the "parametric formulas." The scalars r and Δ are fixed while the variables u and v are parameters.

e. If u_0, v_0 are scalars such that $r^2u_0^2 + \Delta v_0^2 \neq 0$, prove that

$$\frac{r^2u_0^2 - \Delta v_0^2}{r^2u_0^2 + \Delta v_0^2}, \qquad \frac{2r^2u_0 v_0}{r^2u_0^2 + \Delta v_0^2}$$

are the coordinates of a point on the circle $C_r^!$. Prove also that all points of C_r are obtained this way.

f. Prove that the equation of C_r is $rx_1^2 + bx_2^2 = r$, equivalently, $r^2x_1^2 + \Delta x_2^2 = r^2$.

g. If u_0, v_0 are scalars such that $ru_0^2 + \Delta v_0^2 \neq 0$, prove that

$$x_1 = \frac{r^2u_0^2 - \Delta v_0^2}{r^2u_0^2 + \Delta v_0^2}, \qquad x_2 = \frac{2r^2u_0 v_0}{r^2u_0^2 + \Delta v_0^2}$$

is a solution of the equation $r^2x_1^2 + \Delta x_2^2 = r^2$. Prove that all solutions of this equation with values in k are obtained in this way.

The parametric formulas given in Part d of the above exercise parametrize the circle C_r; this was shown in Part e. It was shown in Part g that these formulas give all the solutions of the equation

$$r^2x_1^2 + \Delta x_2^2 = r^2$$

with values in k.

As an example, let $k = \mathbf{Q}$ and assume that V has an orthonormal coordinate system. Then $\Delta = 1$ and V represents the number 1. Hence we may choose $r = 1$ and then C_r is the unit circle C_1 with equation $x_1^2 + x_2^2 = 1$. The parametric formulas in Exercise 286.5d become the standard parametric equations of the unit circle, namely,

$$x_1 = \frac{u^2 - v^2}{u^2 + v^2}, \qquad x_2 = \frac{2uv}{u^2 + v^2}.$$

Consequently, if u_0 and v_0 are rational numbers, not both zero, then

$$\left(\frac{u_0^2 - v_0^2}{u_0^2 + v_0^2}\right)^2 + \left(\frac{2u_0 v_0}{u_0^2 + v_0^2}\right)^2 = 1,$$

equivalently,

$$(u_0^2 - v_0^2)^2 + (2u_0 v_0)^2 = (u_0^2 + v_0^2)^2.$$

This shows that, if u_0 and v_0 are positive integers with $u_0 > v_0$, the triple of positive integers

$$\alpha = u_0^2 - v_0^2, \qquad \beta = 2u_0 v_0, \qquad \gamma = u_0^2 + v_0^2$$

has the property that
$$\alpha^2 + \beta^2 = \gamma^2.$$
A triple of positive integers (α, β, γ) with the property that $\alpha^2 + \beta^2 = \gamma^2$ is called a **Pythagorean triple**. The geometric reason for this name is obvious and the three formulas above which define α, β, and γ show how we can construct Pythagorean triples ad infinitum.

The last theorem of Fermat is equivalent to the statement that, if α, β, and γ are positive integers and $m > 2$, then
$$\alpha^m + \beta^m \neq \gamma^m.$$
Many mathematicians believe that the great Pierre Fermat (1601–1665) could not have had a correct proof of this theorem when he pronounced it, presumably in 1637. However, this belief is based on the flimsy argument that, if Fermat could prove his theorem well over three centuries ago, we should certainly be able to prove it today. It is better not to indulge in such " modern boasting " and to realize that, although Fermat made some wrong conjectures, this lawyer from Toulouse in France never pronounced a theorem which was later proven to be false.

Exercise

6. a. If (α, β, γ) is a Pythagorean triple and d is a positive integer, prove that $(d\alpha, d\beta, d\gamma)$ is again a Pythagorean triple.
   b. Because of Part a, we are only interested in Pythagorean triples (α, β, γ) with the property that the greatest common divisor of α, β, and γ is one. If (α, β, γ) is such a triple, prove that there exist positive integers u_0, v_0 such that either the triple (α, β, γ) or the triple (β, α, γ) is given by the formulas
   $$\alpha = u_0^2 - v_0^2, \qquad \beta = 2u_0 v_0, \qquad \gamma = u_0^2 + v_0^2.$$
   (Hint: Apply Exercise 287.5g to $(\alpha/\gamma)^2 + (\beta/\gamma)^2 = 1$.)
   c. For what values of u_0 and v_0 do the formulas for α, β, γ give the best known Pythagorean triple, (3, 4, 5)?
   d. Find three Pythagorean triples (α, β, γ) with $\beta = 100$.

53. PLANE TRIGONOMETRY

In high school, we learn to associate the six trigonometric functions sine, cosine, tangent, cosecant, secant, and cotangent with angles in the Euclidean plane. The whole justification for angles is that they represent rotations, and trigonometric functions are really functions of rotations. Consequently, we shall not even mention angles, but define the trigonometric functions as functions from the rotation group $O^+(V)$ into k. Furthermore, in order to bring the true foundations of trigonometry to the surface, we shall not restrict ourselves to the Euclidean plane, but impose only those restrictions on k and the geometry of V which are necessary for the immediate argument. Of course, our conditions will never exclude the Euclidean plane and the reader should continually use this plane and his knowledge of ordinary trigonometry to make examples and draw pictures.

We assume throughout this section that $n = 2$ and all lines of V through $\mathbf{0}$ are isometric.

In addition to the Euclidean plane, the negative Euclidean plane has this property; so do other planes over other fields. Nevertheless, the condition that all lines through $\mathbf{0}$ are isometric does restrict the field k severely. For example, no plane with this property exists over $\mathbf{Q}$ or $\mathbf{C}$ (see Exercise 290.7).

Since V is nonsingular, V contains a line which is not a null line. All lines of V are isometric with this one line and hence V contains no null lines. This means, of course, that

V is anisotropic.

The condition that all lines through $\mathbf{0}$ are isometric is much stronger than to say that V is anisotropic. For instance, we have observed many nonisometric lines in anisotropic planes over $\mathbf{Q}$ (Exercise 135.13a).

Exercises

These exercises show how very restrictive the assumption is that all lines through $\mathbf{0}$ are isometric.

1. Assume $k = \mathbf{R}$. Prove that V is either the Euclidean plane or the negative Euclidean plane.
2. Prove that k cannot be algebraically closed and hence cannot be $\mathbf{C}$.
*3. Let A and B be nonzero vectors of V.
   a. Prove that there exists a scalar c such that $A^2 = c^2 B^2$. Prove that $c \neq 0$ and is unique except for sign.
   b. Let c be a scalar such that $A^2 = c^2 B^2$. Prove that there is a unique rotation ρ and a unique symmetry τ such that $\rho A = \tau A = cB$.
   c. Let U and W be lines of V. Prove that there exist two rotations ρ_1, ρ_2 and two symmetries τ_1, τ_2 such that $\rho_i U = \tau_i U = W$ for $i = 1, 2$. Prove, furthermore, that $\rho_2 = -\rho_1$ and $\tau_2 = -\tau_1$.

4. a. Prove that V has a rectangular coordinate system A_1, A_2 with the property that $A_1{}^2 = A_2{}^2$.
   b. Prove that disc V is a square. Of course, this is abuse of language for disc $V = k^{*2}$. (Hint: Compute the discriminant of V relative to a coordinate system with the property described in Part a above.)

*5. a. Let A be a nonzero vector of V and put $A^2 = a$. Prove that V represents all scalars ac^2 where $c \in k$ and no others. In other words, V represents the scalars in the coset ak^{*2} together with the scalar 0. Moral: V represents very few scalars.
   b. Prove that the scalars represented by V are precisely those represented by any one of its lines.
   c. Prove that two nonsingular planes each with the property that all its lines are isometric are themselves isometric if and only if they represent the same scalars. (Hint: Such a plane is an orthogonal sum of two isometric lines and the scalars represented by the plane are those represented by any one of its lines through $\mathbf{0}$.) Observe that this exercise, as well as many other parts of this section, can immediately be extended to higher dimensional nonsingular spaces with the property that all their lines through $\mathbf{0}$ are isometric.
   d. Prove that two nonsingular planes, each with the property that all its lines through $\mathbf{0}$ are isometric, may have the same discriminant without being isometric.

6. Prove that k satisfies the following two conditions. Each condition restricts k severely.
   Condition 6a. If $a_1, \ldots, a_h \in k$, then $a_1{}^2 + \cdots + a_h{}^2 = c^2$ for some $c \in k$. In other words, sums of squares of elements of k are themselves squares.
   Condition 6b. The scalar -1 is not a sum of squares of k. (Hint for 6a: It is sufficient to prove that $a^2 + b^2 = c^2$ for $a, b \in k$. We may assume that $a \neq 0$, $b \neq 0$, and choose a rectangular coordinate system A_1, A_2 of V such that $A_1{}^2 = A_2{}^2$. Now apply Exercise 3a above to the vectors $aA_1 + bA_2$ and A_1.)
   A good name for a field satisfying Condition 6a would be **Pythagorean field**. Fields satisfying Condition 6b are called **formally real fields**. See B. L. van der Waerden, *Modern Algebra*, Vol. 1, 1949, page 225, [22].

*7. a. Prove that $\mathbf{Q}$ does not satisfy Condition 6a, but does satisfy Condition 6b. Hence $k \neq \mathbf{Q}$.
   b. Prove that an algebraically closed field satisfies Condition 6a, but not Condition 6b. Hence k is not algebraically closed. In particular, $k \neq \mathbf{C}$.
   c. Prove that a field of characteristic $p \neq 0$ does not satisfy Condition 6b. Hence k has characteristic 0 and must be infinite.

d. Prove that **R** satisfies both Conditions 6a and 6b. See van der Waerden, [22], p. 225 for the fact that all real closed fields satisfy both Conditions 6a and 6b.

We begin with the cosine function because it is undoubtedly the most important trigonometric function for geometry. The question is: How does one define the cosine of a rotation? Since the cosine of a rotation should be a scalar, we are trying to define a function

$$\text{cosine: } O^+(V) \to k.$$

The definition is based on the following proposition.

Proposition 291.1.   If ρ is a rotation of V, the scalar

$$\frac{A(\rho A)}{A^2}$$

is the same for all nonzero vectors A.

Proof. Let A and B be nonzero vectors of V. In order to show that

$$\frac{A(\rho A)}{A^2} = \frac{B(\rho B)}{B^2},$$

we choose a nonzero scalar c such that $cB = \sigma A$ for some rotation σ of V (Exercise 289.3b). The following computation uses the fact that σ is an isometry (first equality) and $\sigma\rho = \rho\sigma$ (since $O^+(V)$ is commutative);

$$\frac{A(\rho A)}{A^2} = \frac{\sigma A(\sigma\rho A)}{(\sigma A)^2} = \frac{\sigma A(\rho\sigma A)}{(\sigma A)^2} = \frac{cB(\rho(cB))}{(cB)^2} = \frac{B(\rho B)}{B^2}.$$

Done.

Exercise

8. Let ρ be a rotation of V and A be a nonzero vector. If $A \perp \rho A$, prove that $B \perp \rho B$ for all $B \in V$.

Definition 291.1.   The **cosine of a rotation** ρ of V is defined by

$$\cos \rho = \frac{A(\rho A)}{A^2}$$

where A is any nonzero vector of V.

Observe that $\cos \rho$ has been defined without the use of coordinate systems. It is clear that $\cos \rho = 0$ if and only if $A \perp \rho A$ for all vectors A.

Exercise

9. Prove that $\cos 1_V = 1$ and $\cos(-1_V) = -1$. The last equality says that $\cos(180°) = -1$.

If α is an angle in the Euclidean plane, we know from high school trigonometry that $\cos(-\alpha) = \cos \alpha$; in other words, cosine is an "even" function. It is, furthermore, clear that, if ρ is the rotation represented by the angle α, then ρ^{-1} is the rotation represented by the angle $-\alpha$; see Figure 292.1.

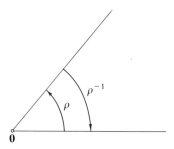

Figure 292.1

We conclude from the following proposition that cosine is an even function.

Proposition 292.1.   If ρ is a rotation of V,
$$\cos(\rho^{-1}) = \cos \rho.$$

Proof. Let A be a nonzero vector of V. Since ρ is an isometry,
$$\frac{A(\rho^{-1}A)}{A^2} = \frac{(\rho A)A}{A^2},$$
equivalently, $\cos(\rho^{-1}) = \cos \rho$. Done.

If α is an angle in the Euclidean plane, we learn in high school trigonometry that $\cos(180° - \alpha) = -\cos \alpha$. If ρ is the rotation represented by α, $-\rho^{-1} = -1_V\rho^{-1}$ is the rotation represented by $180° - \alpha$ (see Figure 293.1).

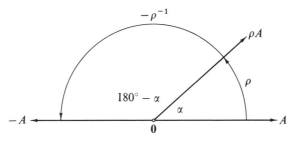

Figure 293.1

Consequently, we would like to show that $\cos(-\rho^{-1}) = -\cos\rho$. This is a consequence of Proposition 292.1 and the following result.

Proposition 293.1.   If ρ is a rotation of V,

$$\cos(-\rho) = -\cos\rho.$$

Proof. Let A be a nonzero vector of V. Then

$$\cos(-\rho) = \frac{A(-\rho A)}{A^2} = -\frac{A(\rho A)}{A^2} = -\cos\rho.$$

Done.

Exercises

10. If V represents a nonzero square of k, prove that the scalars represented by V is the set of squares of k. (Hint: Use Exercise 290.5.)

11. Assume that the field k is ordered. (Readers who are not acquainted with ordered fields should assume that $k = \mathbf{R}$. The field $\mathbf{Q}$, although ordered, cannot be used here.)

   a. Prove that V is either positive definite or negative definite.

   b. If X is a nonzero vector of V, the set of vectors $\{tX \mid t \geq 0\}$ is called the **half line** or the **ray** determined by X. Let A and B be nonzero vectors. We proved in Exercise 289.3a that there exists a nonzero scalar c, unique up to sign, such that $A^2 = c^2 B^2$. Furthermore, there exist unique rotations ρ, ρ' and unique symmetries τ, τ' such that $\rho A = \tau A = cB$ and $\rho' A = \tau' A = (-c)B$. What is the relationship between the rotations ρ and ρ'? the symmetries τ and τ'? Observe that the rotation ρ and the symmetry τ map the ray determined by A onto the ray determined by B (see Figure 294.1) while the rotation ρ' and the symmetry τ' map the ray determined by A onto the ray determined by $-B$ (see Figure 294.2).

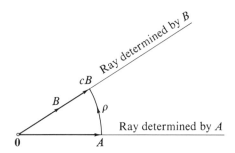

Figure 294.1

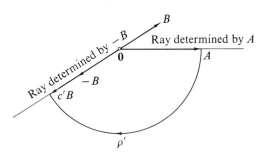

Figure 294.2

*12. Assume that V represents the squares of k and k is ordered. (Since non-zero squares in an ordered field are positive, V is positive definite.) Hence, if $k = \mathbf{R}$, V is the Euclidean plane. If A is a nonzero vector of V, A^2 is a nonzero square in k and we denote by $\sqrt{A^2}$ the positive scalar d such that $d^2 = A^2$.

Let A and B be nonzero vectors of V and ρ the unique rotation such that $\rho A = cB$ for a positive scalar c. Prove that

$$\cos \rho = \frac{AB}{\sqrt{A^2}\sqrt{B^2}}.$$

Indicate where one uses the fact that $c > 0$.

Remark. The above formula for $\cos \rho$ is one of the most important formulas in geometry. It gives the cosine of the rotation ρ which maps the ray determined by A onto the ray determined by B.

We now study the function

$$\text{cosine}: O^+(V) \to k$$

in terms of a rectangular coordinate system A_1, A_2 of V where $A_1{}^2 = A_2{}^2$. Relative to A_1, A_2, the matrix

$$\begin{pmatrix} a & b \\ c & d \end{pmatrix}$$

of a rotation ρ is defined by

$$(\rho A_1, \rho A_2) = (A_1, A_2)\begin{pmatrix} a & b \\ c & d \end{pmatrix},$$

equivalently,

$$A_1 = aA_1 + cA_2, \qquad A_2 = bA_1 + dA_2.$$

Proposition 295.1. Let $\begin{pmatrix} a & b \\ c & d \end{pmatrix}$ be the matrix of a rotation ρ relative to a rectangular coordinate system A_1, A_2 of V where $A_1{}^2 = A_2{}^2$. Then

1. $a = d = \cos \rho$.
2. $b^2 = c^2 = 1 - \cos^2 \rho$.
3. $c = -b$.

Proof. The above equalities among a, b, c, and d are known to us from Exercise 277.11. Nevertheless, we derive them again to stress the power of our assumption that all lines of V through $\mathbf{0}$ are isometric.

We put $A_1{}^2 = A_2{}^2 = \alpha \neq 0$. Then

$$\cos \rho = \frac{A_1(\rho A_1)}{A_1{}^2} = \frac{a\alpha}{\alpha} = a.$$

If A_2 is used to compute $\cos \rho$, one obtains $\cos \rho = d$ and thus $a = d = \cos \rho$.

The matrix of V is

$$\begin{pmatrix} \alpha & 0 \\ 0 & \alpha \end{pmatrix}$$

and the matrix of the isometry ρ is

$$\begin{pmatrix} \cos \rho & b \\ c & \cos \rho \end{pmatrix}.$$

Consequently,

$$\begin{pmatrix} \cos \rho & b \\ c & \cos \rho \end{pmatrix}\begin{pmatrix} \alpha & 0 \\ 0 & \alpha \end{pmatrix}\begin{pmatrix} \cos \rho & c \\ b & \cos \rho \end{pmatrix} = \begin{pmatrix} \alpha & 0 \\ 0 & \alpha \end{pmatrix}.$$

Since $\alpha \neq 0$, this is the same as

$$\begin{pmatrix} \cos \rho & b \\ c & \cos \rho \end{pmatrix}\begin{pmatrix} \cos \rho & c \\ b & \cos \rho \end{pmatrix} = \begin{pmatrix} 1 & 0 \\ 0 & 1 \end{pmatrix},$$

equivalently,

$$\cos^2 \rho + b^2 = 1, \qquad \cos^2 \rho + c^2 = 1, \qquad (b + c) \cos\rho = 0.$$

Statement 2 follows immediately.

There only remains to be shown that $c = -b$. If $\cos \rho \neq 0$, this follows from the equality $(b + c)\cos \rho = 0$. If $\cos \rho = 0$, then $b^2 = c^2 = 1$; hence $b = \pm 1$ and $c = \pm 1$. Furthermore,

$$\det \rho = \cos^2 \rho - bc = -bc = 1$$

which shows that b and c cannot have the same sign. Done.

Since $\cos \rho$ was defined without the use of coordinate systems, the scalar $a = d$ of the matrix of ρ does not depend on the choice of the rectangular coordinate system as long as $A_1{}^2 = A_2{}^2$. The scalars b and c are unique except for sign since $b^2 = 1 - \cos^2 \rho$ and $c^2 = 1 - \cos^2 \rho$. The meaning of these scalars will be given in Proposition 306.1.

Exercises

13. a. Prove that $\cos \rho = 1$ if and only if $\rho = 1_V$.
    b. Prove that $\cos \rho = -1$ if and only if $\rho = -1_V$.
14. If $\rho \in O^+(V)$, $1 - \cos^2 \rho$ is the square of an element in k. Prove that $\rho \neq \pm 1_V$ if and only if $1 - \cos^2 \rho$ is a nonzero square.
15. Assume k is an ordered field.
    a. If $c \in k$, prove that $-1 \leq c \leq 1$ if and only if $0 \leq 1 - c^2$.
    b. If ρ is a rotation, prove that

$$-1 \leq \cos \rho \leq 1.$$

(Hint: Use Part a along with the fact that squares in an ordered field are always positive or zero.)

We now investigate the sine of a rotation ρ. Notation: $\sin \rho$. Of course, we want the equality

$$\sin^2 \rho + \cos^2 \rho = 1,$$

equivalently,

$$\sin^2 \rho = 1 - \cos^2 \rho.$$

We know from Proposition 295.1, that $1 - \cos^2 \rho$ is the square of a scalar b. However, the trouble is that $1 - \cos^2 \rho$ is also the square of $-b$; should we define $\sin \rho = b$ or $\sin \rho = -b$? *For an arbitrary field k, there is no answer to this question and the sine function cannot be defined.* In order to answer the question, the plane must be " oriented " and for this it is necessary that k be an ordered field. Orientation of vector spaces has nothing to do with metrics, but is purely an affine notion, that is, it belongs to linear algebra. We shall develop the concept of orientation here because most courses in linear algebra omit it.

We assume that k is ordered and $\mathscr{W}$ is an m-dimensional vector space over k. (There is no reason to change the letter k because all ordered fields have characteristic zero.) No metric has been assigned to $\mathscr{W}$.

Let $A_1, \ldots, A_m$ and $B_1, \ldots, B_m$ be coordinate systems of $\mathscr{W}$ and denote by P the nonsingular $m \times m$ matrix such that

$$(B_1, \ldots, B_m) = (A_1, \ldots, A_m)P.$$

In other words, if $P = (p_{ij})$ where $p_{ij} \in k$, then

$$B_i = \sum_{j=1}^{m} A_j p_{ji} \qquad \text{for} \quad i = 1, \ldots, m.$$

Since $\det P \neq 0$, either $\det P > 0$ or $\det P < 0$.

The coordinate systems $A_1, \ldots, A_m$ and $B_1, \ldots, B_m$ are said to have the **same orientation** if $\det P > 0$. Notation:

$$(A_1, \ldots, A_m) \sim (B_1, \ldots, B_m).$$

If $\det P < 0$, the coordinate systems are said to have **opposite orientation**. Notation:

$$(A_1, \ldots, A_m) \not\sim (B_1, \ldots, B_m).$$

Exercises

16. Let $m = 2$ and assume A_1, A_2 is a coordinate system of $\mathscr{W}$.
    a. Prove that A_1, A_2 has opposite orientation from each of the three coordinate systems $-A_1, A_2$; $A_1, -A_2$; A_2, A_1.
    b. Prove that A_1, A_2 has the same orientation as each of the three coordinate systems $-A_1, -A_2$; $-A_2, A_1$; $A_2, -A_1$.
*17. a. Prove that $\sim$ is an equivalence relation for the set of coordinate systems of $\mathscr{W}$.
    b. The equivalence relation $\sim$ partitions the set of coordinate systems of $\mathscr{W}$ into disjoint subsets called classes. Prove that there are precisely two such classes. (Hint: Choose a coordinate system $A_1, \ldots, A_m$ of $\mathscr{W}$ and show how one can construct a coordinate

system $B_1, \ldots, B_m$ such that $(A_1, \ldots, A_m) \not\sim (B_1, \ldots, B_m)$; the construction must also work for $m = 1$. Then show that an arbitrary coordinate system $C_1, \ldots, C_m$ is equivalent to $A_1, \ldots, A_m$ or $B_1, \ldots, B_m$.)

According to Exercise 17 above, the set of coordinate systems of $\mathscr{W}$ is partitioned into two classes. There is no law which tells us which of these two classes should be designated as "positively oriented" coordinate systems and which class the "negatively oriented" ones. To "orient" the vector space $\mathscr{W}$ means to decide which class is designated as the positively oriented coordinate systems and which class the negatively oriented ones. Precisely, an **oriented vector space** is a vector space where the coordinate systems of one class have been defined as being positively oriented while the coordinate systems of the other class have been defined as being negatively oriented. Clearly, the vector space $\mathscr{W}$ has two orientations.

Assume that $\mathscr{W}$ is an oriented vector space and $A_1, \ldots, A_m$ and $B_1, \ldots, B_m$ are coordinate systems of $\mathscr{W}$, interrelated by the nonsingular matrix P, that is,

$$(B_1, \ldots, B_m) = (A_1, \ldots, A_m)P.$$

It is immediate that $A_1, \ldots, A_m$ and $B_1, \ldots, B_m$ are both positively oriented or both negatively oriented if and only if $\det P > 0$; also $A_1, \ldots, A_m$ is positively oriented and $B_1, \ldots, B_m$ is negatively oriented (or vice versa) if and only if $\det P < 0$. This finishes the development of orientation and we now return to our plane V, all of whose lines are isometric.

We assume for the remainder of this section that k is an ordered field and the nonsingular plane V has been oriented. Consequently, every coordinate system A_1, A_2 is either positively or negatively oriented and we may speak of the "orientation" of A_1, A_2.

Proposition 298.1. Let ρ be a rotation of V where $\rho \neq \pm 1_V$. If A is a nonzero vector of V, the vectors A and ρA are linearly independent. Further, the orientation of the coordinate system $A, \rho A$ is the same for all nonzero vectors A.

Proof. If the vectors A and ρA were linearly dependent, $\rho A = cA$ and hence $A^2 = c^2 A^2$. Since $A^2 \neq 0$, this would imply that $c = \pm 1$ and hence $\rho = \pm 1_V$ (Exercise 218.7), contrary to hypothesis.

We must still show that, if A and B are nonzero vectors, the coordinate systems $A, \rho A$ and $B, \rho B$ have the same orientation. For this, let c be a nonzero scalar and σ a rotation such that $cB = \sigma A$ (Exercise 289.3b). We denote the matrix of σ relative to the coordinate system $A, \rho A$ by P, that is,

$$(\sigma A, \sigma(\rho A)) = (A, \rho A)P,$$

and recall that $\det P = 1$ since σ is a rotation. Using the fact that $O^+(V)$ is commutative and hence $\sigma\rho = \rho\sigma$, we conclude that

$$(\sigma A, \sigma\rho A) = (\sigma A, \rho\sigma A) = (cB, c(\rho B))$$

from which it follows that

$$(B, \rho B) = (c^{-1}(\sigma A), c^{-1}(\sigma\rho A)) = (A, \rho A)\begin{pmatrix} c^{-1} & 0 \\ 0 & c^{-1} \end{pmatrix}P.$$

Since

$$\det\begin{pmatrix} c^{-1} & 0 \\ 0 & c^{-1} \end{pmatrix}P = (c^{-1})^2 > 0,$$

it follows that $A, \rho A$ and $B, \rho B$ have the same orientation. Done.

It is now clear how one should define positively oriented and negatively oriented rotations. When speaking about plane rotations, it is customary to say counterclockwise instead of positively oriented and clockwise instead of negatively oriented. We follow this custom.

Definition 299.1. Let ρ be a rotation of V where $\rho \neq \pm 1_V$ and let A be a nonzero vector. If the coordinate system $A, \rho A$ is positively oriented, ρ is called a **counterclockwise rotation**. If the coordinate system is negatively oriented, ρ is called a **clockwise rotation**.

The rotations 1_V and -1_V are not given an orientation. *Whenever we speak of a clockwise or counterclockwise rotation, it is understood that the rotation is different from $\pm 1_V$.*

Clockwise and counterclockwise rotations should be viewed as usual. Figure 299.1 shows a counterclockwise rotation on the left and a clockwise rotation on the right.

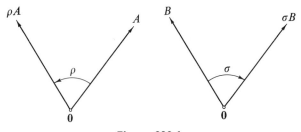

Figure 299.1

Although it is important to visualize clockwise and counterclockwise rotations correctly, the reader should not think that there is anything special about clockwise rotations. Any clockmaker can make a clock run backward and, if all clocks had been constructed that way, we would visualize clockwise and counterclockwise rotations just the opposite way from which we visualize them now.

Exercises

18. Let $\rho \neq \pm 1_V$ be a rotation of V. Prove that ρ and ρ^{-1} have opposite orientations, that is, one is clockwise and the other is counterclockwise. Prove the same for ρ and $-\rho$.

19. This exercise uses the fact that k is ordered, but not that V is oriented. Let A, B be a coordinate system of V.
    a. Prove that rotations preserve orientation. Precisely, if ρ is a rotation, prove that the coordinate systems A, B and ρA, ρB have the same orientation.
    b. Prove that reflections (symmetries) reverse orientation. Precisely, if τ is a symmetry of V, prove that the coordinate systems A, B and τA, τB have opposite orientations.
    c. Prove, in general, that in a nonsingular metric vector space of arbitrary dimension over an ordered field k, rotations preserve orientation while reflections reverse orientation.

All our troubles with defining the function sine: $O^+(V) \to k$ are now over. We must define

$$\sin 1_V = \sin(-1_V) = 0$$

since $\cos 1_V = 1$ and $\cos(-1_V) = -1$ and we want the relation

$$\sin^2 \rho = 1 - \cos^2 \rho$$

for all rotations ρ. If $\rho \neq \pm 1_V$, $1 - \cos^2 \rho$ is a nonzero square of k and we denote by $\sqrt{1 - \cos^2 \rho}$ the *positive* scalar b such that $b^2 = 1 - \cos^2 \rho$. The following definition uses high school trigonometry conventions concerning the sine function.

Definition 300.1.   If ρ is a counterclockwise rotation,

$$\sin \rho = \sqrt{1 - \cos^2 \rho}.$$

If ρ is a clockwise rotation,

$$\sin \rho = -\sqrt{1 - \cos^2 \rho}.$$

Furthermore,

$$\sin 1_V = \sin(-1_V) = 0.$$

Observe that the definition of the sine function does not depend on the choice of a coordinate system. The definition uses only the fact that V is oriented and this does not require that a coordinate system be singled out. It requires only that the coordinate systems of one class be designated as being positively oriented while the coordinate systems of the other class be designated as being negatively oriented.

The whole field of trigonometry is now open to us. The following exercises are an indication of this.

Exercises

20. Prove that $\sin^2 \rho + \cos^2 \rho = 1$ for all $\rho \in O^+(V)$.
21. Prove that $\sin(\rho^{-1}) = -\sin \rho$ for all $\rho \in O^+(V)$. Why should this be interpreted as: The function $\sin: O^+(V) \to k$ is an odd function?
22. Prove that $\sin(-\rho) = -\sin \rho$ for all $\rho \in O^+(V)$.
23. a. Prove that $\sin(-\rho^{-1}) = \sin \rho$ for all $\rho \in O^+(V)$.
    b. If α is an angle in the Euclidean plane, why does Part a of this exercise state that $\sin(180° - \alpha) = \sin \alpha$?
24. a. Prove that there exists a unique counterclockwise rotation α such that $A \perp \alpha A$ for all vectors A. We call α the "90° rotation."
    b. Prove that there exists a unique clockwise rotation β such that $A \perp \beta A$ for all vectors A. We call β the "270° rotation."
    c. Prove that α and β satisfy the following conditions.

$$\alpha^2 = \beta^2 = -1_V, \qquad \alpha^3 = \alpha^{-1} = -\alpha = \beta,$$
$$\alpha^4 = \beta^4 = 1_V, \qquad \beta^3 = -\beta = \alpha.$$

We conclude that α and β are elements of order four of $O^+(V)$ and are inverses of one another.

25. We are already referring to -1_V as the 180° rotation. Let us agree to also call 1_V a 0° rotation. Prove that the following table is correct.

TABLE 301.1

	Cosine	Sine
0° rotation	1	0
90° rotation	0	1
180° rotation	−1	0
270° rotation	0	−1

*26. Let A be a nonzero vector of V. This vector divides V into the usual four quadrants and we show in this exercise how these quadrants can be defined rigorously. The definition depends on the orientation of V, the metric of V, and the vector A.

a. Prove that there is a nonzero vector $B \in \langle A \rangle^*$ such that the coordinate system A, B is positively oriented. Also prove that, if V has this property, the vector B' has this property if and only if $B' = cB$ for a positive scalar c.

We now define the four quadrants. Choose a nonzero vector $B \in \langle A \rangle^*$ such that the coordinate system A, B is positively oriented. The four quadrants are defined by assigning a quadrant to every vector C which does not lie on either the line $\langle A \rangle$ or the line $\langle A \rangle^*$. No quadrant is assigned to vectors on the lines $\langle A \rangle$ and $\langle A \rangle^*$. The quadrant of C depends on the orientations of the coordinate systems A, C and B, C as shown in the following table which conforms to the conventions made in high school trigonometry. A positively oriented coordinate system is indicated by $+$ and a negatively oriented one by $-$.

TABLE 302.1

Quadrant of C	Coordinate system A, C	Coordinate system B, C
First	$+$	$-$
Second	$+$	$+$
Third	$-$	$+$
Fourth	$-$	$-$

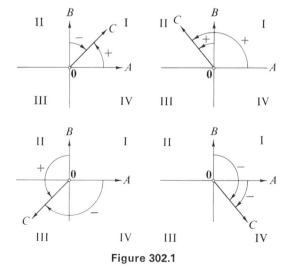

Figure 302.1

Figure 302.1 shows how the four quadrants should be visualized.

b. Prove that the definition of the four quadrants does not depend on the choice of the vector B. If c is a positive scalar, prove also that the vectors A and cA define the same four quadrants.

c. If the orientation of V is reversed, prove that the quadrants are interchanged as follows:

$$\text{First quadrant} \leftrightharpoons \text{Fourth quadrant}$$

$$\text{Second quadrant} \leftrightharpoons \text{Third quadrant}$$

(Hint: If the orientation is reversed, the vector B must be replaced.)

d. If A is replaced by $-A$, prove that the quadrants are interchanged as follows:

$$\text{First quadrant} \leftrightharpoons \text{Third quadrant}$$

$$\text{Second quadrant} \leftrightharpoons \text{Fourth quadrant}$$

e. Prove that the four quadrants could also have been defined in the customary way by using the signs of the coordinates of each vector in the coordinate system A, B. Precisely, if C is a vector not lying on either of the lines $\langle A \rangle$ or $\langle A \rangle^*$, prove that the following table is correct. Here $(+, -)$ means that the first coordinate of C in the coordinate system A, B is positive and the second one is negative, etc.

TABLE 303.1

Quadrant of C	Coordinates of C in the coordinate system relative to A, B
First	$(+, +)$
Second	$(-, +)$
Third	$(-, -)$
Fourth	$(+, -)$

(Hint: If $C = aA + bB$,

$$(A, C) = (A, B)\begin{pmatrix} 1 & a \\ 0 & b \end{pmatrix} \quad \text{and} \quad (B, C) = (A, B)\begin{pmatrix} 0 & a \\ 1 & b \end{pmatrix}.$$

If C lies in the first quadrant, the coordinate systems A, C and A, B have the same orientation and hence

$$\det\begin{pmatrix} 1 & a \\ 0 & b \end{pmatrix} > 0$$

and, similarly, for the other quadrants.)

27. Let ρ be a rotation different from the $0°$ rotation, the $90°$ rotation, the $180°$ rotation, and the $270°$ rotation. Assume that A is a nonzero vector and the four quadrants of V were defined relative to A. (In the hints, B denotes a vector of $\langle A \rangle^*$ such that the coordinate system A, B is positively oriented.)

a. Prove that ρA lies in precisely one of the four quadrants. If ρA lies in the ith quadrant, we say the rotation ρ lies in that quadrant.

b. Prove that the quadrant to which ρ belongs does not depend on the choice of the vector A. (Hint: Let X be a nonzero vector of V. By Exercise 302.26b, one may as well assume $X^2 = A^2$; hence there is a rotation σ such that $X = \sigma A$. Then the coordinate system $\sigma A, \sigma B$ has the same orientation as A, B. Assume $\rho A = aA + bB$. Now use Exercise 303.26e and look at the coordinates of ρX in the coordinate system $\sigma A, \sigma B$.)

c. Prove that the following table is correct. Plus $(+)$ indicates that the corresponding trigonometric function is positive and minus $(-)$ indicates it is negative.

TABLE 304.1

Quadrant of ρ	$\cos \rho$	$\sin \rho$
First	+	+
Second	−	+
Third	−	−
Fourth	+	−

(Hint: $\sin \rho$ is positive if and only if the rotation ρ is counterclockwise, equivalently, if and only if the coordinate system $A, \rho A$ is positively oriented. Hence, for $\sin \rho$, one only has to check that $A, \rho A$ is positively oriented if and only if ρ lies in the first or second quadrant.)

28. This exercise shows that the standard definitions of the cosine and the sine functions in terms of "adjacent," "opposite," and "hypotenuse" are valid.

We assume that V represents the squares of k and hence, if C is a nonzero vector, C^2 is a nonzero square of k. For any nonzero square e of k, $\sqrt{e}$ denotes the positive scalar c such that $c^2 = e$. For any scalar d, the absolute value $|d|$ has the usual meaning, namely, $|d| = d$ if $d \geq 0$ and $|d| = -d$ if $d < 0$.

Let A be the nonzero vector used to define the four quadrants of V and B a vector of $\langle A \rangle^*$ such that the coordinate system A, B is positively oriented. We study a vector C which does not lie on either of the lines $\langle A \rangle$ or $\langle A \rangle^*$. If $C = aA + bB$, a and b are different from zero and the right triangle determined by C (see Figure 305.1) has $\sqrt{C^2}$ as hypotenuse, $|a| \sqrt{A^2}$ as adjacent side, and $|b| \sqrt{B^2}$ as opposite side. Figure 305.1 shows C located in each of the four quadrants; the hypotenuse is denoted by γ, the adjacent side by α, and the opposite side by β.

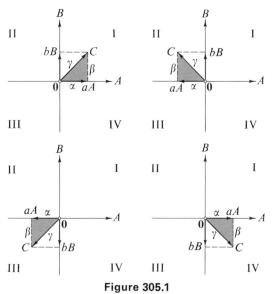

Figure 305.1

We denote by ρ the unique rotation such that $\rho A = cC$ for a positive scalar c.

a. Find the value of α, β, γ in terms of $a, b, \sqrt{A^2}, \sqrt{B^2}$ for each of the four locations of C shown in Figure 305.1.

b. Prove that $\cos \rho = \dfrac{a\sqrt{A^2}}{\sqrt{C^2}}$ and $\sin \rho = \dfrac{b\sqrt{B^2}}{\sqrt{C^2}}$. (Hint: Use Exercise 294.12.)

Remark. Since V represents the squares of k, one can choose A and B such that $A^2 = B^2 = 1$. Then we observe that

$$\cos \rho = \frac{a}{\sqrt{C^2}} \qquad \text{and} \qquad \sin \rho = \frac{b}{\sqrt{C^2}}$$

and the high school trigonometry definitions of cosine and sine have been recovered. If, furthermore, C is chosen on the unit circle, $\cos \rho$ is the "x coordinate" and $\sin \rho$ is the "y coordinate" of C.

The sine function was defined without the choice of a coordinate system. We now study the sine function in terms of a rectangular coordinate system A_1, A_2 of V where $A_1{}^2 = A_2{}^2$. If ρ is a rotation of V, we know from Proposition 295.1 that

$$(\rho A_1, \rho A_2) = (A_1, A_2)\begin{pmatrix} \cos \rho & -c \\ c & \cos \rho \end{pmatrix}.$$

Proposition 306.1. Let ρ be a rotation of V and

$$\begin{pmatrix} \cos \rho & -c \\ c & \cos \rho \end{pmatrix}$$

the matrix of ρ relative to the coordinate system A_1, A_2. If A_1, A_2 is positively oriented, $c = \sin \rho$; if A_1, A_2 is negatively oriented, $c = -\sin \rho$.

Proof. By Proposition 295.1, $c^2 = 1 - \cos^2 \rho$ and hence $c = \pm\sin \rho$. If $\rho = \pm 1_V$, $\sin \rho = 0$ and we are done. Assume $\rho \neq \pm 1_V$, in which case the vectors A_1 and ρA_1 are linearly independent (Proposition 298.1). Furthermore,

$$(A_1, \rho A_1) = (A_1, A_2)\begin{pmatrix} 1 & \cos \rho \\ 0 & c \end{pmatrix}$$

and hence the coordinate systems $A_1, \rho A_1$ and A_1, A_2 have the same orientation if and only if $c > 0$.

Let A_1, A_2 be positively oriented. Then $c > 0$ if and only if ρ is counterclockwise, equivalently, if and only if $\sin \rho > 0$. Consequently, $c = \sin \rho$.

Let A_1, A_2 be negatively oriented. Then $c > 0$ if and only if ρ is clockwise, equivalently, if and only if $\sin \rho < 0$. Consequently, $c = -\sin \rho$. Done.

It is now easy to show that the usual trigonometric formulas for the sum of angles in Euclidean geometry hold in our metric vector space V. These formulas are

$$\cos(\alpha + \beta) = \cos \alpha \cos \beta - \sin \alpha \sin \beta$$
$$\sin(\alpha + \beta) = \sin \alpha \cos \beta + \cos \alpha \sin \beta.$$

If the angle α represents the rotation ρ and the angle β the rotation σ, the angle $\alpha + \beta$ represents the rotation $\rho\sigma$.

Proposition 307.1. If ρ and σ are rotations of V, then

$$\cos \rho\sigma = \cos \rho \cos \sigma - \sin \rho \sin \sigma$$
$$\sin \rho\sigma = \sin \rho \cos \sigma + \cos \rho \sin \sigma.$$

Proof. Choose a positively oriented rectangular coordinate system A_1, A_2 of V such that $A_1{}^2 = A_2{}^2$. It follows from Propositions 295.1 and 306.1 that

$$(\rho A_1, \rho A_2) = (A_1, A_2)\begin{pmatrix} \cos \rho & -\sin \rho \\ \sin \rho & \cos \rho \end{pmatrix}$$

and

$$(\sigma A_1, \sigma A_2) = (A_1, A_2)\begin{pmatrix} \cos \sigma & -\sin \sigma \\ \sin \sigma & \cos \sigma \end{pmatrix}.$$

Consequently, the matrix

$$\begin{pmatrix} \cos \rho\sigma & -\sin \rho\sigma \\ \sin \rho\sigma & \cos \rho\sigma \end{pmatrix}$$

of the rotation $\rho\sigma$ is equal to

$$\begin{pmatrix} \cos \rho & -\sin \rho \\ \sin \rho & \cos \rho \end{pmatrix}\begin{pmatrix} \cos \sigma & -\sin \sigma \\ \sin \sigma & \cos \sigma \end{pmatrix}$$
$$= \begin{pmatrix} \cos \rho \cos \sigma - \sin \rho \sin \sigma & -\cos \rho \sin \sigma - \sin \rho \cos \sigma \\ \sin \rho \cos \sigma + \cos \rho \sin \sigma & -\sin \rho \sin \sigma + \cos \rho \cos \sigma \end{pmatrix}.$$

This proves each formula twice! Done.

Exercises

29. If ρ and σ are rotations of V, verify the following formulas.
   a. $\cos \rho\sigma^{-1} = \cos \rho \cos \sigma + \sin \rho \sin \sigma$.
   b. $\sin \rho\sigma^{-1} = \sin \rho \cos \sigma - \cos \rho \sin \sigma$.
   c. $\cos \rho^2 = \cos^2 \rho - \sin^2 \rho$.
   d. $\sin \rho^2 = 2 \sin \rho \cos \rho$.
   e. $\cos^2 \rho = \frac{1}{2}(1 + \cos \rho^2)$.
   f. $\sin^2 \rho = \frac{1}{2}(1 - \cos \rho^2)$.
   For each of the above formulas, give the corresponding formula for Euclidean angles.

*30. Let V be the Euclidean plane E_2 and hence $k = \mathbf{R}$. As usual, we let i denote the number in $\mathbf{C}$ with the property that $i^2 = -1$. If $a, b \in \mathbf{R}$, the absolute value $|z|$ of the complex number $z = a + bi$ is defined as $|z| = \sqrt{a^2 + b^2}$.
   a. Prove that the set of complex numbers with absolute value one form a group under the operation of multiplication of complex numbers. This group is called the **circle group** since the complex

numbers with absolute value one form the unit circle in the complex plane (see Figure 308.1.)

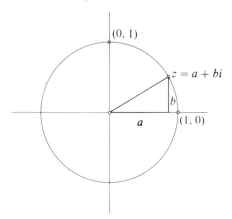

Figure 308.1

b. If a and b are real numbers such that $a^2 + b^2 = 1$, prove there is a unique rotation $\rho \in O^+(E_2)$ such that $\cos \rho = a$ and $\sin \rho = b$. (Hint: For existence, choose a positively oriented orthonormal coordinate system A_1, A_2 of E_2 and study the rotation ρ with the property that $\rho A = aA_1 + bA_2$.)

c. Prove that the mapping

$$\rho \to \cos \rho + i \sin \rho$$

is an isomorphism from the group $O^+(E_2)$ onto the circle group. *Remark.* The above isomorphism establishes the important connection between the group $O^+(E_2)$ which belongs in geometry and the circle group which belongs in algebra. It also establishes the connection between trigonometry and complex numbers, a connection which is often obscured by bringing in the extraneous fact that

$$\cos \theta + i \sin \theta = e^{i\theta}$$

for all real numbers θ. No matter how beautiful this expansion for $e^{i\theta}$ is, it belongs in analysis and is of no use in trigonometry. We underscore this fact in Part d below by deriving the main formula which is often, but unnecessarily, tied up with exponentials.

d. Let ρ be a rotation of E_2 and m a positive integer. Prove that

$$\cos \rho^m + i \sin \rho^m = (\cos \rho + i \sin \rho)^m.$$

(Hint: There is nothing to be proved! It is a formal consequence of the fact that the mapping $\rho \to \cos \rho + i \sin \rho$ is a group homomorphism. Why?) Also, derive the formulas which express $\cos \rho^m$ and $\sin \rho^m$ in terms of $\cos \rho$ and $\sin \rho$ for the special cases that $m = 2, 3, 4$. Give the corresponding formulas for Euclidean angles. Of course, one can use this technique, together with the binomial formula of Newton, to find the formulas which express $\cos \rho^m$ and $\sin \rho^m$ in terms of $\cos \rho$ and $\sin \rho$ for arbitrary m.

e. Here, we use the isomorphism between $O^+(E_2)$ and the circle group to derive a structural property of the Euclidean plane rotation group mentioned on page 275. If s is a positive integer, prove that every element of $O^+(E_2)$ has an sth root; that is, if $\rho \in O^+(E_2)$, prove that there exists a $\sigma \in O^+(E_2)$ such that $\rho = \sigma^s$; then $\sigma = \sqrt[s]{\rho}$. (Hint: Use the fact that $\mathbf{C}$ is algebraically closed to show that every element of the unit circle group has an sth root.) Say what this property means for Euclidean angles.

f. Prove that $O^+(E_2)$ contains a cyclic subgroup of order s for every positive integer s. This exercise shows how important it is to find homomorphisms between the plane rotation group and subgroups of k^*. The technique for finding such homomorphisms in the case of arbitrary geometries (if they exist!) is furnished by Clifford algebras, see Artin, [2].

Here, we stop and leave it to the reader to define the other four trigonometric functions and to develop the remainder of trigonometry. We wanted to demonstrate the fact that trigonometry is part of geometry, more precisely, part of the study of the plane rotation group. From the beginning of this section, angles were never used. Nothing is gained in representing rotations by angles even though this procedure may be sound pedagogy with high school students who are not yet ready to think in terms of rotations.

Remark on Teaching Trigonometry. Every high school curriculum in geometry should contain an elementary treatment of the rotation group $O^+(E_2)$ of the Euclidean plane E_2. Trigonometry is best presented as part of the study of this group. The six trigonometric functions should be defined as functions from $O^+(E_2)$ to the field $\mathbf{R}$ where two elements have to be removed from $O^+(E_2)$ if the function is not the cosine or the sine. The main thing is that the "domain" of the trigonometric functions is not the set of angles, but the group $O^+(E_2)$ or nearly all of $O^+(E_2)$. The transition from the dull configuration of angle (pair of rays with common vertex, ordered pair of rays with common vertex, equivalence class of ordered pairs of rays with common vertex, whatever definition is used) to the rich concept of rotation not only makes geometry more attractive, but, furthermore, results in much simplification and clarification. It is much easier and also much more important to understand the difference between a rotation ρ and its inverse ρ^{-1} than between an angle of $1732°$ and of $-1732°$. Anyway, it is always better to talk about the heart of a matter than about something which merely represents it. The heart is the rotation group $O^+(E_2)$ while the representing is done by the angles.

What about measurement of angles in degrees or radians or any other unit? Is this important for geometry? Again, the real reason for measuring an angle is to measure the rotation it represents. Hence the question is whether it is important to measure rotations. The answer is that geometry proper does not profit much from the measurement of rotations, but that analysis could not exist without it. By analysis we mean calculus and its many offsprings and now elaborate on this point.

The basic reason degrees, radians, or other units for angular measurement can be defined in Euclidean geometry is that there exists a function

$$w: \mathbf{R} \to O^+(E_2)$$

from the real numbers to the Euclidean rotation group which is a continuous homomorphism from the additive group of $\mathbf{R}$ onto the multiplicative group $O^+(E_2)$, that is, $w(\theta + \eta) = w(\theta)w(\eta)$ for real numbers θ and η. We omit the explanation of "continuous" because it involves the topology of both $\mathbf{R}$ and $O^+(E_2)$. We must stress, however, that the function w would have been useless if it had not been continuous. This function is the so-called "winding function" and it is usually introduced in trigonometry courses as a function from $\mathbf{R}$ to the unit circle. The reason the unit circle can be used instead of $O^+(E_2)$ is that this group and the unit circle have the same cardinality (Proposition 283.1). The winding function enables one to define radians in the usual way.

However, the real importance of the winding function for analysis is that it enables one to define new trigonometric functions Cosine, Sine, etc., as functions from $\mathbf{R}$ to $\mathbf{R}$ instead of as functions from $O^+(E_2)$ to $\mathbf{R}$. One simply considers the diagram

$$\mathbf{R} \xrightarrow{w} O^+(E_2) \xrightarrow{\text{cosine}} \mathbf{R}$$

and then defines

$$\text{Cosine}: \mathbf{R} \to \mathbf{R}$$

as the composition of cosine and w. The other trigonometric functions are defined similarly. Needless to say, if the trigonometric function is different from Cosine or Sine, the usual (infinitely many!) points have to be removed from $\mathbf{R}$ so that w does not have one of the two removed elements of $O^+(E_2)$ as value. Analysis is in need of these definitions of the trigonometric functions as functions from $\mathbf{R}$ to $\mathbf{R}$. For instance, how, otherwise, could one define derivatives?

In summary, the measurement of rotations in radians or other units, whether carried out directly or through the intervention of angles, is an absolute necessity for analysis, but is of minor importance for geometry proper.

54. LORENTZ TRANSFORMATIONS

We assume in this section that V is an Artinian plane and M, N is an Artinian coordinate system of V.

If c is a nonzero scalar, the vectors

$$A_1 = M + N \quad \text{and} \quad A_2 = c(M - N)$$

form a rectangular coordinate system of V (since char $k \neq 2$). We choose a rotation ρ of V and ask for the matrix μ of ρ relative to the coordinate system A_1, A_2.

We have already shown (page 273) that, relative to the coordinate system M, N, the matrix of ρ is

$$\begin{pmatrix} a & 0 \\ 0 & a^{-1} \end{pmatrix}$$

for a nonzero scalar a. Since

$$(A_1, A_2) = (M, N) \begin{pmatrix} 1 & c \\ 1 & -c \end{pmatrix},$$

it follows from linear algebra that

$$\mu = \begin{pmatrix} 1 & c \\ 1 & -c \end{pmatrix}^{-1} \begin{pmatrix} a & 0 \\ 0 & a^{-1} \end{pmatrix} \begin{pmatrix} 1 & c \\ 1 & -c \end{pmatrix}.$$

We multiply the matrices and find that

$$\mu = \begin{pmatrix} x & cy \\ c^{-1}y & x \end{pmatrix}$$

where

$$x^2 - y^2 = 1 \quad \text{and} \quad x = \tfrac{1}{2}(a + a^{-1}), \quad y = \tfrac{1}{2}(a - a^{-1}).$$

We may regard the last two equations as a parametrization of the hyperbola $x^2 - y^2 = 1$ in terms of the parameter a. If (x, y) is a point on this hyperbola, the corresponding parameter value is

$$a = x + y \quad \text{and} \quad a^{-1} = x - y.$$

We conclude that the mapping

$$(x, y) \to \begin{pmatrix} x & cy \\ c^{-1}y & x \end{pmatrix}$$

may be interpreted as a one-to-one correspondence between the points of the hyperbola $x^2 - y^2 = 1$ and the rotations of the Artinian plane. It is now but a step to the Lorentz transformations.

Let $k = \mathbf{R}$ and hence V is now the Lorentz plane. If v is the new variable $v = -cy/x$, a slight computation shows that the point (x, y) of the hyperbola $x^2 - y^2 = 1$ satisfies

$$x = \frac{1}{\pm\sqrt{1 - (v^2/c^2)}} \quad \text{and} \quad y = \frac{-vc^{-1}}{\pm\sqrt{1 - (v^2/c^2)}}.$$

Consequently,

$$\mu = \begin{pmatrix} \dfrac{1}{\pm\sqrt{1 - (v^2/c^2)}} & \dfrac{-v}{\pm\sqrt{1 - (v^2/c^2)}} \\[3mm] \dfrac{-vc^{-2}}{\pm\sqrt{1 - (v^2/c^2)}} & \dfrac{1}{\pm\sqrt{1 - (v^2/c^2)}} \end{pmatrix}$$

which is precisely the matrix of the Lorentz transformations as used in relativity theory (see P. G. Bergmann, *Introduction to the Theory of Relativity*, Equations (4.13) on page 35, [5]).

It is clear that the Lorentz transformations describe the rotations of the Lorentz plane in the special rectangular coordinate system $M + N$, $c(M - N)$ where c is the speed of light. We warn the reader that the famous Lorentz group is not the group of rotations of the Lorentz plane, but is the commutator subgroup of the rotation group of Minkowski space. The purpose of the following exercises is to develop more technique with Artinian planes.

Exercises

1. Assume that τ is a symmetry of V. The matrix of τ, relative to the coordinate system M, N, is

$$\begin{pmatrix} 0 & b \\ b^{-1} & 0 \end{pmatrix}$$

for some nonzero scalar b. Prove that
   a. The vector $A = bM - N$ is not isotropic and τ is the symmetry τ_A.
   b. $\langle A \rangle^* = \langle bM + N \rangle$.
   c. The line $\langle bM + N \rangle$ is the fixed space of τ and the line $\langle bM - N \rangle$ consists of the vectors inverted by τ.
2. Let σ be an isometry of V with the property that $\sigma M = 2N$.
   a. Prove that σ is a symmetry of V.
   b. Find all fixed vectors of σ whose square is 1.
   c. Find all vectors which are inverted by σ and have square -1.

3. Assume ρ is a rotation of V. The matrix of ρ relative to the coordinate system M, N is

$$\begin{pmatrix} a & 0 \\ 0 & a^{-1} \end{pmatrix}$$

for some nonzero scalar a. By the Cartan–Dieudonné theorem, ρ is the product of two symmetries. Find scalars s, t, u, v (as functions of a) such that the vectors $A = sM + tN$ and $B = uM + vN$ are nonisotropic and $\rho = \tau_A \tau_B$. (Hint: Use the simple matrix representation of τ_A and τ_B relative to the coordinate system M, N.)

4. Let c be a nonzero scalar and consider the rectangular coordinate system

$$A_1 = M + N \qquad \text{and} \qquad A_2 = c(M - N)$$

of the beginning of this section.

a. Relative to this coordinate system, prove that the matrix of a symmetry has the form

$$\begin{pmatrix} cy & x \\ x & c^{-1}y \end{pmatrix}$$

where $x^2 - y^2 = 1$. (Hint: The matrix of the symmetry $A_1 \rightleftharpoons A_2$ relative to the coordinate system A_1, A_2 is

$$\begin{pmatrix} 0 & 1 \\ 1 & 0 \end{pmatrix}.$$

Hence, if M is the matrix of a symmetry,

$$M\begin{pmatrix} 0 & 1 \\ 1 & 0 \end{pmatrix}$$

is the matrix of a rotation. Now use the results obtained on page 311.)

b. Prove that the mapping

$$(x, y) \rightarrow \begin{pmatrix} cy & x \\ x & c^{-1}y \end{pmatrix}$$

may be interpreted as a one-to-one correspondence between the points of the hyperbola $x^2 - y^2 = 1$ and the set of reflections (symmetries) of the Artinian plane.

5. For many theoretical questions concerning the Artinian plane, the Artinian coordinate system M, N is best, even though it is not a rectangular coordinate system. For questions in relativity theory, the rectangular coordinate system $M + N$, $c(M - N)$ is best suited. In this

exercise, we show how another rectangular coordinate system is often used in theoretical investigations. We showed, in Exercise 189.8, that the vectors

$$A_1 = \tfrac{1}{2}M + N \qquad \text{and} \qquad A_2 = \tfrac{1}{2}M - N$$

form a rectangular coordinate system relative to which the matrix of V is

$$\begin{pmatrix} 1 & 0 \\ 0 & -1 \end{pmatrix}.$$

We also saw that the equation of the unit circle of V in the coordinate system A_1, A_2 is $x^2 - y^2 = 1$ (Exercise 190.9). Henceforth, this curve will be called the unit circle instead of the hyperbola $x^2 - y^2 = 1$.

a. Prove that, relative to the coordinate system A_1, A_2, the matrix of a rotation has the form

$$\begin{pmatrix} x & y \\ y & x \end{pmatrix}$$

where $x^2 - y^2 = 1$. (Hint: The matrix of a given rotation relative to the coordinate system M, N is

$$\begin{pmatrix} a & 0 \\ 0 & a^{-1} \end{pmatrix}$$

while, furthermore,

$$(A_1, A_2) = (M, N)\begin{pmatrix} \tfrac{1}{2} & \tfrac{1}{2} \\ 1 & -1 \end{pmatrix}.$$

The method which gave the matrix μ on page 311 can again be used here.)

b. Prove that the mapping

$$(x, y) \to \begin{pmatrix} x & y \\ y & x \end{pmatrix}$$

may be interpreted as a one-to-one correspondence between the points on the unit circle $x^2 - y^2 = 1$ and the rotations of the Artinian plane. This fact that the unit circle and the rotation group $O^+(V)$ have the same cardinality is also contained in Proposition 283.1.

c. Prove that, relative to the coordinate system A_1, A_2, the matrix of a symmetry has the form

$$\begin{pmatrix} y & x \\ x & y \end{pmatrix}$$

where $x^2 - y^2 = 1$.

d. Prove that the mapping

$$(x, y) \rightarrow \begin{pmatrix} y & x \\ x & y \end{pmatrix}$$

may be interpreted as a one-to-one correspondence between the points of the unit circle $x^2 - y^2 = 1$ and the reflections of the Artinian plane.

6. Let $\rho \neq 1_V$ be a rotation of V.
   a. Prove that the linear transformation $\rho - 1_V$ is a linear automorphism of V. Why is this result also true for non-Artinian planes?
   b. Prove that M and N are characteristic vectors of $\rho - 1_V$.
   c. If X is a nonisotropic vector of V, prove that $(\rho - 1_V)X$ is also nonisotropic. (Hint: This follows from Parts a and b by linear algebra alone.)

7. Let τ be a symmetry of V.
   a. Prove that the linear transformation $\tau - 1_V$ of V into itself is singular. Why is this also correct if τ is an arbitrary involution of type $< n$ of an arbitrary nonsingular space?
   b. Prove that $\tau - 1_V$ maps all of V on the nonsingular line which consists of the vectors inverted by τ. Why is this also correct for non-Artinian planes?
   c. Prove that $\tau - 1_V$ maps the line of fixed vectors of τ onto $\mathbf{0}$ and all other vectors of V on nonisotropic vectors. In particular, the vectors $(\tau - 1_V)M$ and $(\tau - 1_V)N$ are not isotropic.

55. ROTATIONS AND REFLECTIONS IN THREE-SPACE

We now investigate three-dimensional geometry, traditionally called "solid geometry." These next three sections describe the various kinds of isometries which occur in three-space.

We assume throughout this section that $n = 3$.

By the Cartan–Dieudonné theorem, every isometry of V is either 1_V, a symmetry, or the product of two or three symmetries. This already shows that a rotation is either 1_V or the product of two and not less than two symmetries while a reflection is either a symmetry or the product of three and not less than three symmetries. In Euclidean three-space, we are accustomed to the fact that a rotation different from 1_V rotates the space around an axis of rotation. The following proposition shows that this is true for all spaces. As always, the fixed space of an isometry is the linear subspace of V consisting of the vectors left fixed by the isometry.

Proposition 316.1.   A rotation different from 1_V has a line as fixed space.

Proof. Let $\rho \neq 1_V$ be a rotation. Then ρ is the product of two symmetries and hence leaves a nonzero vector A fixed (Corollary 258.1). We must show that the fixed space of ρ is the line $\langle A \rangle$, equivalently, if the vector B does not lie on $\langle A \rangle$, $\rho B \neq B$. Suppose that $B \notin \langle A \rangle$, but $\rho B = B$. Then ρ would be a rotation which leaves the plane $\langle A, B \rangle$ pointwise fixed. However, the planes are now hyperplanes and hence the only rotation which leaves a plane pointwise fixed is 1_V (Exercise 262.13). This contradicts the hypothesis that $\rho \neq 1_V$, consequently, $\rho B \neq B$. Done.

A line which is left pointwise fixed by an isometry is called an **axis** of that isometry; this terminology conforms with our use of the word axis on page 244. The rotation 1_V has all lines of V as axis while every other rotation has a unique line as axis. An axis can be either a nonsingular line or a null line.

It has already been pointed out that there are four kinds of isometries, depending on the number of symmetries necessary to factor them. The four kinds are characterized by their fixed spaces as follows.

Class 1. $\{1_V\}$. The fixed space is V.
Class 2. Symmetries. The fixed space is a nonsingular plane.
Class 3. Rotations $\neq 1_V$. The fixed space is a line. These isometries are products of two and not less than two symmetries.
Class 4. Reflections which are not symmetries. These isometries are products of three and not less than three symmetries. We shall show that the fixed space of such an isometry is $\{0\}$. Readers who have done Exercise 262.14 know this already, but we shall find an independent proof of this fact by making a careful study of the isometries which leave a nonzero vector fixed.

An isometry σ leaves a nonzero vector A fixed if and only if the line $\langle A \rangle$ is an axis of σ. Let us denote the set of isometries which have a line l as axis by $O(V; l)$. Hence $\sigma \in O(V; l)$ if and only if the fixed space of σ contains l; this fixed space may be larger than l, as is the case when $\sigma = 1_V$. We recall from Exercise 252.4 that $O(V; l)$ is a subgroup of $O^+(V)$.

Similarly, the set of rotations of V with the line l as axis is denoted by $O^+(V; l)$. Clearly, $O^+(V; l)$ is a subgroup of $O^+(V)$.

We assume for the remainder of this section that l is a nonsingular line.

In this case, the groups $O(V; l)$ and $O^+(V; l)$ have been investigated in Exercise 252.5 with no restriction on the dimension of V or the nonsingular axis. It was found that one puts $V = l \oplus l^*$ and then each $\sigma \in O(V; l)$ has the form $\sigma = 1_l \oplus \sigma'$ where σ' is the restriction of σ to the nonsingular hyperplane l^*. In our case, l^* is a nonsingular plane. The isometry σ is completely determined by l and its restriction σ' to l^* while the restriction mappings

$$\text{res}: O(V; l) \rightarrow O(l^*)$$

and

$$\text{res}: O^+(V; l) \rightarrow O^+(l^*)$$

are isomorphisms. Here $O(l^*)$ and $O^+(l^*)$ are, respectively, the orthogonal group and the rotation group of the nonsingular plane l^*. We now state these results as a proposition.

Proposition 317.1. The group $O(V; l)$ is isomorphic to the orthogonal group $O(l^*)$ of the nonsingular plane l^*. The group $O^+(V; l)$ is isomorphic to the rotation group $O^+(l^*)$ of l^*.

For example, in Euclidean three-space, all lines are nonsingular and all planes are Euclidean. Hence all groups $O(V; l)$ are isomorphic to the orthogonal group of the Euclidean plane and all groups $O^+(V; l)$ are isomorphic to the circle group. If $\sigma \in O^+(V; l)$, the restriction σ' of σ to l^* is a rotation of the Euclidean plane l^*. If we use the language of angles, σ' is determined by its angle of rotation and this angle is also called the angle of σ. Since σ is completely determined by l and σ', we obtain the usual statement that a rotation in Euclidean three-space is determined by its axis and its angle.

The situation is radically different in the three-space over **R** with matrix

$$\begin{pmatrix} 1 & 0 & 0 \\ 0 & 1 & 0 \\ 0 & 0 & -1 \end{pmatrix}.$$

(The underlying coordinate system is only mentioned when it is relevant for the discussion.) In addition to containing singular planes, V now contains two types of nonsingular planes and the structure of the groups $O(V; l)$ and $O^+(V; l)$ depends heavily on the choice of the nonsingular line l. We know from Exercise 164.7 that, if a line lies inside the light cone, its orthogonal complement is a Euclidean plane and hence the structure of the groups $O(V; l)$ and $O^+(V; l)$ does not change as l moves inside the light cone. If a line m lies outside the light cone, its orthogonal complement is a Lorentz plane and, again, the structure of the groups $O(V; m)$ and $O^+(V; m)$ does not depend on which line m outside the light cone is used. Nevertheless, the groups $O(V; l)$ and $O(V; m)$ are not isomorphic; the groups $O^+(V; l)$ and $O^+(V; m)$ are also not isomorphic as is shown in Exercise 1 below.

Exercises

*1. Let V be the three-space over $\mathbf{R}$ with the matrix

$$\begin{pmatrix} 1 & 0 & 0 \\ 0 & 1 & 0 \\ 0 & 0 & -1 \end{pmatrix}.$$

a. If l is a line inside the light cone, prove that the group $O^+(V; l)$ is isomorphic to the circle group.

b. If m is a line outside the light cone, prove that the group $O^+(V; m)$ is isomorphic to the multiplicative group $\mathbf{R}^*$.

c. Prove that the groups $O^+(V; l)$ and $O^+(V; m)$ are not isomorphic. (Hint: The problem is to show that $\mathbf{R}^*$ is not isomorphic to the circle group. What is the order of the complex number i of the circle group? Does $\mathbf{R}^*$ have an element of that order?)

*2. If $\sigma \in O(V; l)$, prove that σ is the product of at most two symmetries.

56. NULL AXES IN THREE-SPACE

We continue the investigation of the groups $O(V; l)$ and $O^+(V; l)$, but now assume that l is a null line in three-space. If l and m are both null lines, the Witt theorem guarantees the existence of an isometry σ of V such that $\sigma l = m$. From this alone it follows easily that $O(V; l) \simeq O(V; m)$ and $O^+(V; l) \simeq O^+(V; m)$ (see Exercise 326.12). We shall study the structure of the groups $O(V; l)$ and $O^+(V; l)$ directly and obtain the isomorphisms in this way. Furthermore, this will bring out the surprising fact that, when l is a line on the light cone, the structure of the groups $O(V; l)$ and $O^+(V; l)$ has nothing to do with the geometry of V, but is completely determined by the field k alone.

We assume in this whole section that $n = 3$ and l is a null line of V.

Consequently, $l \subset l^*$ and l^* is a singular plane with l as radical. Observe that l^* is not a null plane since its radical l is not l^*. For reasons explained in Exercise 151.4b, l^* should be visualized as the tangent plane of the light cone along the generator l. Every isometry of l^* transforms the radical l onto itself, but may not leave l pointwise fixed. It is clear that those isometries of l^* which leave l pointwise fixed form a subgroup of the orthogonal group $O(l^*)$ of l^*; we denote this subgroup by $O(l^*; l)$.

An isometry $\sigma \in O(V; l)$ transforms l^* onto itself and the restriction σ' of σ to l^* belongs to $O(l^*; l)$. In other words, the restriction mapping $\sigma \to \sigma'$ maps $O(V; l)$ into $O(l^*; l)$.

Proposition 319.1.   The restriction mapping

$$\text{res}: O(V; l) \to O(l^*; l)$$

is an isomorphism.

Proof. We leave it to the reader to prove that res is a group homomorphism. If $\alpha \in O(l^*; l)$, α can be extended to an isometry σ of V by the Witt theorem; obviously, $\sigma \in O(V; l)$ and res $\sigma = \alpha$ which shows that res is onto. There still remains to show that the kernel of res is 1_V. For this, let $\sigma \in O(V; l)$ and assume that res $\sigma = 1_{l^*}$. Then σ leaves the singular hyperplane l^* pointwise fixed and hence $\sigma = 1_V$ (Exercise 262.13). Done.

According to the last proposition, the groups $O(V; l)$ and $O(l^*; l)$ have the same structure and we now concentrate on the latter group. This group consists of the isometries of the singular plane l^* which leave the radical of that plane pointwise fixed. We study this situation in the exercises below.

Exercises

In Exercises 1–11, P stands for a singular plane whose radical is a line m. N is a fixed nonzero vector in m and A is a fixed vector of P which does not belong to m. It follows that $m = \langle N \rangle$, $N^2 = 0$ and $N \perp A$ since $m = \text{rad } P$. Consequently, N, A is a rectangular coordinate system of P. The isomorphism between the ring of 2×2 matrices and the ring of linear transformations of P is with respect to this fixed coordinate system.

1. Prove that $A^2 \neq 0$.
2. An isometry of P must carry the radical m onto itself, but may not leave m pointwise fixed.
   a. Prove that those isometries of P which leave m pointwise fixed form a group. We denote this group by $O(P; m)$. It is obvious that $O(P; m)$ is a subgroup of the orthogonal group of P.
   b. Let $\alpha \in O(P; m)$ and put $\alpha A = xN + yA$. Prove that $y = \pm 1$.
   c. Since

$$(\alpha N, \alpha A) = (N, A)\begin{pmatrix} 1 & x \\ 0 & \pm 1 \end{pmatrix},$$

the matrix of α is either

$$\begin{pmatrix} 1 & x \\ 0 & 1 \end{pmatrix} \quad \text{or} \quad \begin{pmatrix} 1 & x \\ 0 & -1 \end{pmatrix}.$$

Prove that, for every scalar x, including 0, both these matrices determine an isometry of P which belongs to $O(P; m)$. The isometry defined by

$$\begin{pmatrix} 1 & x \\ 0 & 1 \end{pmatrix}$$

is denoted by $\alpha_x^{\ +}$ and the isometry defined by

$$\begin{pmatrix} 1 & x \\ 0 & -1 \end{pmatrix}$$

is denoted by $\alpha_x^{\ -}$. Consequently, the group $O(P; m)$ is the disjoint union of the two subsets

$$\{\alpha_x^{\ +} \mid x \in k\} \quad \text{and} \quad \{\alpha_x^{\ -} \mid x \in k\}.$$

*3. a. Prove that the set of matrices

$$\psi = \left\{ \begin{pmatrix} 1 & x \\ 0 & \pm 1 \end{pmatrix} \mid x \in k \right\}$$

is a group under multiplication of matrices.
   b. Prove that the mapping

$$\begin{pmatrix} 1 & x \\ 0 & 1 \end{pmatrix} \rightarrow \alpha_x^{\ +} \quad \text{and} \quad \begin{pmatrix} 1 & x \\ 0 & -1 \end{pmatrix} \rightarrow \alpha_x^{\ -}$$

is an isomorphism from the group ψ onto the group $O(P; m)$.
*4. We saw in the previous exercise that $O(P; m) \simeq \psi$ and we now study the latter group.
   a. Prove that the matrices

$$\left\{ \begin{pmatrix} 1 & x \\ 0 & 1 \end{pmatrix} \mid x \in k \right\}$$

form a subgroup of ψ. We denote this group by ψ^+.
   b. Prove that the mapping

$$x \rightarrow \begin{pmatrix} 1 & x \\ 0 & 1 \end{pmatrix}$$

is an isomorphism from the additive group k^+ onto the multiplicative group ψ^+.

c. Prove that ψ^+ is a commutative, normal subgroup of ψ of index two. (Hint: Consider the determinant mapping

$$\det: \psi \to \{1, -1\}.$$

Observe that the argument uses the fact that char $k \neq 2$.)

5. In the previous exercise, we saw that the group ψ contains a normal, commutative subgroup of index two. We denote the isomorphism from ψ onto $O(P; m)$ by ϕ and transfer our knowledge about ψ to $O(P; m)$ by means of ϕ. Observe that ϕ maps ψ^+ onto the set of isometries $\{\alpha_x{}^+ \mid x \in k\}$. We denote this set of isometries by $O^+(P; m)$.

a. Prove that $O^+(P; m)$ is a normal, commutative subgroup of $O(P; m)$ of index two. The following diagram shows the relationships among the groups under consideration.

$$
\begin{array}{ccc}
\psi & \overset{\phi}{\longrightarrow} & O(P; m) \\
| & & | \\
\psi^+ & \overset{\phi}{\longrightarrow} & O^+(P; m)
\end{array}
$$

Observe that the subgroup $O^+(P; m)$ has two cosets, the group $O^+(P; m)$ itself and the coset $\{\alpha_x{}^- \mid x \in k\}$.

b. Prove that the multiplicative group $O^+(P; m)$ is isomorphic to the additive group k^+. Observe that the structure of the group $O^+(P; m)$ is completely determined by k alone and hence is the same for all singular planes over k which have a line as radical.

*6. a. Prove that the structure of the group $O(P; m)$ is also completely determined by the field k alone and hence is also the same for all singular planes which have a line as radical. Precisely, if P' is a singular plane with a line m' as radical, prove that $O(P; m) \simeq O(P'; m')$ and the structure of these groups depends only on k.

b. Readers who know about semidirect products of groups (see M. Hall, Jr., *The Theory of Groups*, pp. 88–90, [12]) can go further and say precisely what the structure of the group $O(P; m)$ is by using the isomorphism $k^+ \simeq O^+(P; m)$. (Hint: Study the exact sequence

$$1 \longrightarrow k^+ \longrightarrow O(P; m) \overset{\det}{\longrightarrow} \{1, -1\} \longrightarrow 1$$

where det denotes the determinant function. Conclude that $O(P; m)$ is the semidirect product of the additive group k^+ with the cyclic group of order two relative to the automorphism $x \to -x$ of k^+. Of course, this again shows that the structure of $O(P; m)$ does not depend on P.)

*7. Since the plane P is singular, there is an isometry of P with determinant a for every $a \in k^*$ (Exercise 216.3). We now show that the determinants of isometries which leave m pointwise fixed are very restricted.

a. Prove that an isometry of $O(P; m)$ has determinant ± 1.

b. Prove that the isometries of $O(P; m)$ with determinant 1 form the group $O^+(P; m)$. Also show that $O^+(P; m)$ consists of the rotations of P which leave m pointwise fixed (see Exercise 230.10 for the definitions of rotation and reflection of singular spaces).

c. According to Exercise 5b above, the subgroup $O^+(P; m)$ of $O(P; m)$ has the two cosets

$$\{\alpha_x^+ \mid x \in k\} \quad \text{and} \quad \{\alpha_x^- \mid x \in k\}.$$

We denote the latter group by $O^-(P; m)$. Prove that $O^-(P; m)$ consists of the isometries of $O(P; m)$ with determinant -1. Also show that this coset consists of the reflections of P which leave m pointwise fixed. Readers who have done Exercise 230.10g can go further and prove that all the reflections of P are symmetries, even those which do not leave m pointwise fixed.

*8. We saw in the previous exercise that isometries in the group

$$O^+(P; m) = \{\alpha_x^+ \mid x \in k\}$$

have determinant 1 and are the rotations of P while isometries in the coset

$$O^-(P; m) = \{\alpha_x^- \mid x \in k\}$$

have determinant -1 and are reflections of P. The isometries in these two cosets can also be distinguished by the vectors they invert. For the sake of completeness, we shall also discuss the fixed spaces of these isometries. Let $x \in k$.

a. Prove that m is the fixed space of α_x^-. Also show that $\alpha_0^+ = 1_P$ and, if $x \neq 0$, m is the fixed space of α_x^+. Observe that the fixed spaces do not distinguish between α_x^+ and α_x^- when $x \neq 0$.

b. Prove that α_x^+ inverts no nonzero vector.

c. Prove that the space of vectors inverted by α_x^- is the line $\langle xN - 2A \rangle$. Also show that this line is nonsingular.

d. Prove that $O^+(P; m)$ consists of the isometries of P which leave m pointwise fixed and invert no vector.

e. Prove that the coset $O^-(P; m)$ consists of the isometries of P which leave m pointwise fixed and invert a nonzero vector. Prove that every vector inverted by α_x^- is nonisotropic.

*9. If $x \in k$, prove that $\alpha_x^+ = \alpha_0^- \alpha_x^-$. We conclude that every rotation of P which leaves m pointwise fixed is the product of two reflections which leave m pointwise fixed.

*10. If $x \in k$, prove that $\alpha_x{}^+ = (\alpha_{x/2}^+)^2$. Consequently, every element of $O^+(P; m)$ has a square root in $O^+(P; m)$.

11. This exercise is not needed for the sequel. In Exercises 1–8 above, we discussed only those isometries of P which leave m pointwise fixed. We suggest the reader extend the discussion to the full orthogonal group $O(P)$. (Hint: $P = m \oplus \langle A \rangle$ even though $\langle A \rangle$ is not the orthogonal complement of m. An isometry σ of P must transform m onto itself and hence the matrix of σ is

$$\begin{pmatrix} z & x \\ 0 & y \end{pmatrix}$$

for scalars x, y, z. Prove again that $y = \pm 1$ and that x can be any scalar while z can be any nonzero scalar.) Observe that, consequently, $\det \sigma$ can be any nonzero scalar. Prove that σ is a rotation if and only if $y = 1$. Proceed by answering the following questions. Is the rotation group $O^+(P)$ commutative as in the case of nonsingular planes? Is $O^+(P)$ a normal subgroup of $O(P)$ of index two? Are all reflections symmetries? Does the Witt theorem hold? Does the Cartan–Dieudonné theorem hold? And so on!

We return to the null line l of V and the isomorphism

$$\text{res: } O(V; l) \to O(l^*; l)$$

of Proposition 319.1. If m is another null line, one obtains the isomorphism $O(V; m) \simeq O(m^*; m)$. Since the groups $O(l^*; l)$ and $O(m^*; m)$ are isomorphic (Exercise 321.6a), we conclude that

$$O(V; l) \simeq O(V; m).$$

In other words, the structure of the group $O(V; l)$ does not depend on the choice of the null line l. Furthermore, the structure of the group $O(l^*; l)$ is determined by the field k and hence the same is true for the group $O(V; l)$.

Let ext: $O(l^*; l) \to O(V; l)$ denote the isomorphism which is the inverse of the isomorphism res. Of course, ext stands for "extension" and, if $\alpha \in O(l^*; l)$, ext α is the unique isometry in $O(V; l)$ which acts on the plane l^* the same way as α. Consequently, if an isometry σ of V leaves l pointwise fixed, and, if $\sigma A = \alpha A$ for even one vector A of l^* such that $A \notin l$, then $\sigma = $ ext α. We now transfer the properties of the group $O(l^*; l)$ to the group $O(V; l)$ by the isomorphism ext.

The group $O(l^*; l)$ contains the subgroup $O^+(l^*; l)$ of index two and hence $O^+(l^*; l)$ has two cosets, namely, $O^+(l^*; l)$ itself and one other coset which was denoted by $O^-(l^*; l)$. We recall that the coset $O^-(l^*; l)$ consists of the isometries of l^* which leave l pointwise fixed and invert a nonisotropic vector (Exercise 322.8e). Furthermore, if $\alpha \in O^+(l^*; l)$, then $\alpha = \beta\gamma$ where $\beta, \gamma \in O^-(l^*; l)$ (Exercise 322.9).

Proposition 324.1. If $\alpha \in O^+(l^*; l)$, then ext α is a rotation of V. If $\alpha \in O^-(l^*; l)$, ext α is a symmetry of V.

Proof. Assume that $\alpha \in O^-(l^*; l)$. Choose a nonisotropic vector $B \in l^*$ such that $\alpha B = -B$. The symmetry τ_B of V also inverts B; since B is orthogonal to the vectors of l, τ_B leaves l pointwise fixed. Hence $\tau_B = $ ext α.

Now assume $\alpha \in O^+(l^*; l)$. Put $\alpha = \beta\gamma$ where $\beta, \gamma \in O^-(l^*; l)$. Then

$$\text{ext } \alpha = (\text{ext } \beta)(\text{ext } \gamma)$$

and, since ext β and ext γ are symmetries of V, ext α is a rotation. Done.

According to the above proposition, the isomorphism ext maps the group $O^+(l^*; l)$ onto rotations of $O(V; l)$, that is, onto the subgroup $O^+(V; l)$ of $O(V; l)$. In other words, the tower

$$O^+(l^*; l) \subset O(l^*; l)$$

is transported by ext onto the tower

$$O^+(V; l) \subset O(V; l)$$

as shown in the following diagram.

$$O(l^*; l) \xrightarrow{\text{ext}} O(V; l)$$
$$|\qquad\qquad\qquad |$$
$$O^+(l^*; l) \xrightarrow{\text{ext}} O^+(V; l)$$

Figure 324.1

We now draw conclusions from the fact that the algebraic properties of the tower

$$O^+(l^*; l) \subset O(l^*; l)$$

are preserved under the isomorphism ext.

Proposition 325.1. $O^+(V; l)$ is a commutative, normal subgroup of $O(V; l)$ of index two. Furthermore,

$$O^+(V; l) \simeq k^+.$$

Proof. $O^+(l^*; l)$ is a commutative, normal subgroup of $O(l^*; l)$ of index two and $O^+(l^*; l) \simeq k^+$ (Exercises 320.4b and 321.4c). By means of the isomorphism ext, these same properties hold for the tower

$$O^+(V; l) \subset O(V; l).$$

Done.

Observe that the structure of $O^+(V; l)$ is completely determined by the field k alone and does not depend on the choice of the null line l.

Proposition 325.2. If $\rho \in O^+(V; l)$, there exists a $\sigma \in O^+(V; l)$ such that $\rho = \sigma^2$.

Proof. Every element of $O^+(l^*; l)$ has a square root in $O^+(l^*; l)$ (Exercise 323.10). The isomorphism

$$\text{ext}: O^+(l^*; l) \to O^+(V; l)$$

shows that the same is true in $O^+(V; l)$. Done.

According to Proposition 325.1, the subgroup $O^+(V; l)$ of $O(V; l)$ has two cosets, the group $O^+(V; l)$ itself and one other coset which we denote by $O^-(V; l)$.

Proposition 325.3. The group $O^-(V; l)$ consists of symmetries of V. If $\rho \in O^+(V; l)$, $\rho = \tau_1 \tau_2$ where τ_1 and τ_2 are symmetries in $O^-(V; l)$.

Proof. It follows from Proposition 324.1 that ext maps the coset $O^-(l^*; l)$ onto the coset $O^-(V; l)$ and, by the same proposition, that $O^-(V; l)$ consists of symmetries. Finally, let $\rho \in O^+(V; l)$; then $\rho = \text{ext } \alpha$ for some

$$\alpha \in O^+(l^*; l).$$

Put $\alpha = \beta\gamma$ where $\beta, \gamma \in O^-(l^*; l)$; then

$$\rho = (\text{ext } \alpha) = (\text{ext } \beta)(\text{ext } \gamma)$$

where ext β and ext γ are symmetries in $O^-(V; l)$. Done.

An isometry which leaves only the origin fixed must be a reflection which cannot be written as the product of less than three symmetries. We remarked on page 316 that the converse is true, that is, a reflection which is not the product of less than three symmetries leaves only the origin fixed. We are now ready to prove this.

Proposition 326.1. An isometry of V leaves a nonzero vector fixed if and only if it is the product of two or less symmetries.

Proof. Let σ be an isometry of V. We have only to show that, if $\sigma A = A$ where $A \neq \mathbf{0}$, then σ is the product of two or less symmetries. If $A^2 \neq 0$, σ has the nonsingular line $\langle A \rangle$ as axis and the proposition follows from Exercise 318.2. If $A^2 = 0$, σ has the null line $\langle A \rangle$ as axis. Then σ is either a symmetry or a rotation (Proposition 325.2) and hence can be factored into two or less symmetries. Done.

Exercises

*12. The structure of the groups $O(V; l)$ and $O^+(V; l)$ does not depend upon the choice of the null line l because all null lines are isometric. This exercise shows that this fact has nothing to do with the dimension of V or l.

    Let W be a nonsingular metric vector space of arbitrary (finite) dimension. Assume that T and U are isometric subspaces of W which may or may not be singular. We put

$$O(W; T) = \{\sigma \mid \sigma \in O(W), \quad \sigma \mid T = 1_T\}$$

and

$$O^+(W; T) = \{\rho \mid \rho \in O^+(W), \quad \rho \mid T = 1_T\}$$

and, similarly, for $O(W; U)$ and $O^+(W; U)$.
    a. Prove that $O(W; T) \simeq O(W; U)$. (Hint: By the Witt theorem, there is an isometry α of W such that $\alpha T = U$. For each $\sigma \in O(W; T)$, consider the isometry $\alpha \sigma \alpha^{-1}$.)
    b. Prove that $O^+(W; T) \simeq O^+(W; U)$.

13. The fact that res: $O(V; l) \to O(l^*; l)$ is an isomorphism shows that an isometry σ of l^* which leaves l pointwise fixed can be extended in only one way to an isometry of V. Actually, all isometries of l^*, even the ones which do not leave l pointwise fixed, can be extended to only

one isometry of V. The present exercise shows that this depends only on the fact that l^* is a singular hyperplane of V.

Let W be a nonsingular metric vector space of arbitrary (finite) dimension. Suppose that H is a singular hyperplane of W.

a. Prove that H has a line as radical.

b. If σ is an isometry of H, prove that σ can be extended to one and only one isometry of W. (Hint: Apply Exercise 262.13.)

14. Assume A_1, A_2, A_3 is a rectangular coordinate system of V relative to which the matrix of V is

$$\begin{pmatrix} 1 & 0 & 0 \\ 0 & 1 & 0 \\ 0 & 0 & -1 \end{pmatrix}.$$

Put $N = A_1 + A_2$ and $l = \langle N \rangle$.

a. Prove that l is a null line of V.

b. Prove that $l^* = \langle N, A_1 \rangle$.

c. Consider the isometry σ of l^* where $\sigma N = N$ and $\sigma A_1 = 3N - A_1$. Find a nonisotropic vector B in V such that the symmetry τ_B is the extension of σ to V.

15. a. Use our assumption that V is a nonsingular three-space and l is a null line of V and prove that

$$V = W \oplus Art_2$$

where W is a nonsingular line. (Hint: The nonsingular completions of l are Artinian planes.)

b. Prove that there are as many nonsingular three-spaces containing a null line (up to isometry) as there are elements in the group k^*/k^{*2}.

c. If $k = \mathbf{R}$ or k is finite, prove that there are precisely two nonsingular three-spaces containing a null line (up to isometry).

Remark. This exercise demonstrates again the power of the geometric approach. In algebraic language, we have proved the following: Consider the set of nonsingular quadratic forms in three variables which have a nontrivial solution. There are as many inequivalent such forms as there are elements in the group k^*/k^{*2}. ("Inequivalent" refers to forms which cannot be transformed into one another by a nonsingular linear transformation of variables.) A strictly algebraic proof of this result would be rather difficult and not at all instructive.

57. REFLECTIONS IN THREE-SPACE

We assume throughout this section that n = 3.

The isometries of V which leave a nonzero vector fixed have just been investigated. These are the isometries which can be factored into two or less symmetries (Proposition 326.1). We now turn to the isometries which leave only the origin fixed, equivalently, to the isometries which can be factored into three but not less than three symmetries. Clearly, they are all reflections.

Exercises

1. Prove that the reflections of V consist of symmetries, together with those reflections which leave only the origin fixed. Observe that -1_V is a reflection of the second kind.

2. Let σ be an isometry of V. Prove that
   a. The fixed space of σ is the space of vectors inverted by $-\sigma$.
   b. The space of vectors inverted by σ is the fixed space of $-\sigma$.

*3. Prove that ρ is a rotation of V which inverts no nonzero vector if and only if $-\rho$ is a reflection which leaves only the origin fixed.

*4. a. Prove that the involutions of V consist of $\pm 1_V$, the symmetries, and the 180° rotations.
   b. Prove that ρ is a 180° rotation if and only if $-\rho$ is a symmetry.

According to Exercise 3 above, the reflections of V which leave only the origin fixed form the set of isometries

$$\{-\rho \,|\, \rho \in O^+(V), \quad \rho \text{ inverts no nonzero vector}\}.$$

We must find the rotations of V which invert no nonzero vector.

Proposition 328.1. An isometry of V leaves a nonzero vector fixed and inverts a nonzero vector if and only if that isometry is either a symmetry or a 180° rotation.

Proof. We have already seen that symmetries and 180° rotations leave nonzero vectors fixed and invert nonzero vectors. Conversely, assume that σ is an isometry of V and A, B are nonzero vectors such that $\sigma A = A$ and $\sigma B = -B$. Then A, B are linearly independent and hence $\langle A, B \rangle$ is a plane. Furthermore,

$$\sigma^2 A = \sigma(\sigma A) = \sigma A = A$$
$$\sigma^2 B = \sigma(\sigma B) = \sigma(-B)\sigma = B;$$

hence σ^2 leaves the plane $\langle A, B \rangle$ pointwise fixed. Since σ^2 is a rotation and $\langle A, B \rangle$ is a hyperplane of V, $\sigma^2 = 1_V$ (Exercise 262.13) and, therefore, σ^2 is an involution. By Exercise 4 above, $\pm 1_V$ are the only involutions besides symmetries and $180°$ rotations. But 1_V inverts no nonzero vector and -1_V leaves no nonzero vector fixed, so σ must be either a symmetry or a $180°$ rotation. Done.

We are mainly interested in the following corollary.

Corollary 329.1. The only rotations which invert a nonzero vector are $180°$ rotations.

Proof. A $180°$ rotation inverts a whole plane of vectors. Conversely, let ρ be a rotation of V which inverts a nonzero vector. Since ρ also leaves a nonzero vector fixed, ρ is a $180°$ rotation by Proposition 328.1. Done.

We can now describe the rotations of V which invert no nonzero vector. By the above corollary, these are all the rotations except for the $180°$ rotations. For every nonsingular line, there is precisely one $180°$ rotation with that line as axis and, by definition, a null line is never the axis of a $180°$ rotation. Consequently, the rotations which invert no nonzero vector fall into the following two types.

Type 1.   For each nonsingular line l, all rotations in the group $O^+(V; l)$ except for the $180°$ rotation with l as axis. These are the rotations $1_l \oplus \sigma'$ where σ' is any rotation of the nonsingular plane l^* different from the $180°$ rotation -1_{l^*}. Observe that the rotation 1_V is of this type.

Type 2.   For each null line l, all rotations in the group $O^+(V; l)$. These are the rotations ext α where $\alpha \in O^+(l^*; l)$. Only the rotation 1_V belongs to both types.

We have seen that the reflections $-\rho$ where ρ is a rotation of Type 1 or 2 are precisely the reflections which leave only the origin fixed. Hence these reflections fall into two types which have only the reflection -1_V in common.

Type 1'.   The reflections $-\rho$ where ρ is a rotation of Type 1. In addition to the reflection -1_V, this gives the following reflections for each nonsingular line l: All reflections which invert only the vectors of l with the exception of the symmetry with respect to the nonsingular plane l^*.

Type 2'.   The reflections $-\rho$ where ρ is a rotation of Type 2. For each null line, we obtain the reflection -1_V and those reflections which invert the vectors of that line and no others.

Exercises

5. Prove that -1_V is the only reflection which is both Type 1' and Type 2'.
6. Let l be a nonsingular line in V. Prove that the reflections of Type 1' corresponding to l are the reflections of the form $-1_l \oplus \sigma'$ where σ' is any rotation of the nonsingular plane l^* different from 1_{l^*}.
7. Assume l is a null line of V. Prove that the reflections of Type 2' corresponding to l are the reflections $-\text{ext } \alpha$ where $\alpha \in O^+(l^*; l)$.
8. Prove that the isometries of V fall into the following two classes.
   Class 1. Isometries which leave a nonzero vector fixed.
   Class 2. Isometries which invert a nonzero vector.
   Prove that the isometries common to these two classes are the symmetries and the 180° rotations. Observe that every isometry in three-space either leaves a nonzero vector fixed or inverts a nonzero vector or both.
9. Assume that V has a rectangular coordinate system A_1, A_2, A_3 relative to which the matrix of V is

$$\begin{pmatrix} 1 & 0 & 0 \\ 0 & 1 & 0 \\ 0 & 0 & -1 \end{pmatrix}.$$

All matrix representations of isometries are to be relative to this coordinate system.
   a. Find a matrix which represents a reflection $\sigma \neq -1_V$ such that σ leaves no nonzero vector fixed and where the axis of the rotation $-\sigma$ is the nonsingular line $\langle A_3 \rangle$.
   b. Answer the same question as in Part a but with the axis of the rotation $-\sigma$ equal to the null line $\langle A_2 + A_3 \rangle$.

Remark on High School Teaching. The description of the various isometries of Euclidean three-space is an excellent topic for discussion in a high school course in solid geometry. The description is a simple one since no null axes occur. Combined with some work on the Witt theorem for Euclidean three-space, an attractive unit in solid geometry can be built.

It is unfortunate that solid geometry has fallen into disrepute in many high school circles. The reason for this is twofold. First, for many years, solid geometry meant plugging numbers in unproven mensuration formulas. Of course, such activity is neither geometry nor algebra. It is, at best, arithmetic of the deadliest kind. Second, even today, a discussion of solid geometry often centers around the properties of lines and planes. Although such discussions are usually logically sound, they cannot shake off the stigma of dullness. The color of solid geometry comes from the isometries of three-space and they can very well be discussed in the eleventh and twelfth grades. The authors support the study of solid geometry in high school.

58. THE CARTAN–DIEUDONNÉ THEOREM FOR ROTATIONS

It is dangerous to base one's geometric intuition on plane geometry alone. Many theorems in geometry are valid only for dimension three or larger and are false in the plane. The present section and most of the remainder of this chapter deal with such theorems. We begin with a highly geometric problem, namely, the factorization of rotations into 180° rotations.

The dimension of our nonsingular metric vector space V is again arbitrary. The Cartan–Dieudonné theorem states that every isometry is the product of not more than n symmetries. Symmetries are involutions of Type 1. What about involutions of Type 2, the 180° rotations? Of course, the product of 180° rotations is always a rotation. Could it be possible that every rotation is the product of 180° rotations? The following exercises show that this is false for dimensions one and two, but is correct for dimension three.

Exercises

1. Prove that V has a 180° rotation if and only if $n \geq 2$.
2. Let $n = 1$. Prove that V has precisely one rotation. Since V has no 180° rotations, this lonely rotation is not the product of 180° rotations.
3. Let $n = 2$.
   a. Prove that V has only one 180° rotation.
   b. Prove that the only rotations which are products of 180° rotations are 1_V and -1_V.
   c. Prove that every rotation is a product of 180° rotations if and only if V is the exceptional plane.
*4. Let $n = 3$. Prove that every rotation is the product of two or less 180° rotations. (Hint: Use the fact from Exercise 328.4b that ρ is a 180° rotation if and only if $-\rho$ is a symmetry.)

We want to show that, for $n \geq 3$, every rotation is the product of n or less 180° rotations. This result could be called the Cartan–Dieudonné theorem for rotations. The following exercises prepare for the proof.

Exercises

*5. Let U be a nonsingular subspace of V.
    a. Assume that ρ is an involution of U of type r and σ is an involution of U^* of type s. Prove that $\rho \oplus \sigma$ is an involution of V of type $r + s$.
    b. If σ is a $180°$ rotation of U^*, prove that $1_U \oplus \sigma$ is a $180°$ rotation of V.

6. Let $B_1, \ldots, B_r$ be nonisotropic vectors of V and consider the symmetries $\tau_{B_1}, \ldots, \tau_{B_r}$. Prove that the isometry $\tau_{B_1} \cdots \tau_{B_r}$ leaves the subspace $\langle B_1, \ldots, B_r \rangle^*$ pointwise fixed. (Hint: The symmetry τ_{B_i} leaves the hyperplane $\langle B_i \rangle^*$ pointwise fixed and

$$\langle B_1, \ldots, B_r \rangle^* = \langle B_1 \rangle^* \cap \cdots \cap \langle B_r \rangle^*.)$$

*7. a. Prove that V contains nonsingular subspaces of dimension $0, 1, \ldots, n$.
    b. Let W be a metric vector space of dimension m whose radical has dimension s. Prove that W contains nonsingular subspaces of dimension $0, 1, \ldots, m - s$.

We are now ready to prove the main result of this section.

Theorem 332.1. (the Cartan–Dieudonné theorem for rotations). If $n \geq 3$, every rotation is the product of an even number $r \leq n$ rotations.

Proof. Every rotation ρ of V is the product $\tau_1 \cdots \tau_r$ of an even number $r \leq n$ of symmetries $\tau_1, \cdots, \tau_r$. By taking these symmetries two at a time, we have

$$\rho = (\tau_1 \tau_2) \cdots (\tau_{r-1} \tau_r)$$

and conclude that it is sufficient to show the following. If τ_1 and τ_2 are distinct symmetries of V, the rotation $\tau_1 \tau_2$ is the product of two $180°$ rotations.
    Choose nonisotropic vectors A, B in V such that $\tau_1 = \tau_A$ and $\tau_2 = \tau_B$. Since $\tau_1 \neq \tau_2$, A and B are linearly independent (Exercise 220.12) and we denote the plane $\langle A, B \rangle$ by P. This plane contains nonisotropic vectors and hence is not a null plane. This is the same as saying that P is either nonsingular or its radical is a line. The $(n - 2)$-dimensional space P^* has the same radical as P and hence P^* is either nonsingular or has a line as radical. Now $n - 2 \geq 1$ since $n \geq 3$; it follows from Exercise 332.7b that P^* contains a nonsingular subspace W of dimension $n - 3$.

The rotation $\tau_1 \tau_2$ leaves P^* pointwise fixed and hence $\tau_1 \tau_2$ leaves W pointwise fixed. Consequently,

$$\tau_1 \tau_2 = 1_W \oplus \rho$$

where $\rho \neq 1_{W^*}$ is a rotation of the nonsingular three-space W^*. We showed in Exercise 331.4 that $\rho = \alpha\beta$ where α and β are $180°$ rotations of W^*. Thus

$$\tau_1 \tau_2 = 1_W \oplus \rho = 1_W \oplus \alpha\beta = (1_W \oplus \alpha)(1_W \oplus \beta)$$

and we know that $1_W \oplus \alpha$ and $1_W \oplus \beta$ are $180°$ rotations (Exercise 332.5b). Done.

59. THE COMMUTATOR SUBGROUP OF A GROUP

We found it very interesting that the plane rotation group is always commutative (Theorem 272.1). Of course, the one-dimensional rotation group consists of 1_V and hence is also commutative. Even so, in nature, operations usually do not commute, so we refuse to believe that rotation groups of metric vector spaces are always commutative. Indeed, we shall see that the rotation group $O^+(V)$ is never commutative if $n \geq 3$. Since $O^+(V)$ is a subgroup of $O(V)$, this also shows that $O(V)$ is never commutative if $n \geq 3$.

The amount by which a group fails to be commutative is measured by the size of its "commutator subgroup." The remainder of this section contains no geometry but develops the theory of the commutator subgroup for readers who are not acquainted with this concept. Other readers need study only Propositions 336.1 and 336.2.

Throughout this section, G denotes a group written multiplicatively with 1 as unit element.

If $a, b \in G$, the **commutator** of the ordered pair (a, b) is the group element $a^{-1}b^{-1}ab$. Observe that a and b commute if and only if $a^{-1}b^{-1}ab = 1$. Indeed, one thinks of the commutator of (a, b) as a measure for the amount by which the group elements a and b fail to commute. By a "commutator of G," one means a commutator of a pair of elements (a, b) of G. (Some authors define the commutator of (a, b) as $aba^{-1}b^{-1}$. It is immaterial for group theory which definition is adopted. We chose the definition best suited to left cosets.)

Exercise

1. Let $a, b \in G$ and let $c = a^{-1}b^{-1}ab$, that is, c is the commutator of the ordered pair (a, b).
   a. Prove that c^{-1} is the commutator of the ordered pair (b, a).

b. If $g \in G$, prove that gcg^{-1} is the commutator of (gag^{-1}, gbg^{-1}).
c. The product of two commutators of G is, in general, not a commu-tator. Nevertheless, prove that, if $d \in G$ and c' is the commutator of (b, d), then cc' is the commutator of $(a^{-1}ba, a^{-1}d)$.

Since the product of two commutators of G may not be a commutator, the set of commutators

$$C = \{a^{-1}b^{-1}ab \,|\, a,b \in G\}$$

may not be a subgroup of G. In order to study the subgroup generated by C, we study first the subgroup generated by an arbitrary nonempty subset S of G.

Let $\langle S \rangle$ denote the set of products $s_1 \cdots s_h$ where either s_i or s_i^{-1} belongs to S for $i = 1, \ldots, h$; s_i may be equal to s_j for $i \neq j$ and h can be any positive integer. It is easy to show that $\langle S \rangle$ is a subgroup of G (see Exercise 2 below) and this group is called the **group generated by** S. The set S is called a **normal set** if, for all $s \in S$ and $g \in G$, $gsg^{-1} \in S$.

Exercises

*2. a. Prove that $\langle S \rangle$ is a subgroup of G which contains S.
   b. Prove that $\langle S \rangle$ is the intersection of all the subgroups of G which contain S.
   c. Prove that $\langle S \rangle$ is the smallest subgroup of G which contains S, that is, if H is a subgroup of G which contains S, then $\langle S \rangle \subset H$.
   d. Prove that S is a set of generators of G (see page 251) if and only if $\langle S \rangle = G$.

*3. Assume that S is closed under inverses, that is, if $s \in S$, then $s^{-1} \in S$. Prove that the group $\langle S \rangle$ consists of the products $s_1 \cdots s_h$ where $s_i \in S$ and h can be any positive integer.

4. a. Prove that the set C of commutators of G is closed under inverses.
   b. Prove the following more general result. The set of commutators

   $$\{a^{-1}b^{-1}ab \,|\, a, b \in S\}$$

   is closed under inverses. No assumptions are made about S except that S is nonempty.

*5. Assume that S contains only elements of finite order; in other words, for each $s \in S$, $s^i = 1$ for some $i \geq 1$. Prove that the group $\langle S \rangle$ con-sists of the products $s_1 \cdots s_h$ where $s_i \in S$ and h can be any integer. This happens, for instance, when G is finite. (Hint: Prove the result directly or apply Exercise 251.2.)

6. Assume that S is a normal set.
   a. Prove that $\langle S \rangle$ is a normal subgroup of G.
   b. Prove that the set of commutators $\{a^{-1}b^{-1}ab \,|\, a, b \in S\}$ is a normal set which is closed under inverses.

7. Assume the elements of S commute pairwise, that is, $st = ts$ for all $s, t \in S$. Prove that the group $\langle S \rangle$ is commutative. (*Warning:* This result is trivial if

$$\langle S \rangle = \{s_1 \cdots s_h \,|\, s_i \in S, \quad i = 1, \ldots, h\},$$

   but, otherwise, needs a little argument.)

8. Assume that S is a set of generators of G and H is a normal subgroup of G. Prove that the cosets $\{sH \,|\, s \in S\}$ form a set of generators for the quotient group G/H.

*9. a. We recall that an element $g \in G$ is called an involution if $g^2 = 1$. By saying that a group is generated by involutions, one means that G has a set of generators which consists of involutions. There are many groups which are generated by involutions, but contain elements which are not involutions. Prove that the following groups are of this kind.
   i. The symmetric group S_m where $m \geq 3$. (S_m denotes the group of permutations of a set with m elements.)
   ii. The orthogonal group $O(V)$ when $n \geq 2$ and V is not the exceptional plane.
   iii. The rotation group $O^+(V)$ when $n \geq 3$.
   b. However, if G is commutative and is generated by involutions, prove that all elements of G are involutions.

10. We denote the set of squares $\{s^2 \,|\, s \in G\}$ of G by G^2. If H is a normal subgroup of G, prove that all elements of the quotient group G/H are involutions if and only if $G^2 \subset H$.

The set of commutators C of G is normal and is closed under inverses, therefore, it generates a normal subgroup of G which consists of the products of finitely many commutators. This group is called the **commutator subgroup of** G and is denoted by G'.

Exercises

*11. Let H be a normal subgroup of G. Prove that the quotient group G/H is commutative if and only if $G' \subset H$.

12. a. Prove that G/G' if a commutative group.
    b. Prove that G is commutative if and only if $G' = \{1\}$.

c. If H is a subgroup which contains G', prove that H is normal. (Hint: If $h \in H$ and $g \in G$, one has to prove that $hgh^{-1} \in H$. Use the fact that the commutator $hgh^{-1}g^{-1}$ of (g^{-1}, h^{-1}) belongs to G'.)

It is clear from Exercise 11 above that G/G' is the "largest" quotient group of G which is commutative. This group G/G' is called the **commutator factor group** of G. The larger G', the smaller G/G'; for this reason, one considers G to be less commutative the larger its commutator subgroup.

The following propositions will be used in the next section.

Proposition 336.1. Let S be a set of generators of G. If S is normal, G' consists of the products of finitely many commutators $a^{-1}b^{-1}ab$ where $a, b \in S$.

Proof. The set

$$T = \{a^{-1}b^{-1}ab \,|\, a, b \in S\}$$

is closed under inverses (Exercise 334.3); hence all we have to show is that $\langle T \rangle = G'$. The set T is normal and hence $\langle T \rangle$ is a normal subgroup of G (Exercise 335.6a). It is trivial that $\langle T \rangle \subset G'$; in order to show that $G' \subset \langle T \rangle$, we prove that the quotient group $G/\langle T \rangle$ is commutative. The group $G/\langle T \rangle$ is generated by the cosets $\{s\langle T \rangle \,|\, s \in S\}$; therefore, all we have to show is that

$$(a\langle T \rangle)(b\langle T \rangle) = (b\langle T \rangle)(a\langle T \rangle)$$

for $a, b \in S$. The above equality is the same as

$$ab\langle T \rangle = ba\langle T \rangle,$$

equivalently,

$$a^{-1}b^{-1}ab \in \langle T \rangle.$$

However, this is trivial since $a^{-1}b^{-1}ab \in T$. Done.

Proposition 336.2. Let S be a set of generators of G. Then $S^2 \subset G'$ if and only if $G^2 \subset G'$. In particular, if all elements of S are involutions, $G^2 \subset G'$.

Proof. The last sentence needs no proof and neither does the implication "$G^2 \subset G'$ implies $S^2 \subset G'$." Hence we assume that $S^2 \subset G'$ and must prove that $G^2 \subset G'$, equivalently, that G/G' consists of involutions (Exercise 335.9). The group G/G' is commutative and is generated by the cosets $\{sG' \mid s \in S\}$, whence we only have to show that each of these cosets is an involution of G/G'. To say that the coset sG' is an involution means that $(sG')^2 = G'$, equivalently, $s^2G' = G'$. But this last equality is true if and only if $s^2 \in G'$ which is given for each $s \in S$. Done.

60. THE COMMUTATOR SUBGROUP OF THE ORTHOGONAL GROUP

The results of this section hold for all dimensions and are not restricted to $n \geq 3$. We are interested in determining how noncommutative the orthogonal group $O(V)$ can be. It is on the borderline whether this is a geometric question or an algebraic one. Many of the most interesting problems in mathematics straddle the main branches of our science.

Of course, the commutativity of $O(V)$ is investigated by studying its commutator subgroup which we denote by $\Omega(V)$.

Proposition 337.1. The commutator subgroup $\Omega(V)$ of $O(V)$ consists of rotations, that is, $\Omega(V) \subset O^+(V)$.

Proof. $O^+(V)$ is a normal subgroup of $O(V)$ and the quotient group $O(V)/O^+(V)$ consists of two elements. Since the group with two elements is commutative, $\Omega(V) \subset O^+(V)$. Done.

We shall prove in Theorem 338.1 that $\Omega(V)$ consists of very special rotations. To prepare for this, we recall that the symmetries of V form a set of generators of $O(V)$ (the Cartan–Dieudonné theorem). It is shown in the exercises below that this set of generators is normal.

Exercises

*1. Let $\mathscr{W}$ be a vector space (without metric) over a field ℓ which may have characteristic two. (We need the results only when $\mathscr{W}$ is finite dimensional, but the reader may observe that the finite dimensionality of $\mathscr{W}$ is never used. As always, whenever it is necessary for the proofs that $\mathscr{W}$ have finite dimension, this will be mentioned explicitly.)

Let $\sigma: \mathscr{W} \to \mathscr{W}$ be a linear transformation of $\mathscr{W}$ which may be singular. We again let F denote the fixed space of σ and I the space of

vectors inverted by σ. Finally, assume that $\alpha: \mathscr{W} \to \mathscr{W}$ is a linear auto-morphism of $\mathscr{W}$; we study the linear transformation $\alpha\sigma\alpha^{-1}: \mathscr{W} \to \mathscr{W}$.
a. Prove that αF is the space of vectors left fixed by $\alpha\sigma\alpha^{-1}$.
b. Prove that αI is the space of vectors inverted by $\alpha\sigma\alpha^{-1}$.
c. If α and σ commute, prove that $\alpha F = F$ and $\alpha I = I$. This means that α leaves the spaces F and I invariant, but not necessarily pointwise fixed.

2. Let U be a nonsingular subspace of V and consider the involution $\sigma = -1_U \oplus 1_{U*}$. If α is an isometry of V, prove that $\alpha\sigma\alpha^{-1}$ is the in-volution corresponding to the nonsingular subspace αU, that is, prove that

$$\alpha\sigma\alpha^{-1} = -1_{\alpha U} \oplus 1_{(\alpha U)*}.$$

(Hint: From the previous exercise, it is easy to determine the spaces F and I for $\alpha\sigma\alpha^{-1}$. This exercise does not use the fact that α is an isometry; where then is this fact used?)

*3. If σ is an involution and α an arbitrary isometry of V, prove that $\alpha\sigma\alpha^{-1}$ is an involution of the same type as σ. Conclude that the involutions of V of a fixed type p form a normal set of $O(V)$ for $p = 0, 1, \ldots, n$.

By Exercise 3 above, the symmetries (involutions of type one) form a normal subset of $O(V)$. If τ_1 and τ_2 are symmetries, $\tau_1\tau_2$ is a rotation and the commutator

$$\tau_1^{-1}\tau_2^{-1}\tau_1\tau_2 = \tau_1\tau_2\tau_1\tau_2 = (\tau_1\tau_2)^2$$

is the square of $\tau_1\tau_2$. We call a rotation which is the product of two sym-metries a **birotation**; the squares of birotations belong to $\Omega(V)$ and, there-fore, products of finitely many squares of birotations belong to $\Omega(V)$. The next theorem says they are all of $\Omega(V)$. It gives a concrete description of the elements of the commutator subgroup of $O(V)$. Observe that, if τ is any symmetry, $1_V = \tau\tau$ and hence 1_V is a birotation.

Theorem 338.1. $\Omega(V)$ consists of the products of finitely many squares of birotations. If $\sigma \in O(V)$, then $\sigma^2 \in \Omega(V)$, equivalently, the commutator factor group $O(V)/\Omega(V)$ consists of involutions.

Proof. Since the set of symmetries is a normal set of generators of $O(V)$, it follows from Proposition 336.1 that $\Omega(V)$ consists of the products of finitely many squares of birotations. Because symmetries are involutions, the last sentence of our theorem follows from Exercise 335.9 and Proposition 336.2. Done.

Exercises

4. Let $n = 1$. Prove that the commutator subgroup of both $O(V)$ and $O^+(V)$ is equal to $\{1_V\}$.

5. Let $n = 2$.
   a. Prove that the commutator subgroup of $O^+(V)$ is $\{1_V\}$.
   b. Prove that $\Omega(V) = \{1_V\}$ if and only if V is the exceptional plane. In Exercise 8 below, we shall determine $\Omega(V)$ for all planes.

6. Let G be a group written multiplicatively. In general, its set of squares G^2 is not a group.
   a. Prove that G^2 is a normal set which is closed under inverses.
   b. Prove that the products of finitely many squares of G form a normal subgroup of G.
   c. If G is commutative, prove that G^2 is a subgroup of G.

*7. a. Prove that the products of finitely many squares of $O(V)$ form the commutator subgroup $\Omega(V)$.
   b. Prove that the products of finitely many squares of $O^+(V)$ form the commutator subgroup $\Omega(V)$.

*8. Let $n = 2$. $O(V)^2$ and $O^+(V)^2$ denote the set of squares of, respectively, $O(V)$ and $O^+(V)$.
   a. Prove that $\Omega(V) = O^+(V)^2 = O(V)^2$. (Hint: For the equality $\Omega(V) = O^+(V)^2$, use Exercise 7b above and the fact that $O^+(V)$ is commutative.)
   b. Give examples of noncommutative groups whose squares form a group. Why is the orthogonal group of the exceptional plane not such an example? Observe that, in the examples provided by plane geometry, the squares form a commutative group.

*9. If V is the Euclidean plane, prove that $\Omega(V) = O^+(V)$. (Hint: Use Exercise 309.30e where it was shown that every element of $O^+(V)$ has a square root.)

10. If V is the negative Euclidean plane, prove that $\Omega(V) = O^+(V)$. (Hint: Use Exercise 224.22.)

11. Let V be the Lorentz plane and M, N an Artinian coordinate system of V. We have seen that, relative to this coordinate system, the rotations of V are represented by matrices

$$\begin{pmatrix} a & 0 \\ 0 & a^{-1} \end{pmatrix}$$

where a can be any nonzero number.
   a. Which real numbers a represent elements of $\Omega(V)$?
   b. Prove that $\Omega(V)$ is isomorphic to the multiplicative group of the positive real numbers.

12. If V is the Artinian plane and k is algebraically closed, prove that $\Omega(V) \simeq k^*$.

*13. a. Prove that the following inclusions always hold.

$$O^+(V)^2 \subset O(V)^2 \subset \Omega(V) \subset O^+(V) \subset O(V).$$

Observe that the first two sets in this tower, $O^+(V)^2$ and $O(V)^2$, are, in general, not groups.

b. If the elements of $O^+(V)$ are squares (of elements of $O^+(V)$), prove that

$$O(V)^2 = \Omega(V) = O^+(V).$$

This gives a second solution to Exercise 9 above which avoids the remark that, in the plane, $\Omega(V) = O^+(V)^2$.

14. Let V be Euclidean three-space.
a. Prove that every rotation is the square of a rotation.
b. Prove that $\Omega(V) = O^+(V) = O(V)^2$. Observe that the squares of the elements of $O(V)$ form $O^+(V)$. We shall see in Exercise 16 below that $O(V)$ is not commutative and in Section 61 that $O^+(V)$ is also not commutative.

15. The present exercise interrelates commutativity and orthogonality. Let A and B be nonisotropic vectors of V which are linearly independent. We proved in Exercise 259.7 that $A \perp B$ implies $\tau_A \tau_B = \tau_B \tau_A$. Now prove that, if $\tau_A \tau_B = \tau_B \tau_A$, then $A \perp B$. (Hint: Since τ_A and τ_B commute, $\tau_A \tau_B \tau_A^{-1} = \tau_B$ and hence $\tau_A \langle B \rangle = \langle B \rangle$ by Exercise 337.1. Consider the possibilities for $\tau_A B$.)

*16. Let $n \geq 3$. Prove that $O(V)$ is never commutative, equivalently, that $\Omega(V) \neq \{1_V\}$. (Hint: Find two nonisotropic vectors of V which are linearly independent and not orthogonal. The argument works just as well when $n = 2$ if V is not the exceptional plane. Why does the argument break down for the exceptional plane?)

61. THE COMMUTATOR SUBGROUP OF THE ROTATION GROUP

We denote the commutator subgroup of the rotation group $O^+(V)$ by $\Omega^+(V)$. Since the rotation groups of the line and the plane are commutative, $\Omega^+(V) = \{1_V\}$ for $n = 1$ and 2. The situation changes radically for higher dimensions. When $n \geq 3$, we shall see that $\Omega^+(V) = \Omega(V)$ from which it follows that $\Omega^+(V) \neq \{1_V\}$ (Exercise 340.16) and hence $O^+(V)$ is not commutative.

The following exercise will be used in the proof of our next theorem.

Exercise

*1. Let G be a group and H a subgroup of G. Prove that $H' \subset G'$. (Primes denote commutator subgroups.)

We have already seen many examples which show that the condition $n \geq 3$ cannot be omitted from the hypothesis of the following theorem.

Theorem 341.1. Let $n \geq 3$. Then $\Omega^+(V)$ consists of the products of finitely many squares $(\rho\sigma)^2$ where ρ and σ are 180° rotations. If $\alpha \in O^+(V)$, then $\alpha^2 \in \Omega^+(V)$ which is equivalent to saying that the commutator factor group $O^+(V)/\Omega^+(V)$ consists of involutions. Finally, $\Omega^+(V) = \Omega(V)$.

Proof. Since $n \geq 3$, every rotation is the product of a finite number of 180° rotations (Theorem 332.1). Consequently, the set S of 180° rotations is a set of generators of $O^+(V)$; moreover, S is a normal set of $O(V)$ (Exercise 338.3) and hence all the more of $O^+(V)$. It follows that $\Omega^+(V)$ consists of the product of finitely many commutators $\rho^{-1}\sigma^{-1}\rho\sigma$ where ρ and σ are 180° rotations (Proposition 336.1). Since ρ and σ are involutions

$$\rho^{-1}\sigma^{-1}\rho\sigma = \rho\sigma\rho\sigma = (\rho\sigma)^2;$$

thus we have shown that $\Omega^+(V)$ consists of the products of finitely many squares $(\rho\sigma)^2$ where ρ and σ are 180° rotations. Furthermore, since S consists of involutions, $O^+(V)^2 \subset \Omega^+(V)$ (Proposition 336.2) and there only remains to be shown that $\Omega^+(V) = \Omega(V)$. Since $O^+(V) \subset O(V)$, we conclude from Exercise 1 above that $\Omega^+(V) \subset \Omega(V)$. We now show $\Omega(V) \subset \Omega^+(V)$. An element of $\Omega(V)$ is a product of squares of rotations (Exercise 339.7b) and we have just shown that the square of every rotation belongs to $\Omega^+(V)$. Done.

The following diagrams show the inclusion relations among the orthogonal group, the rotation group, the commutator subgroups, and the sets of

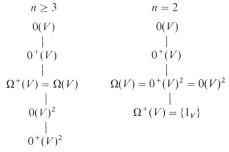

$n \geq 3$	$n = 2$
$0(V)$	$0(V)$
$\mid$	$\mid$
$0^+(V)$	$0^+(V)$
$\mid$	$\mid$
$\Omega^+(V) = \Omega(V)$	$\Omega(V) = 0^+(V)^2 = 0(V)^2$
$\mid$	$\mid$
$0(V)^2$	$\Omega^+(V) = \{1_V\}$
$\mid$	
$0^+(V)^2$	

Figure 341.1

squares $O^+(V)^2$ and $O(V)^2$ for $n = 2$ and $n \geq 3$. Exercises 339.8a and 340.13 can be used to check the diagrams.

Exercises

2. Let $n = 1$; we know that $O(V) = \{1_V, -1_V\}$. Prove that
$$O^+(V) = \Omega(V) = \Omega^+(V) = O(V)^2 = O^+(V)^2 = \{1_V\}.$$

3. If $n = 2$, prove that $\Omega(V) = \Omega^+(V)$ if and only if V is the exceptional plane.

4. If $n \geq 3$, prove that neither $O^+(V)$ nor $O(V)$ is commutative. (The argument is indicated at the beginning of this section.)

When $n \geq 3$, $\Omega(V)$ is the common commutator subgroup of $O(V)$ and $O^+(V)$ and, when $n = 1$ or 2, $\Omega^+(V) = \{1_V\}$. Therefore, there is no further need for the symbol $\Omega^+(V)$.

62. THE ISOMETRIES $\pm 1_V$

Let G be a group written multiplicatively. If S is a nonempty subset of G, the **centralizer of** S consists of the elements of G which commute with each element of S. Denoting the centralizer by $Z(S)$, we have
$$Z(S) = \{g \mid g \in G, \quad gs = sg \quad \text{for all } s \in S\}.$$
The centralizer of G is called the **center of** G and hence this center is
$$\{g \mid g \in G, \quad gs = sg \quad \text{for all } s \in G\}.$$

Exercises

1. a. If S is a nonempty subset of G, prove that $Z(S)$ is a subgroup of G.
   b. Let G be the symmetric group on three objects. Thus G consists of the six permutations of these objects. Let t be one of the three transpositions of G. Prove that $Z(t) = \{1, t\}$ and that $\{1, t\}$ is not a normal subgroup of G. This example shows that $Z(S)$ is, in general, not a normal subgroup of G.
2. a. Prove that g belongs to the center of G if and only if $gsg^{-1} = s$ for all $s \in G$.
   b. Prove that the center of G is a commutative normal subgroup of G.
   c. Prove that G is commutative if and only if the center of G is equal to G.

3. a. Prove that the group $\{1_V, -1_V\}$ is a subgroup of the center of $O(V)$.
   b. If n is even, prove that $\{1_V, -1_V\}$ is also a subgroup of the center of $O^+(V)$. Of course, if n is odd, -1_V does not even belong to $O^+(V)$.

In addition to the commutator subgroup, the center of a group is also a measure of the commutativity of that group. If the center of G is G, the group is commutative and, in general, the smaller the center, the less commutative the group is considered to be. Even so, the commutator subgroup seems to be a more informative measure of commutativity than the center.

We shall see in the next section that, when $n \geq 3$, the three groups $O(V)$, $O^+(V)$, and $\Omega(V)$ are highly noncommutative in the sense that their centers are very small. A basic tool in the study of these centers is criteria which tell when an isometry of V is equal to $\pm 1_V$; we derive these criteria in the remainder of this section.

Let $\mathcal{W}$ be a vector space without metric over a field $\not{k}$ which may have characteristic two. If r is a nonzero scalar, the linear automorphism (nonsingular linear transformation) $M(0, r)$ of $\mathcal{W}$, defined by

$$M(0, r)A = rA$$

for all $A \in \mathcal{W}$, is called a **magnification** of $\mathcal{W}$. If the dimension of $\mathcal{W}$ is finite, $M(0, r)$ is the magnification with center 0 and ratio r of Chapter 1. The next lemma gives a criterion for telling when a linear automorphism of $\mathcal{W}$ is a magnification. This lemma belongs in linear algebra and has nothing to do with metrics.

Lemma 343.1. A linear automorphism of $\mathcal{W}$ is a magnification if and only if it leaves all lines of $\mathcal{W}$ through 0 invariant.

Proof. Magnifications obviously leave all lines of $\mathcal{W}$ through 0 invariant. Hence we assume σ is a linear automorphism of $\mathcal{W}$ leaving all lines through 0 invariant. If A is a nonzero vector $\sigma\langle A \rangle = \langle A \rangle$, therefore, $\sigma A = rA$ for some nonzero scalar r. We shall prove that $\sigma = M(0, r)$ by choosing a second nonzero vector B in $\mathcal{W}$ and proving that $\sigma B = rB$.

Case 1. A and B are linearly dependent. Then $B = bA$ and hence

$$\sigma B = \sigma(bA) = b\sigma A = b(rA) = r(bA) = rB.$$

Case 2. A and B are linearly independent. Since the lines $\langle B \rangle$ and $\langle A + B \rangle$ are invariant under σ, there are scalars s and t such that $\sigma B = sB$ and $\sigma(A + B) = t(A + B)$. Let us compute $\sigma(A + B)$ in two ways,

$$\sigma(A + B) = t(A + B) = tA + tB$$
$$\sigma(A + B) = \sigma A + \sigma B = rA + sB.$$

The linear independence of A and B implies that $t = r$ and $t = s$ whence $r = s$. Done.

We now come to the first criterion for telling when an isometry of V is $\pm 1_V$.

Lemma 344.1.   An isometry of V is equal to $\pm 1_V$ if and only if it leaves all lines through $\mathbf{0}$ invariant.

Proof. The isometries $\pm 1_V$ are magnifications of V and hence leave all lines through $\mathbf{0}$ invariant. We assume σ is an isometry of V which leaves all lines through $\mathbf{0}$ invariant. Then σ is a magnification; but the only isometries which are magnifications are $\pm 1_V$ (Exercise 218.6a). Done.

Exercise

4. This exercise extends Lemma 344.1 to singular spaces which are not null spaces. Let W be a metric vector space which may be singular, but is not a null space. Prove that an isometry of W is equal to $\pm 1_W$ if and only if it leaves all lines through $\mathbf{0}$ invariant. (Hint: Apply Exercise 218.6b.)

In case $n \geq 3$ and V has null lines, Lemma 344.1 can be improved. An isometry is then equal to $\pm 1_V$ if it leaves only the lines of the light cone invariant. The following exercise will be used in the proof of this result.

Exercise

*5. Let $\mathscr{W}$ be a vector space without a metric over a field k which may have characteristic two. Assume that $W = U \oplus W$ and σ is a linear transformation of $\mathscr{W}$ which leaves the subspaces U and W invariant. If $A \in U$ and $B \in W$, prove that
$$\sigma(A + B) = c(A + B)$$
for $c \in k$ if and only if $\sigma A = cA$ and $\sigma B = cB$.

Lemma 344.2   Assume that $n \geq 3$ and V has null lines. Then an isometry is equal to $\pm 1_V$ if and only if it leaves all null lines of V invariant.

Proof. We only have to show that, if an isometry σ of V leaves all null lines invariant, then $\sigma = \pm 1_V$. Since V has null lines, V contains an Artinian plane P (why?) and we put $V = P \oplus P^*$. By assumption, σ leaves the two null lines of V invariant and hence σ leaves P invariant; therefore, σ also leaves P^* invariant.

Let us show that σ leaves all lines of P^* invariant. Select $A \in P^*$; we must show that $\sigma A = cA$ for some scalar c. Since an Artinian plane represents all scalars, there is a vector $B \in P$ such that $B^2 = -A^2$. Then

$$(A + B)^2 = A^2 + B^2 = 0$$

which implies that $\sigma(A + B) = c(A + B)$ for $c \in k$. From Exercise 5 above, we conclude that $\sigma A = c$ and, therefore, σ leaves all lines of P^* invariant. Thus far, we have not used the assumption $n \geq 3$.

As always, we denote the restriction of σ to P and P^* by $\sigma | P$ and $\sigma | P^*$. The isometry $\sigma | P^*$ leaves all lines of P^* invariant; therefore, $\sigma | P^* = \pm 1_{P*}$ (Lemma 344.1).

We now prove the following. If $\sigma | P^* = 1_{P*}$, then $\sigma | P = 1_P$; if $\sigma | P^* = -1_{P*}$, then $\sigma | P = -1_P$. The idea of the proof is simple. Since σ leaves the two null lines of P invariant, $\sigma | P$ is a rotation of P. Hence, if σ leaves a nonzero vector of P fixed, $\sigma | P = 1_P$ (Exercise 269.1) and, if σ inverts a nonzero vector of P, $\sigma | P = -1_P$ (Exercise 269.2).

Assume that $\sigma | P^* = 1_{P*}$. Since $n \geq 3$, $\dim P^* \geq 1$ and there exists a nonisotropic vector $A \in P^*$. Again, let $B \in P$ such that $B^2 = -A^2$. Then $B^2 \neq 0$ and hence $B \neq \mathbf{0}$; we shall show that $\sigma B = B$. Now $(A + B)^2 = 0$ implies $\sigma(A + B) = c(A + B)$; as before,

$$\sigma A = cA \quad \text{and} \quad \sigma B = cB.$$

But $A \in P^*$ and $\sigma | P^* = 1_{P*}$ implies that $\sigma A = A$. Hence $c = 1$ and, therefore, $\sigma B = B$.

If $\sigma | P^* = -1_{P*}$, one reasons in precisely the same way, but now concludes from $\sigma A = -A$ that $c = -1$ and $\sigma B = -B$.

Finally, we have shown that either

$$\sigma = 1_P \oplus 1_{P*} = 1_V \quad \text{or} \quad \sigma = -1_P \oplus -1_{P*} = -1_V.$$

Done.

The condition $n \geq 3$ cannot be removed from Lemma 344.2. Let V be an Artinian plane. We have seen that every rotation of V leaves the two null lines invariant. Moreover, V is the exceptional plane if and only if $O^+(V) = \{1_V, -1_V\}$. Therefore, if V is not the exceptional plane, there are rotations different from $\pm 1_V$ which leave the two null lines invariant.

One immediately asks whether an isometry which leaves only the lines through **0** not on the light cone invariant must also be $\pm 1_V$. The answer is affirmative and the condition $n \geq 3$ is now superfluous. However, the exceptional plane has to be excluded. We add this result for the sake of completeness; it will be used only in Exercises 349.5 and 350.6. The condition that V has null lines has been omitted from the following lemma because, if V is anisotropic, this lemma is contained in Lemma 344.1.

Lemma 346.1. Assume that V is not the exceptional plane. An isometry of V is equal to $\pm 1_V$ if and only if it leaves all nonsingular lines invariant.

Proof. The only isometries of a nonsingular line are $\pm 1_V$, so we may assume $n \geq 2$. The isometries $\pm 1_V$ obviously leave all nonsingular lines invariant; thus we must prove that an isometry σ of V which leaves all nonsingular lines invariant must be $\pm 1_V$.

Case 1. n = 2. We first show σ cannot be a symmetry. A symmetry leaves only two lines invariant, its fixed space and the space of vectors it inverts. Since V is not the exceptional plane, there are more than two nonsingular lines and hence at least one nonsingular line is not invariant by a symmetry, contrary to assumption. Consequently, σ is a rotation. If A is a nonisotropic vector, $\sigma\langle A \rangle = \langle A \rangle$, equivalently, $\sigma A = \pm A$. If $\sigma A = A$, then $\sigma = 1_V$ and, if $\sigma A = -A$, $\sigma = -1_V$. Observe that the argument is valid whether or not V has null lines.

Case 2. n \geq 3. We assume that V has null lines, the anisotropic case being covered by Lemma 344.1. By the same lemma (or Lemma 344.2), it is sufficient to show that σ also leaves the null lines invariant. We select a nonzero isotropic vector N of V and show that $\sigma N = bN$ for some scalar b.

The hyperplane $\langle N \rangle^*$ has dimension at least two (because $n \geq 3$) and has the line $\langle N \rangle$ as radical. Consequently, $\langle N \rangle^*$ is not a null space and, therefore, contains a nonisotropic vector A. But the vector $N + A$ is also nonisotropic since

$$(N + A)^2 = N^2 + 2NA + A^2 = A^2 \neq 0.$$

By hypothesis, there are scalars a and b such that $\sigma A = aA$ and $\sigma(N + A) = b(N + A)$; we conclude that

$$bN + bA = \sigma N + \sigma A = \sigma N + aA$$

and hence

$$\sigma N = bN + (b - a)A.$$

However, σN must be an isotropic vector; thus

$$(bN + (b - a)A)^2 = (b - a)^2 A^2 = 0.$$

Since $A^2 \neq 0$, $b = a$, whence $\sigma N = bN$. Done.

Exercise

6. The exceptional plane has the two rotations $\pm 1_V$ and two symmetries. Prove that both symmetries leave the nonsingular lines through **0** invariant and hence Lemma 346.1 does not hold for the exceptional plane. (Hint: Use the setup and results of Exercise 280.5c.)

63. CENTERS OF $O(V)$, $O^+(V)$, AND $\Omega(V)$

It is the purpose of this section to show that, if $n \geq 3$, the centers of the three groups $O(V)$, $O^+(V)$, and $\Omega(V)$ are very small. The center of $O(V)$ will turn out to be the group $\{1_V, -1_V\}$ while the center of $O^+(V)$ is the same group if n is even and $\{1_V\}$ if n is odd. The center of $\Omega(V)$ is also either $\{1_V, -1_V\}$ or $\{1_V\}$; but, if n is even, there is no satisfactory geometric criterion which separates these two cases. The results of the following exercises will be used in the proofs.

Exercises

1. Let U be a nonsingular subspace of V. Assume that σ is an isometry of V which leaves U and hence U^* invariant.
    a. Prove that the fixed space of σ is the orthogonal sum of the fixed spaces of $\sigma | U$ and $\sigma | U^*$. (Hint: When showing that the fixed space of σ is contained in the orthogonal sum of the fixed spaces of $\sigma | U$ and $\sigma | U^*$, use Exercise 344.5. The opposite inclusion is straightforward.)
    b. Prove that the space of vectors inverted by σ is the orthogonal sum of the spaces of vectors inverted by $\sigma | U$ and $\sigma | U^*$.
*2. a. If U is a nonsingular subspace of V, prove that U is contained in a nonsingular subspace of dimension d of V for all d satisfying $\dim U \leq d \leq n$.
    b. If U is an arbitrary subspace of V, prove that U is contained in a nonsingular subspace of dimension d of V if and only if

    $$\dim U + \dim(\text{rad } U) \leq d \leq n.$$

    (Hint: A nonsingular completion of U has dimension

    $$\dim U + \dim(\text{rad } U).)$$

    c. The following special case of Part b will be used in the proof of Theorem 348.1. If $n \geq 3$ and l is a null line of V, prove that V contains a nonsingular three-space which contains l.

3. Let $\mathscr{W}$ be a vector space without metric over a field $\mathscr{k}$ which may have characteristic two. Assume dim $\mathscr{W} \geq 3$.
   a. Prove that each line through **0** is the intersection of two planes through **0**.
   b. If $\sigma\colon \mathscr{W} \to \mathscr{W}$ is a linear transformation which leaves the planes through **0** invariant, prove that σ also leaves all lines through **0** invariant. Observe that this result is false when dim $\mathscr{W} = 2$. (The proof is strictly set theoretic and does not use the linearity of σ.)

As before, $O^+(V)^2$ denotes the set of squares of rotations of V. We now determine the centralizer of the set $O^+(V)^2$ in $O(V)$. All results on the centers of the groups $O(V)$, $O^+(V)$, and $\Omega(V)$ then follow easily.

Theorem 348.1.   If $n \geq 3$, the centralizer in $O(V)$ of the set $O^+(V)^2$ is the group $\{1_V, -1_V\}$.

Proof. Since $\pm 1_V$ belong to the center of $O(V)$, they belong to the centralizer of every subset of $O(V)$. Consequently, we must show that an isometry σ which belongs to the centralizer of $O^+(V)^2$ is $\pm 1_V$.

Case 1. V has null lines. Since $n \geq 3$, it is sufficient to show that σ leaves all null lines of V invariant (Lemma 344.2). Every null line l is contained in a nonsingular three-space $U \subset V$ (Exercise 2c above). There exist rotations of U which have l as axis and each such rotation is the square ρ^2 of a rotation ρ of U (Proposition 325.2). Furthermore, $V = U \oplus U^*$ and the fixed space of the rotation $\rho^2 \oplus 1_{U*}$ is $l \oplus U^*$. Since

$$\rho^2 \oplus 1_{U*} = (\rho \oplus 1_{U*})^2 \in O^+(V)^2,$$

σ commutes with $\rho^2 \oplus 1_{U*}$; it follows from Exercise 338.1c that

$$\sigma(l \oplus U^*) = l \oplus U^*.$$

However, the radical of the singular space $l \oplus U^*$ is l (Exercise 145.12a). Since σ must map the radical of $l \oplus U^*$ into itself, we conclude that $\sigma l = l$. Thus $\sigma = \pm 1_V$.

Case 2. V is anisotropic. It is sufficient to show that all lines of V are invariant. Since $n \geq 3$, we need only prove that σ leaves all planes through **0** invariant. Let P be a plane of V. Since P is anisotropic, it certainly is not the exceptional plane; thus there is a rotation ρ of P which is not an involution (Theorem 282.1). Then $\rho^2 \neq 1_P$ and ρ^2 leaves only the origin fixed. We put $V = P \oplus P^*$ and consider the rotation

$$\rho^2 \oplus 1_{P*} = (\rho \oplus 1_{P*})^2$$

whose fixed space is P^*. The isometry σ commutes with this rotation and hence $\sigma P^* = P^*$; therefore, also $\sigma P = P$. Done.

Since

$$O^+(V)^2 \subset O^+(V) \subset O(V),$$

the following theorem is almost an obvious corollary of the above theorem and, therefore, indicates the importance of Theorem 348.1.

Theorem 349.1. Let $n \geq 3$. Then

1. The center of $O(V)$ is the group $\{1_V, -1_V\}$.
2. The center of $O^+(V)$ is $\{1_V, -1_V\}$ if n is even and $\{1_V\}$ if n is odd.
3. The center of $\Omega(V)$ is $\{1_V\}$ if n is odd. If n is even, the center of $\Omega(V)$ is contained in $\{1_V, -1_V\}$.

Proof. Since

$$O^+(V)^2 \subset \Omega(V) \subset O^+(V) \subset O(V),$$

it follows from Theorem 348.1 that the center of each of the groups $\Omega(V)$, $O^+(V)$, and $O(V)$ is either $\{1_V\}$ or $\{1_V, -1_V\}$; this already proves Statement 3. We have observed that $\{1_V, -1_V\}$ is contained in the center of $O(V)$ hence this center is $\{1_V, -1_V\}$. It is clear that the center of $O^+(V)$ is $\{1_V, -1_V\}$ if $-1_V \in O^+(V)$ and $\{1_V\}$, otherwise. Done.

If n is even, there is no good geometric criterion which decides whether $-1_V \in \Omega(V)$; thus Statement 3 is far from satisfactory.

Exercises

4. Let $n \geq 3$.
   a. If S is a subset of $O(V)$ which contains $O^+(V)^2$, prove that the centralizer of S in $O(V)$ is the group $\{1_V, -1_V\}$.
   b. Prove that the centralizer of either of the groups $\Omega(V)$ or $O^+(V)$ in $O(V)$ is $\{1_V, -1_V\}$.
   A knowledge of the centralizer of $O^+(V)$ in $O(V)$ is only necessary to obtain the center of $\Omega(V)$. The centers of $O(V)$ and $O^+(V)$ can be obtained much more easily without using the set $O^+(V)^2$, as is shown in the next two exercises.

*5. Without using Theorems 348.1 and 349.1, prove that $\{1_V, -1_V\}$ is the center of $O(V)$ for all $n \geq 1$, except for the exceptional plane. (Hint: It is only necessary to show that the center of $O(V)$ is a subset of $\{1_V, -1_V\}$. If an isometry σ belongs to the center of $O(V)$, it commutes with all symmetries of V. Hence, if A is a nonisotropic vector of V, σ commutes with the symmetry τ_A. This implies that $\sigma\langle A \rangle = \langle A \rangle$. Now apply Lemma 346.1.) If V is the exceptional plane, we have already shown that $O(V)$ is commutative and is the Klein four group (Exercise 281.4). Hence the center is also the Klein four group.

*6. a. Let $n \geq 3$. Without using Theorems 348.1 and 349.1, prove that the center of $O^+(V)$ is $\{1_V, -1_V\}$ if n is even and $\{1_V\}$ if n is odd. (Hint: It is only necessary to show that the center of $O^+(V)$ is contained in $\{1_V, -1_V\}$. If a rotation ρ belongs to the center of $O^+(V)$, it commutes with all 180° rotations. Conclude that ρ leaves all nonsingular planes of V invariant. Since $n \geq 3$, it follows that ρ leaves all nonsingular lines invariant.)
   b. If $n = 2$, prove that the center of $O^+(V)$ is $O^+(V)$.
   c. If $n = 1$, prove that the center of $O^+(V)$ is $\{1_V\}$.
7. a. If $n = 2$, prove that $\Omega(V)$ is the group of squares of $O^+(V)$. In this case, one may take the attitude that the commutator subgroup $\Omega(V)$ is known.
   b. If $n = 1$, prove that the center of $\Omega(V)$ is $\{1_V\}$.
8. In Case 1 of the proof of Theorem 348.1, we used the fact that, if l is a null line in three-space, there is a rotation whose square has l as fixed space. It is shown in this exercise that such a rotation may not exist if l is nonsingular.

   Let $n = 3$ and $k = \mathbf{Z}_3$. Assume that A_1, A_2, A_3 is a rectangular coordinate system relative to which the matrix of V is

$$\begin{pmatrix} 1 & 0 & 0 \\ 0 & 1 & 0 \\ 0 & 0 & -1 \end{pmatrix}.$$

   a. Prove that there are only two rotations of V with the nonsingular line $\langle A_1 \rangle$ as axis, namely, 1_V and the 180° rotation with $\langle A_1 \rangle$ as fixed space.
   b. Prove that there exists no rotation ρ of V whose square ρ^2 has $\langle A_1 \rangle$ as fixed space. (Hint: First show that ρ would also have $\langle A_1 \rangle$ as fixed space.)

We finish this section by determining the centralizer of $O^+(V)^2$ in $O(V)$, also when $n < 3$. We denote this centralizer by Z and hence Z is $\{1_V, -1_V\}$ if $n \geq 3$. If $n = 1$, $O^+(V) = O^+(V)^2 = \{1_V\}$; therefore, Z is again $\{1_V, -1_V\}$.

The situation changes radically when $n = 2$. As shown in the exercises below, Z is never equal to $\{1_V, -1_V\}$. There are precisely three planes for which $Z = O(V)$; for all other planes $Z = O^+(V)$.

Exercises

*9. Let G be a multiplicative group and H a subgroup of G of index two. Assume that J is a subgroup of G which contains H. Prove that either $J = H$ or $J = G$. Thus, if J contains one element of G which is not in H, then $J = G$.

In the remaining exercises of this section, $n = 2$ and Z denotes the centralizer of $O^+(V)^2$ in $O(V)$.

10. a. Prove that either $Z = O^+(V)$ or $Z = O(V)$. (Hint: $O^+(V)$ is commutative and hence $Z \supset O^+(V)$.)
   b. Prove that $Z = O(V)$ as soon as there is even one symmetry which commutes with all squares of rotations.
*11. Prove that $Z = O(V)$ if and only if $\rho^4 = 1_V$ for all rotations ρ. (Hint for the if part: If $\rho^4 = 1_V$ for all $\rho \in O^+(V)$, all squares of rotations are involutions. Consequently, $O^+(V)^2 \subset \{1_V, -1_V\}$. Hint for the only if part: If $Z = O(V)$, ρ^2 commutes with all symmetries for every rotation ρ. Therefore, the rotation ρ^2 leaves a nonsingular line invariant and hence $\rho^2 = \pm 1_V$.)
12. In order to study the Artinian planes for which $Z = O(V)$, we first study a strictly field theoretic question. Find all fields k with the property that $a^4 = 1$ for every nonzero element a. (Hint: The nonzero elements of k are roots of the polynomial $X^4 - 1$ and hence k has at most four nonzero elements. Consequently, k has 2, 3, 4, or 5 elements. Use the fact from modern algebra that there is precisely one field (within isomorphism) with 2, 3, 4, or 5 elements. Show, however, that only three of these fields have the required property.)
13. Let V be the Artinian plane over k. Prove that $Z = O(V)$ if and only if k has either three or five elements. Thus there are precisely two Artinian planes for which $Z = O(V)$. Of course, when k has three elements, V is the exceptional plane. (Hint: For Artinian planes, $O^+(V) \simeq k^*$, hence $Z = O(V)$ if and only if $a^4 = 1$ for every nonzero scalar a.)
14. Prove that there is precisely one anisotropic plane for which $Z = O(V)$, namely, the anisotropic plane over the field with three elements. Here we use the fact that there is precisely one anisotropic plane (within isometry) whose field has three elements; see Artin, [2], page

143. (Hint: Assume that V is anisotropic and k has three elements. Then V has four nonsingular lines and hence four symmetries and four rotations. Use Exercise 11 above to conclude that $Z = O(V)$. Conversely, assume that V is anisotropic and $Z = O(V)$. In order to show that k has three elements, we prove that V has precisely four rotations. If $\rho \in O^+(V)$, $\rho^4 = 1_V$ and hence the order of an element of $O^+(V)$ can only be one, two, or four. Now V is not the exceptional plane, equivalently, $O^+(V)$ does not consist of only involutions. It follows that $O^+(V)$ contains a cyclic subgroup H of order four while H contains a cyclic subgroup of order two. Since $O^+(V)$ contains at most one cyclic subgroup of given order (Theorem 275.1), it is easy to conclude that $O^+(V) = H$.)

15. Prove that Z is never equal to $\{1_V, -1_V\}$.

*16. a. Prove that $Z = O(V)$ if and only if $O^+(V)^2 \subset \{1_V, -1_V\}$. Consequently, $O^+(V)^2 \subset \{1_V, -1_V\}$ if and only if V is one of the following three planes: The exceptional plane, the Artinian plane over the field $\mathbf{Z}_5$ or the anisotropic plane over $\mathbf{Z}_3$.

   b. If V is the exceptional plane, prove that $O^+(V)^2 = \{1_V\}$.

   c. If V is the Artinian plane over $\mathbf{Z}_5$ or the anisotropic plane over $\mathbf{Z}_3$, prove that $O^+(V)^2 = \{1_V, -1_V\}$. (Hint: In both cases, $O^+(V)$ has four elements and hence is the cyclic group with four elements.)

64. LINEAR REPRESENTATIONS OF THE GROUPS $O(V)$, $O^+(V)$, AND $\Omega(V)$

Let G be a group, written multiplicatively, with unit element 1 and $\mathcal{W}$ a vector space of finite dimension m over k. A **linear representation of** G is an action of G on $\mathcal{W}$ which respects both the multiplication in G and the addition and scalar multiplication in $\mathcal{W}$. Precise statements of these terms are now given. The group G is said to *act* on $\mathcal{W}$ if, for each $\sigma \in G$ and $A \in \mathcal{W}$, a unique vector $\sigma A \in \mathcal{W}$ has been defined. We denote this *action* of G on $\mathcal{W}$ by $(G, \mathcal{W})$. This action satisfies the following four axioms. For all $\rho, \sigma \in G$, $A, B \in \mathcal{W}$ and $t \in k$,

1. $(\rho\sigma)A = \rho(\sigma A)$.
2. $1A = A$.
3. $\sigma(A + B) = \sigma A + \sigma B$.
4. $\sigma(tA) = t(\sigma A)$.

The first two axioms say that the action respects the multiplication in G, while the latter two axioms say the action respects addition and scalar multiplication in $\mathcal{W}$. Moreover, Axioms 3 and 4 are equivalent to saying that the mapping $A \to \sigma A$ is a linear transformation of $\mathcal{W}$ for each $\sigma \in G$. Therefore, it follows that $\sigma 0 = 0$.

A **linear representation of** G then consists of an action $(G, \mathcal{W})$ of G on $\mathcal{W}$ which satisfies Axioms 1–4. Examples of linear representations are the actions $(O(V), V)$, $(O^+(V), V)$, and $(\Omega(V), V)$ where, in each case, the vector σA is obtained by letting the isometry σ act as usual on the vector A (see Exercise 1 below). We refer to these actions as the "natural actions" or the "natural representations" of the respective groups $O(V)$, $O^+(V)$, and $\Omega(V)$.

Exercises

*1. Prove that the natural representations of the groups $O(V)$, $O^+(V)$, and $\Omega(V)$ are linear representations of these groups.

*2. Let $(G, \mathcal{W})$ be a linear representation of G. To each $\sigma \in G$ is associated the linear transformation $L(\sigma): \mathcal{W} \to \mathcal{W}$ defined by $L(\sigma)A = \sigma A$ for all $A \in \mathcal{W}$.

 a. Prove that $L(\sigma)$ is nonsingular for each $\sigma \in G$.

 b. The result in Part a says that $L(\sigma)$ is an element of the general linear group $GL(m, \mathit{k})$. Prove that the mapping $\sigma \to L(\sigma)$ is a homomorphism from G into $GL(m, \mathit{k})$. We refer to the group homomorphism $\sigma \to L(\sigma)$ as the "homomorphism associated with the linear representation $(G, \mathcal{W})$."

3. The three groups $O(V)$, $O^+(V)$, and $\Omega(V)$ are subgroups of the general linear group $GL(n, k)$. The three inclusion mappings

$$O(V) \to GL(n, k), \qquad O^+(V) \to GL(n, k), \qquad \Omega(V) \to GL(n, k)$$

are group homomorphisms. Prove that these inclusion mappings are the homomorphisms associated with the natural representations of the groups $O(V)$, $O^+(V)$, and $\Omega(V)$, respectively. Observe that these homomorphisms are all one-to-one.

*4. Let $L: G \to GL(m, \mathit{k})$ be an arbitrary group homomorphism. Define an action $(G, \mathcal{W})$ by

$$\sigma A = L(\sigma)A$$

for $\sigma \in G$ and $A \in \mathcal{W}$.

 a. Prove that $(G, \mathcal{W})$ is a linear representation of G.

 b. Prove that the homomorphism from G into $GL(m, \mathit{k})$ associated with the linear representation $(G, \mathcal{W})$ (see Exercise 2 above) is the given homomorphism L.

5. Prove that the notions of linear representation $(G, \mathcal{W})$ and of group homomorphism $G \to GL(m, \mathit{k})$ are equivalent notions. (Hint: Prove that Exercises 2 and 4 above set up a one-to-one correspondence

between the set of group homomorphisms from G into $GL(m, \text{\it k})$ and the set of linear representations $(G, \mathscr{W})$.

6. A linear representation $(G, \mathscr{W})$ is called **faithful** if the unit element is the only element of G which leaves all vectors of $\mathscr{W}$ fixed. Prove that a linear representation $(G, \mathscr{W})$ is faithful if and only if the corresponding homomorphism from G into $GL(m, \text{\it k})$ is one-to-one. In this case, G may be regarded as a subgroup of $GL(m, \text{\it k})$.

7. Prove that the natural representations of the groups $O(V)$, $O^+(V)$, and $\Omega(V)$ are all faithful.

8. Let $(G, \mathscr{W})$ be a linear representation of the group G and let H be a subgroup of G. If $\sigma \in H$ and $A \in \mathscr{W}$, the vector σA is given by the action of G on $\mathscr{W}$; we denote this resulting action of H on $\mathscr{W}$ by $(H, \mathscr{W})$. One says that $(H, \mathscr{W})$ is obtained from $(G, \mathscr{W})$ by "restricting the action of G to H" or by "restricting the representation of G to H."

   a. Prove that $(H, \mathscr{W})$ is a linear representation of H.

   b. If $(G, \mathscr{W})$ is a faithful representation of G, prove that $(H, \mathscr{W})$ is a faithful representation of H.

9. a. Prove that the natural representation of $O^+(V)$ is obtained by restricting the natural representation of $O(V)$ to $O^+(V)$.

   b. Prove that the natural representation of $\Omega(V)$ is obtained by restricting the natural representations of either $O(V)$ or $O^+(V)$ to $\Omega(V)$.

If $(G, \mathscr{W})$ is a linear representation of G, a linear subspace U of $\mathscr{W}$ is called an **invariant subspace of** $(G, \mathscr{W})$ if $\sigma A \in U$ for each $\sigma \in G$ and $A \in U$. Clearly, $\mathscr{W}$ and $\{0\}$ are always invariant subspaces. If $\mathscr{W} \neq \{0\}$ and if $\mathscr{W}$ and $\{0\}$ are the only invariant subspaces, the linear representation $(G, \mathscr{W})$ is called **simple** or **irreducible.** We prefer the term "simple."

Exercises

10. Let $(G, \mathscr{W})$ be a linear representation of G.

   a. If $m = 1$, prove that $(G, \mathscr{W})$ is simple. Examples are the natural representations of $O(V)$, $O^+(V)$, and $\Omega(V)$ when V is a line.

   b. If $m \geq 2$ and G consists of only the unit element, prove that $(G, \mathscr{W})$ is not simple.

*11. Let $(G, \mathscr{W})$ be a linear representation of G and H a subgroup of G. Assume that $(H, \mathscr{W})$ is the linear representation of H obtained by restricting the action of G to H.

a. If U is an invariant subspace of the representation $(G, \mathcal{W})$, prove that U is also an invariant subspace of the representation $(H, \mathcal{W})$.
b. If $(H, \mathcal{W})$ is a simple representation of H, prove that $(G, \mathcal{W})$ is a simple representation of G.
c. The converse of Part b is false, that is, $(G, \mathcal{W})$ may be simple while $(H, \mathcal{W})$ is not simple. Prove that this happens when $(G, \mathcal{W})$ is simple, $m \geq 2$, and H consists of the unit element of $G.$. A much more interesting example with $H \neq \{1\}$ is mentioned after Proposition 357.1.

The theory of linear representations is probably the most powerful tool for the investigation of groups. The theory concentrates on finding the various simple representations $(G, \mathcal{W})$ of a given group where the dimension of $\mathcal{W}$ is finite. If $n \geq 3$, the metric vector space V offers a simple representation for each of the classical groups $O(V)$, $O^+(V)$, and $\Omega(V)$; it would be a pity not to prove this.

In Euclidean spaces, no line or plane plays a special role and, from this fact alone, we expect the natural representations of the groups $O(V)$, $O^+(V)$, and $\Omega(V)$ to be simple if V is Euclidean three-space. The next theorem states that this is correct for all nonsingular metric vector spaces of dimension at least three. Some exercises precede the theorem.

Exercises

*12. Let U be a nonsingular subspace of V and consider the involution $\sigma = -1_U \oplus 1_{U*}$. Prove that every isometry of V which leaves U invariant commutes with σ. (Hint: If σ leaves U invariant, it also leaves U^* invariant and hence $\sigma = \beta \oplus \gamma$ where β is an isometry of U and γ is an isometry of U^*.)

*13. Assume that V contains a null line l.
    a. Prove that $n \geq 2$ and V contains a null line different from l. (Hint: Consider a nonsingular completion of l.)
    b. If l and m are two distinct null lines of V and n is at most three, prove that the plane which contains l and m is an Artinian plane. (Hint: Prove that a plane with two distinct null lines cannot have a line as radical. Hence, such a plane is either a null plane or is Artinian; why can V not contain null planes?)

14. Let $n = 3$ and assume that V has a null line l. Let ρ be a rotation of V with l as fixed space. Prove that ρ leaves no null line of V invariant except l. (Hint: If ρ leaves a null line $m \neq l$ invariant, show that ρ leaves the Artinian plane which contains l and m pointwise fixed, contradicting the fact that l is the fixed space of ρ.)

*15. Let $n \geq 3$ and U be a null space in V. Assume that l is a line in U and W is a nonsingular three-space in V which contains l. Prove that $U \cap W = l$.

Theorem 356.1. If $n \geq 3$, the natural representation of the groups $O(V)$, $O^+(V)$, and $\Omega(V)$ are simple.

Proof. Let U be a subspace of V where $U \neq \{0\}$ and $U \neq V$. Since

$$O^+(V)^2 \subset \Omega(V) \subset O^+(V) \subset O(V),$$

we have only to show that U is not left invariant by all the elements of the set $O^+(V)^2$. In fact, we shall show that there exists a rotation ρ of V and a vector $A \in U$ such that $\rho^2 A \notin U$.

Case 1. U is nonsingular. Then $V = U \oplus U^*$ and we consider the involution $\sigma = -1_U \oplus 1_{U^*}$. If all the elements of $O^+(V)^2$ left U invariant, σ would commute with all these elements (Exercise 12 above) and hence $\sigma = \pm 1_V$. This means that $U = \{0\}$ or $U = V$, contrary to assumption.

Case 2. U is a null space. Choose a nonzero vector A in U. We shall produce a rotation ρ of V such that $\rho^2 A \notin U$. The null line $\langle A \rangle$ is contained in a nonsingular three-space W (Exercise 347.2c). By Exercise 355.13a, there is a null line in W different from $\langle A \rangle$; let α be a rotation of W which has l as fixed space. Then α does not leave the line $\langle A \rangle$ invariant (Exercise 355.14) and hence $\alpha \langle A \rangle$ is a line of W different from $\langle A \rangle$. Since $\langle A \rangle = U \cap W$ (Exercise 15 above), the line $\alpha \langle A \rangle$ cannot belong to U, that is, $\alpha A \notin U$.

By Proposition 325.2, there is a rotation β of W such that $\alpha = \beta^2$. Since W is nonsingular, $V = W \oplus W^*$ and we put $\rho = \beta \oplus 1_{W*}$. Then ρ is a rotation of V and

$$\rho^2 = \beta^2 \oplus 1_{W*} = \alpha \oplus 1_{W*}.$$

Since ρ^2 and α act the same way on the vectors of W, $\rho^2 A \notin U$.

Case 3. U is an arbitrary singular subspace of V. By Case 2, there exists a rotation ρ of V such that ρ^2 does not leave rad U invariant. Then ρ^2 could not leave U invariant since every isometry of V which leaves U invariant must also leave rad U invariant. Done.

There remains the investigation of the natural representations of the classical groups $O(V)$, $O^+(V)$, and $\Omega(V)$ for $n = 1$ and 2. It was shown in Exercise 354.11 that all representations are simple if $n = 1$ and we now turn to the two-dimensional case.

Proposition 357.1. Let V be an anisotropic plane. Then the natural representations of $O(V)$ and $O^+(V)$ are simple. If $k \neq \mathbf{Z}_3$, the natural representation of $\Omega(V)$ is also simple. If $k = \mathbf{Z}_3$, $\Omega(V)$ leaves all lines of V through **0** invariant; therefore, its natural representation is not simple.

Proof. V is not the exceptional plane and hence there is a rotation $\rho \neq \pm 1_V$. If l is a line through **0**, it is nonsingular and not left invariant by ρ. Since $\rho \in O^+(V) \subset O(V)$, this shows that the natural representations of both $O(V)$ and $O^+(V)$ are simple.

We recall that $\Omega(V) = O^+(V)^2$ (Exercise 339.8a). Therefore, if $k = \mathbf{Z}_3$, $\Omega(V) = \{1_V, -1_V\}$ (Exercise 352.16c). In that case, $\Omega(V)$ leaves all lines of V through **0** invariant. If $k \neq \mathbf{Z}_3$, $\Omega(V) \nsubseteq \{1_V, -1_V\}$, equivalently, $\Omega(V)$ contains a rotation $\rho \neq \pm 1_V$. We then conclude, as in the case for the groups $O(V)$ and $O^+(V)$, that the natural representation of $\Omega(V)$ is simple. Done.

The field $\mathbf{Z}_3$ also plays a special role when V is Artinian.

Proposition 357.2. Let V be an Artinian plane. The natural representations of $O^+(V)$ and $\Omega(V)$ are not simple. The natural representation of $O(V)$ is simple, unless V is the exceptional plane. In that case, $O(V)$ has two invariant lines and hence its natural representation is not simple.

Proof. Every rotation leaves the two null lines of V invariant; this shows that the natural representation of $O^+(V)$ is not simple. It follows that the natural representation of $\Omega(V)$ is also not simple because it is just the action of $O^+(V)$ restricted to $\Omega(V)$ (Exercise 355.11b).

We now turn to $O(V)$. Assume, first, that V is not the exceptional plane. Then V has a rotation $\rho \neq \pm 1_V$. This rotation leaves the two null lines invariant, but no other line. Furthermore, any symmetry interchanges the two null lines and thus does not leave them invariant. Since both ρ and τ belong to $O(V)$, the natural representation of $O(V)$ is simple.

Finally, let V be the exceptional plane. Then $O(V)$ consists of the two rotations $\pm 1_V$ and two symmetries. The two rotations leave all lines invariant while the two symmetries leave the two nonsingular lines invariant. Consequently, the two nonsingular lines are left invariant by all isometries of V, therefore, implying that the natural representation of $O(V)$ is not simple. Done.

If V is an Artinian plane which is not the exceptional plane, the natural representation of $O(V)$ is simple, but the natural representation of $O^+(V)$ is not simple. Since the action of $O^+(V)$ is obtained by restricting the action of

$O(V)$ to $O^+(V)$, this shows again that a simple representation need not be simple when restricted.

Here, we finish our investigation of the classical groups $O(V)$, $O^+(V)$, and $\Omega(V)$. Deeper results can only be obtained by introducing the theory of Clifford algebras. See Artin, [2]. This theory allows one to say a great deal about the precise structure of the groups in question for $n \le 4$ and to derive strong theorems about the general nature of these groups when $n > 4$. Even with the use of Clifford algebras and further techniques of more and more algebraic character, many problems concerning the classical groups remain unsolved today. Somewhat unexpectedly, the main difficulties arise when V is anisotropic.

65. SIMILARITIES

We assume that $n \ge 1$ throughout this section.

Similarities are not as important as isometries, but without them a person does not have a well-rounded view of geometry. Isometries preserve distance and hence angular measure while a similarity (similarity transformation) preserves only angular measure and not necessarily distance. The definition which says that two triangles are similar if their corresponding angles are congruent is equivalent to saying that two triangles are similar if there is a similarity which maps one triangle onto the other. We shall return to the relationship between similar triangles and the notion of similarity in Section 70 in Chapter 3. Among the several equivalent ways of defining similarities of nonsingular metric vector spaces, we now choose the one we like best. It is the definition which gives immediately that similar triangles have proportional sides (see Exercise 425.24).

Definition 358.1. A similarity γ of V is a linear automorphism of V for which there exists a nonzero scalar q such that

$$(\gamma A)(\gamma B) = q(AB)$$

for all $A, B \in V$. The scalar q is called the **square ratio** of the similarity.

If $A = B$ in the above definition, one has

$$(\gamma A)^2 = qA^2$$

and hence q is the scalar by which the squares of vectors are multiplied. Observe that q can be computed from the action of γ on just one nonisotropic vector A since

$$q = \frac{(\gamma A)^2}{A^2}.$$

If V is Euclidean space, the square ratio q of a similarity γ must be a positive real number. Namely, if A is a nonzero vector, both A^2 and $(\gamma A)^2$ are then positive real numbers; therefore, so is $(\gamma A)^2/A^2$. Furthermore, the length of A is then the positive number $\sqrt{A^2}$ and hence

$$\text{length } \gamma A = \sqrt{q} \text{ length } A.$$

This shows that $\sqrt{q}$ is the real number by which lengths are multiplied; for this reason, $\sqrt{q}$ is the true ratio of the similarity. However, for arbitrary geometries, $\sqrt{q}$ does not exist in k, even if $k = \mathbf{R}$ (see Exercise 363.17c).

It is clear that isometries are those similarities with square ratio 1. Further examples of similarities are the magnifications $M(0, r)$ where we recall that $r \neq 0$ and

$$M(0, r)A = rA$$

for all $A \in V$.

If $A, B \in V$,

$$(M(0, r)A)(M(0, r)B) = (rA)(rB) = r^2(AB);$$

thus a magnification of ratio r is a similarity of square ratio r^2. We shall see more examples of similarities in the exercises.

Two similarities γ and μ are multiplied as linear transformations, that is, $\gamma\mu$ means "first μ and then γ." The inverse γ^{-1} is the inverse of γ as a linear automorphism of V. The following exercises bring out the fact that, under this multiplication, the similarities of V form a group and the scalars which occur as square ratios are restricted by the geometry of V. The group of similarities will be denoted by $S(V)$.

Exercises

1. If γ and μ are similarities of V with square ratios q and r, respectively, prove that $\gamma\mu$ is a similarity with square ratio qr.

*2. a. If α is an arbitrary linear transformation of V (possibly singular) and $M(0, r)$ is a magnification, prove that

$$M(0, r)\alpha = \alpha M(0, r).$$

b. If σ is an isometry and $M(0, r)$ is a magnification of V, prove that $M(0, r)\sigma$ is a similarity with square ratio r^2. Observe that this result gives us many more examples of similarities than we had before. We shall prove later that, in Euclidean and many other geometries, these are the only similarities; that is, in those geometries, every similarity is the product of an isometry and a magnification (see Theorem 366.1).

3. If γ is a similarity of V with square ratio q, prove that γ^{-1} is a similarity with square ratio q^{-1}.

4. Prove that the similarities of V form a subgroup of the general linear group $GL(n, k)$; also show that this subgroup contains $O(V)$. Thus we have the tower of groups

$$O(V) \subset S(V) \subset GL(n, k).$$

*5. Prove that the mapping

$$\gamma \to \text{square ratio of } \gamma$$

is a homomorphism from the group $S(V)$ into the multiplicative group k^*. We shall see in Exercise 7 below that this homomorphism is, in general, not onto k^*.

6. Prove that $O(V)$ is a normal subgroup of $S(V)$ which contains the commutator subgroup of $S(V)$. (Hint: Find the kernel of the homomorphism given in Exercise 5 above.)

*7. In general, the square ratio of a similarity cannot be chosen arbitrarily. The purpose of this exercise is to see how the geometry of V severely limits the scalars which can occur as square ratios.

   a. If γ is a similarity of V with square ratio q, prove that $q = a/b$ where a and b are both nonzero scalars represented by V. We shall see in Exercise 361.13 that the converse is false; it may happen that a scalar q is the quotient of two nonzero scalars which are both represented by V, but there is no similarity which has q as square ratio. On the other hand, we now show that there are always many scalars which are square ratios of similarities.

   b. Prove that every square of a nonzero scalar occurs as a square ratio of a similarity.

   c. We denote the set of scalars which are square ratios of similarities by $\mathscr{S}$; in Part b it was shown that $\mathscr{S} \supset k^{*2}$. Prove that $\mathscr{S}$ is a subgroup of k^*. Hence we have the tower of groups

$$k^{*2} \subset \mathscr{S} \subset k^*.$$

   (Hint: Consider the homomorphism $\gamma \to$ square ratio of γ given in Exercise 5 above.)

8. a. Let V be Euclidean n-space. Prove that $\mathscr{S} = \mathbf{R}^{*2}$. (Hint: In order to show that $\mathscr{S} \subset \mathbf{R}^{*2}$, use the fact that every element of $\mathscr{S}$ is the quotient of two nonzero scalars represented by V.)

   b. If V is negative Euclidean n-space, prove that $\mathscr{S} = \mathbf{R}^{*2}$. Observe that V represents only nonpositive real numbers, thus it can happen that no scalar which is represented by V can be the square ratio of a similarity. We shall see in Exercise 365.23c that $\mathscr{S} = \mathbf{R}^{*2}$ for all geometries over $\mathbf{R}$ except when V is Artinian.

*9. Here we study the tower $k^{*2} \subset \mathscr{S} \subset k^*$ further and give sufficient (but not necessary) conditions in order that $\mathscr{S} = k^{*2}$ or $\mathscr{S} = k^*$.

    a. Assume that the quotient of every pair of nonzero scalars which are represented by V is a square in k^*. Prove that $\mathscr{S} = k^{*2}$. Also prove that this condition is fulfilled when V is Euclidean or negative Euclidean n-space. Exercise 361.13 shows that $\mathscr{S}$ may equal k^{*2} without the condition being satisfied.

    b. Assume that every element of k has a square root. Prove that $\mathscr{S} = k^*$. This happens, for instance, when $k = \mathbf{C}$ or, more generally, when k is algebraically closed. Exercise 363.17c shows that $\mathscr{S}$ may equal k^* without the condition being fulfilled.

10. Let $n = 1$. Prove that

$$S(V) = GL(1, k) \simeq k^*$$

and hence $S(V)$ is the group of magnifications. Furthermore, prove that $\mathscr{S} = k^{*2}$ and the condition in Exercise 9a above is satisfied.

*11. Here, we relate the notion of similarity to that of changing the metric of V by a scalar. Let q be a nonzero scalar and define the new nonsingular metric vector space V_0 by

$$A \circ B = q(AB).$$

The vectors of V_0 are the same as those of V; only the inner product has been changed. The space V_0 was first discussed in Exercise 166.13.

    a. Let γ be a linear automorphism of V. Prove that γ is a similarity of V with square ratio q if and only if γ is an isometry from V_0 to V.

    b. If n is odd, the result of Part a gives an extra condition which the elements of $\mathscr{S}$ must satisfy, namely, if $q \in \mathscr{S}$, prove that $q^n \in k^{*2}$. Observe that, if $q \in \mathscr{S}$, we are in the situation of Exercise 166.13h. If n is even, $q^n \in k^{*2}$ does not represent a restriction on the scalar q.

12. Let n be odd and assume $k = \mathbf{Q}$. Prove that $\mathscr{S} = \mathbf{Q}^{*2}$. (Hint: A rational number q whose odd power q^n is a square must be a square itself.)

*13. Let $k = \mathbf{Q}$.

    a. Assume that $n = 3$ and V has a rectangular coordinate system relative to which the matrix of V is

$$\begin{pmatrix} 1 & 0 & 0 \\ 0 & 1 & 0 \\ 0 & 0 & -1 \end{pmatrix}.$$

Prove that $\mathscr{S} = \mathbf{Q}^{*2}$, but the condition of Exercise 361.9a is not satisfied; that is, find a rational number q which is the quotient of two nonzero rational numbers, both represented by V while there is no similarity having q as square ratio.

b. More generally, let n be odd and assume that V has a nonzero isotropic vector. Do the same as in Part a.

A linear automorphism γ of V is said to **preserve orthogonality** if $A \perp B$ implies $\gamma A \perp \gamma B$ for all $A, B \in V$. It follows immediately from the definition of similarity that similarities preserve orthogonality. The interesting thing is that the converse is true, namely, if a linear automorphism preserves orthogonality, it is a similarity (see Theorem 362.1). The following exercises will be used in the proof of this theorem.

Exercises

*14. Let $A_1, \ldots, A_n$ be a rectangular coordinate system of V. Assume that γ is a linear automorphism of V and put $B_i = \gamma A_i$ for $i = 1, \ldots, n$. Then $B_1, \ldots, B_n$ is a coordinate system of V which may not be rectangular. Prove that γ is a similarity if and only if the following two conditions hold:

1. $B_1, \ldots, B_n$ is a rectangular coordinate system of V.
2. There exists a scalar q such that $B_i{}^2 = qA_i{}^2$ for $i = 1, \ldots, n$. (q is necessarily nonzero.)

Also prove that, under these conditions, q is the square ratio of γ.

*15. Let $(a_1, \ldots, a_n)$ and $(b_1, \ldots, b_n)$ be n-tuples of scalars. Prove that there exists a scalar q such that $b_i = qa_i$ for $i = 1, \ldots, n$ if and only if every solution of the equation $a_1x_1 + \cdots + a_n x_n = 0$ is a solution of $b_1x_1 + \cdots + b_n x_n = 0$. (Hint: Use only the most elementary theory of linear equations.)

The following theorem shows that similarities may be defined as linear automorphisms which preserve orthogonality.

Theorem 362.1.   A linear automorphism of V is a similarity if and only if it preserves orthogonality.

Proof. It is only necessary to prove that, if a linear automorphism γ preserves orthogonality, it is a similarity. For this, we choose a rectangular coordinate system $A_1, \ldots, A_n$ of V and put $B_i = \gamma A_i$ for $i = 1, \ldots, n$ and show that the two conditions of Exercise 14 above are satisfied. Condition 1 is trivially true since γ preserves orthogonality by hypothesis. Condition 2 is equivalent to saying that, if $c_1 A_1{}^2 + \cdots + c_n A_n{}^2 = 0$, then $c_1 B_1{}^2 + \cdots + c_n B_n{}^2 = 0$ (Exercise 15 above). The first equality says that the vector $c_1 A_1 + \cdots + c_n A_n$ is orthogonal to the vector $A_1 + \cdots + A_n$. Since γ preserves orthogonality, it follows that the vector $c_1 B_1 + \cdots + c_n B_n$ is orthogonal to the vector $B_1 + \cdots + B_n$, equivalently, that $c_1 B_1{}^2 + \cdots + c_n B_n{}^2 = 0$. Done.

We let

$$J = \{A^2/B^2 \mid A \quad \text{and} \quad B \text{ nonisotropic vectors of } V\}.$$

In other words, J is the set of scalars which are quotients of nonzero scalars both of which are represented by V. We proved in Exercise 360.7a that $\mathscr{S} \subset J$. Of course, whenever $\mathscr{S} = k^*$, then $\mathscr{S} = J = k^*$.

Exercises

16. Exercise 361.13 shows that J can be different from $\mathscr{S}$ when $n = 3$.
    a. If $n = 1$, prove that $\mathscr{S} = J$. (Hint: Show that $J = k^{*2}$ and use the relation $k^{*2} \subset \mathscr{S} \subset J$.)
    b. If $n = 2$, prove that $\mathscr{S} = J$. (Hint: Let $q = B_1{}^2/A_1{}^2$ where A_1 and B_1 are nonisotropic vectors. In order to show that there is a similarity with q as square ratio, choose vectors A_2 and B_2 such that A_1, A_2 and B_1, B_2 are two rectangular coordinate systems of V. Then $A_1{}^2 A_2{}^2 = c^2 B_1{}^2 B_2{}^2$ where $c \in k^*$, equivalently, $A_2{}^2 = c^2 q B_2{}^2$. Next, show that the coordinate systems A_1, A_2 and $B_1, cq B_2$ satisfy the two conditions of Exercise 362.14.)

*17. Let $n = 2$.
    a. If V represents 1, prove that every scalar which is represented by V is the square ratio of a similarity.
    b. If V is the Artinian plane, prove that $\mathscr{S} = k^*$.
    c. Let V be the Lorentz plane. Find a real number q which is the square ratio of a similarity while $\sqrt{q} \notin \mathbf{R}$. The Lorentz plane also shows that $\mathscr{S}$ may be equal to k^* while not every element of k has a square root.

18. In order to show that $\mathscr{S} = k^*$ for all Artinian spaces, we extend the notion of orthogonal sum of isometries to that of orthogonal sum of similarities with the same square ratio. Let $V = U_1 \oplus \cdots \oplus U_h$ where $U_1, \ldots, U_h$ are nonsingular subspaces of V. Let $\gamma_i \colon U_i \to U_i$ be a similarity of U_i for $i = 1, \ldots, h$ and assume that q is the common

square ratio of the similarities $\gamma_1, \ldots, \gamma_h$. Define $\gamma_1 \oplus \cdots \oplus \gamma_h$ as the "direct sum" of the linear transformations $\gamma_1, \ldots, \gamma_h$; that is, if $B_i \in U_i$ for $i = 1, \ldots, h$,

$$(\gamma_1 \oplus \cdots \oplus \gamma_h)(B_1 + \cdots + B_h) = \gamma B_1 + \cdots + \gamma_h B_h.$$

Prove that $\gamma_1 \oplus \cdots \oplus \gamma_h$ is a similarity of V with square ratio q. This similarity is called the **orthogonal sum** of the similarities $\gamma_1, \ldots, \gamma_h$.

19. If V is an Artinian space, prove that $\mathscr{S} = k^*$.
20. When n is odd, it may happen that $\mathscr{S} \neq J$, seemingly because $q^n \in k^{*2}$ for all $q \in \mathscr{S}$ (Exercises 361.11b and 361.13). It is reasonable to conjecture that $\mathscr{S} = J$ whenever n is even because this is true for planes and all Artinian spaces. We now show this conjecture is already false for $n = 4$.

Let V be Minkowski space and A_1, A_2, A_3, A_4 a rectangular coordinate system relative to which the matrix of V is

$$\begin{pmatrix} 1 & 0 & 0 & 0 \\ 0 & 1 & 0 & 0 \\ 0 & 0 & 1 & 0 \\ 0 & 0 & 0 & -1 \end{pmatrix}.$$

Prove that $-1 \in J$ but $-1 \notin \mathscr{S}$. (Hint: To show that $-1 \notin \mathscr{S}$, suppose γ were a similarity of V with -1 as square ratio. Then $\gamma A_1, \gamma A_2, \gamma A_3, \gamma A_4$ would be a rectangular coordinate system of V. Reach a contradiction by considering the signature of V relative to $\gamma A_1, \gamma A_2, \gamma A_3, \gamma A_4$. The signature was defined on page 172.)

21. If V has a nonzero isotropic vector, prove that $J = k^*$.
22. Let k be an ordered field.
    a. If V has a similarity with a negative square ratio, prove that n is even and the signature of V is $n/2$. (Hint: Let $A_1, \ldots, A_n$ be a rectangular coordinate system. By Theorem 173.1, the signature of V is the number of these coordinate vectors whose squares are positive. If γ is a similarity with negative square ratio, compute the signature of V also using the rectangular coordinate system $\gamma A_1, \ldots, \gamma A_n$.)
    b. Assume that k is ordered and every positive element of k has a square root. Prove that V has a similarity with a negative square ratio if and only if V is Artinian.
*23. Assume $k = \mathbf{R}$ and let us find the group $\mathscr{S}$ for each geometry.
    a. Prove that either $\mathscr{S} = \mathbf{R}^{*2}$ or $\mathscr{S} = \mathbf{R}^*$. (Hint: $\mathbf{R}^{*2} \subset \mathscr{S} \subset \mathbf{R}^*$ and $\mathbf{R}^{*2}$ is a subgroup of $\mathbf{R}^*$ of index two; now apply Exercise 351.9).
    b. Prove that $\mathscr{S} = \mathbf{R}^*$ if and only if V has a similarity with square ratio -1.

   c. Prove that $\mathscr{S} = \mathbf{R}^*$ if and only if V is Artinian and that $\mathscr{S} = \mathbf{R}^{*2}$ in all other cases.

24. Assume $k = \mathbf{R}$ and let us determine the set J for each geometry.

   a. Prove that $J = \mathbf{R}^{*2}$ if and only if V is either Euclidean or negative Euclidean space and that $J = \mathbf{R}^*$ in all other cases.

   b. If V is Euclidean or negative Euclidean space, prove that $\mathscr{S} = J = \mathbf{R}^{*2}$. If V is Artinian, prove that $\mathscr{S} = J = \mathbf{R}^*$. In all other cases, prove that $\mathscr{S} = \mathbf{R}^{*2}$ and $J = \mathbf{R}^*$.

*25. Let $n = 2$ and A_1, A_2 be a rectangular coordinate system of V. We study the linear automorphism defined by

$$\gamma A_1 = A_2 \quad \text{and} \quad \gamma A_2 = A_1.$$

   a. Prove that γ is a similarity if and only if $A_2{}^2 = \pm A_1{}^2$.

   b. If $A_2{}^2 = A_1{}^2$, prove that γ is a symmetry.

   c. If $A_2{}^2 = -A_1{}^2$, prove that γ is a similarity with square ratio -1 and V is Artinian.

   d. If $A_2{}^2 = A_1{}^2$, show, by examples, that V may or may not be Artinian, depending on the choice of k.

We learned in Exercise 359.2 that the product of an isometry σ and a magnification $M(0, r)$ is a similarity with square ratio r^2; furthermore, the transformations $M(0, r)$ and σ commute. Now we determine in which metric vector spaces every similarity is the product of a magnification and an isometry.

Theorem 365.1.   A similarity of V is the product of a magnification and an isometry if and only if the square ratio of that similarity is a square in k^*.

Proof. One way has been done; there only remains to be shown that, if γ is a similarity of V with square ratio r^2, there exists an isometry σ of V such that $\gamma = M(0, r)\sigma$. For this, we put

$$\sigma = M(0, r^{-1})\gamma$$

and ask the reader to verify that $\gamma = M(0, r)\sigma$. Since σ is the product of a similarity with square ratio r^{-2} and a similarity with square ratio r^2, σ is a similarity with square ratio 1. Thus σ is an isometry. Done.

The following theorem gives the best criterion for determining when a similarity of V is the product of a magnification and an isometry.

Theorem 366.1. Every similarity of V is the product of a magnification and an isometry if and only if $\mathscr{S} = k*^2$.

Proof. We always have $k*^2 \subset \mathscr{S}$ and it follows from the previous theorem that every similarity of V is the product of a magnification and an isometry if and only if $\mathscr{S} \subset k*^2$. Done.

For example, let $k = \mathbf{R}$. Then every similarity is the product of a magnification and an isometry if and only if V is not an Artinian space (Exercise 365.23c). This includes Euclidean spaces, negative Euclidean spaces, Minkowski spaces, and all odd dimensional spaces. However, if V is Artinian, a similarity can only be factored into a magnification and an isometry if the square ratio of that similarity is positive.

Exercises

26. The factorization $\gamma = M(0, r)\sigma$ of a similarity γ into a magnification $M(0, r)$ and an isometry σ is not unique, although nearly so. Let $r, r' \in k^*$ and $\sigma, \sigma' \in O(V)$.
    a. Prove that

    $$M(0, r)\sigma = M(0, r')\sigma'$$

    if and only if one of the following two conditions is satisfied.
    1. $r' = r$ and $\sigma' = \sigma$.
    2. $r' = -r$ and $\sigma' = -\sigma$.
    b. If a similarity γ is the product of a magnification $M(0, r)$ and an isometry σ, prove that there are precisely the four factorizations

    $$\gamma = M(0, r)\sigma = \sigma M(0, r) = M(0, -r)(-\sigma) = (-\sigma)M(0, -r).$$

27. Assume that k is ordered and $\mathscr{S} = k*^2$. (Readers who are not well versed in ordered fields should assume that $k = \mathbf{R}$ and V is not an Artinian space.)
    a. Prove that V is not an Artinian space. (Hint: Use the fact that k is ordered to conclude that $k*^2 \neq k*$.)
    b. If γ is a similarity of V, prove that there exists a unique positive scalar r and a unique isometry σ such that

    $$\gamma = M(0, r)\sigma = \sigma M(0, r).$$

    c. According to Part b, if r is a positive scalar, $M(0, -r)$ is the product of a magnification $M(0, r')$ with $r' > 0$ and an isometry. Find r' and the isometry.

28. This exercise should be omitted by readers who already know about the direct products of groups. Assume C and D are two groups, both written multiplicatively. Consider the ordered pairs (c, d) where $c \in C$ and $d \in D$ and define a product for them as follows:

$$(c, d)(c', d') = (cc', dd')$$

where $c, c' \in C$ and $d, d' \in D$.

a. Prove that the set of ordered pairs

$$\{(c, d) \mid c \in C, \quad d \in D\}$$

is a group under the above multiplication. This group is called the **direct product** of the groups C and D and is denoted by $C \times D$.

b. Prove that the direct product $C \times D$ is a commutative group if and only if both groups C and D are commutative. (Hint: For the "only if" part, assume C is not commutative. Then there exist elements $c, c' \in C$ such that $cc' \neq c'c$. Show that the pairs $(c, 1)$ and $(c', 1)$ where 1 is the unit element of D do not commute.)

c. If C and D are finite groups, prove that $C \times D$ is a finite group and that its order is the product of the orders of C and D.

d. Prove that the groups $C \times D$ and $D \times C$ are isomorphic.

29. Assume that k is ordered and $\mathscr{S} = k^{*2}$. We denote the multiplicative group of positive elements of k by P; thus $k^{*2} \subset P$. (Again, readers who are not well versed in ordered fields should assume that $k = \mathbf{R}$ and V is not an Artinian space.)

a. Prove that the group of similarities $S(V)$ is isomorphic to the direct product $P \times O(V)$. (Hint: For each $\gamma \in S(V)$, there is a unique ordered pair (r, σ) where r is a positive scalar and σ is an isometry such that $\gamma = M(0, r)\sigma$. Study the mapping $\gamma \to (r, \sigma)$ from $S(V)$ to $P \times O(V)$.)

b. Prove that $S(V)$ is commutative if and only if $n = 1$.

30. Assume that k is ordered and V is the Artinian plane. Let A_1, A_2 be a rectangular coordinate system such that $A_1{}^2 = 1$ and $A_2{}^2 = -1$. Prove that the linear automorphism which interchanges A_1 and A_2 is a similarity which is not the product of a magnification and an isometry. (Hint: This similarity was studied in Exercise 365.25.)

We assume in the remainder of this section that k is ordered and $\mathscr{S} = k^{*2}$. Readers who are not well versed in ordered fields should assume that $k = \mathbf{R}$ and V is not an Artinian space.

If γ is a similarity of V, there exists a unique positive scalar r and a unique isometry σ such that

$$\gamma = M(0, r)\sigma = \sigma M(0, r).$$

The similarity γ is called a **direct similarity** if σ is a rotation and an **opposite similarity** if σ is a reflection.

Exercises

31. Prove that rotations are direct similarities and reflections are opposite similarities.

32. Let r be a positive scalar.
    a. Prove that the magnification $M(0, r)$ is a direct similarity.
    b. Prove that the magnification $M(0, -r)$ is a direct similarity if n is even and an opposite similarity if n is odd.

33. a. Prove that the direct similarities of V form a group which is isomorphic to the direct product $P \times O^+(V)$ where P again denotes the multiplicative group of positive elements of k. We denote this group by $S^+(V)$.
    b. Prove that $S^+(V)$ is commutative if and only if n is 1 or 2.
    c. Prove that $O^+(V)$ is a normal subgroup of $S^+(V)$ which contains the commutator subgroup of $S^+(V)$. (Hint: Study the mapping $\gamma \to$ square ratio of γ.)

34. a. If n is even, prove that direct (opposite) similarities could just as well have been defined as products of rotations (reflections) with magnifications $M(0, r)$ where $r < 0$.
    b. If n is odd, prove that direct (opposite) similarities are products of reflections (rotations) with magnifications $M(0, r)$ where $r < 0$.

35. Assume that $n = 2$ and every positive element of k has a square root in k.
    a. Prove that V is either the positive definite plane or a negative definite plane. (Hint: There are only three nonisometric planes; eliminate the "undesirable" one.)
    b. If A and B are nonzero vectors of V, prove that there exists a unique direct similarity γ and a unique opposite similarity γ' such that $\gamma A = B$ and $\gamma' A = B$. (Hint: First show that there is a positive element t such that $B^2 = (tA)^2$, then use the fact that there is a unique rotation ρ and a unique reflection σ such that

$$\rho(tA) = B \quad \text{and} \quad \sigma(tA) = B$$

(Exercise 271.4).)

36. Assume that $A_1, \ldots, A_n$ is an arbitrary coordinate system of V and γ is a similarity. Prove that the coordinate systems $A_1, \ldots, A_n$ and $\gamma A_1, \ldots, \gamma A_n$ have the same orientation if γ is a direct similarity and

have opposite orientation if γ is an opposite similarity. (See pages 297–298 for the notion of orientation of coordinate systems and vector spaces.) Consequently, if V is oriented, direct similarities leave the orientation of V intact while opposite similarities reverse the orientation.

chapter 3
metric
affine
spaces

Introduction. We now study truly metric affine geometry. For this, we return to the affine space (X, V, k) of Chapter 1 and assume, in addition, that V is a metric vector space. The inner product of V induces a metric on X and it is this space X, with the vector space V acting on it, which is the true concept of metric affine space.

The step from the metric vector space V to the metric affine space (X, V, k) is, admittedly, a small one. In fact, it is so small that many people never make it. It certainly seems sufficient to know metric vector spaces for the applications of metric geometry to group theory. Nevertheless, if only for the purpose of gaining correct geometric insight, one should carry out the transition from metric vector spaces to metric affine spaces, at least once.

In order to give the reader the opportunity to carry out this transition largely by himself, more material than usual is relegated to the exercises. We hope the reader will not need the hints (some are actually short solutions) which are included in many of the exercises.

66. SQUARE DISTANCE

The affine spaces considered in this chapter are of a more restricted nature than those of Chapter 1. We must limit the scalar domain k of the affine space (X, V, k) in order to make V into a metric vector space.

We assume in this entire chapter that k is a field of characteristic different from 2.

The dimension of the affine space (X, V, k) is the dimension of V and is again denoted by n. In the following definition, V is allowed to be a singular metric vector space.

Definition 372.1.   A **metric affine space** is an affine space (X, V, k) where V is a metric vector space.

If $n = 1$, (X, V, k) is called a metric affine line; if $n = 2$, (X, V, k) is called a metric affine plane, etc. All metric notions of the vector space V can be carried, in a rather obvious way, from V to the affine space X.

Definition 372.2.   Two affine subspaces of X are called **orthogonal (perpendicular)** if their direction spaces are orthogonal.

If $x, y \in X$ and U, W are linear subspaces of V, the affine subspaces $S(x, U)$ and $S(y, W)$ are orthogonal if $U \perp W$. Notation: $S(x, U) \perp S(y, W)$. The transfer of several other metric notions from V to X, such as isometry and the square of vectors, will be carried out in the text and in exercises. We strongly urge the reader to do this for himself before continuing the development of this chapter.

If the metric vector space V is Euclidean n-space, the metric affine space $(X, V, \mathbf{R})$ coincides with the classical concept of Euclidean n-dimensional space. More explicitly, the axioms underlying the construction of $(X, V, \mathbf{R})$ are equivalent to those of Euclid and Hilbert for Euclidean n-space. There is no controversy as to what Euclidean geometry is, but only to what is the best set of axioms for Euclidean geometry. In the opinion of the authors and many others, the axioms of Euclid and Hilbert are unmanageable for technical geometry, in spite of their geometric appeal. Those who oppose the definition of Euclidean n-space as the metric affine space $(X, V, \mathbf{R})$ say that the axioms underlying $(X, V, \mathbf{R})$ are too algebraic and conceal the true geometric nature of Euclidean space. The authors' answer is that one should not close one's eyes to the algebraic structures which lie embedded in Euclidean geometry. On the contrary, we should become accustomed to them and use them whenever convenient. It is particularly convenient to use them in the axiomatic characterization of Euclidean space.

In Chapter 2, we referred to the Euclidean metric vector space V as "Euclidean space." From a strict logical point of view, this was not justified. True Euclidean space is the metric affine space $(X, V, \mathbf{R})$ where X is not a vector space and has no preferred point. The Euclidean vector space V has the origin as preferred point and its points (vectors) can be added and can be multiplied by real numbers. Henceforth, we shall only call $(X, V, \mathbf{R})$ Euclidean space and shall refer to V as the Euclidean *vector* space.

The same remarks can be made about all other geometries we have studied. For instance, when we say "the Lorentz plane," we shall now mean the metric affine plane $(X, V, \mathbf{R})$ where V is the Lorentz vector space; similar remarks hold for Minkowski space, Artinian space, the exceptional plane, and so on. This takes care of the fact that a metric affine space is never a vector space, as should be the case.

Let (X, V, k) be a metric affine space and let us see how a metric can be introduced in the set of points X. If x and y are points in X, the unique vector A of V such that $Ax = y$ is again denoted by $\overrightarrow{x, y}$.

Definition 373.1.   The **square distance** between the points x and y of X is the scalar $\overrightarrow{x, y}^2$.

The reason for calling $\overrightarrow{x, y}^2$ the square distance instead of the distance should be clear. If $(X, V, \mathbf{R})$ is Euclidean space, $\overrightarrow{x, y}^2 \geq 0$ and the Euclidean distance between the points x and y is the nonnegative square root $\sqrt{\overrightarrow{x,y}^2}$. In this case, the square distance is the square of the Euclidean distance. It would have been preferable to always work with the distance itself rather than with the square distance, but this is rarely possible. For instance, in the Lorentz plane, $\overrightarrow{x, y}^2$ may be negative and, therefore, there is no real number whose square is $\overrightarrow{x, y}^2$. This is the reason one must work with the square distance $\overrightarrow{x, y}^2$ in arbitrary geometries which are not Euclidean.

Let c be a point of X. We recall that X can be given the structure of an n-dimensional vector space over k with origin c (Definition 59.1), namely, the tangent space $X(c)$ of X at c. We showed that there are linear isomorphisms

$$g: V \to X(c) \quad \text{and} \quad f: X(c) \to V$$

which are inverses of one another; if $A \in V$ and $x \in X$,

$$g(A) = Ac \quad \text{and} \quad f(x) = \overrightarrow{c, x}.$$

The vectors of $X(c)$ are the points of X; if b is another point of X, the vector spaces $X(c)$ and $X(b)$ consist of the same set of vectors and are isomorphic. Nevertheless, these vector spaces cannot be identified, as was shown on pages 60 and 61.

The metric of the vector space $X(c)$ is defined as follows.

Definition 374.1. If $x, y \in X$, the **inner product** xy is defined by

$$xy = (\overrightarrow{c, x})(\overrightarrow{c, y}).$$

Since $\overrightarrow{c, x} = f(x)$ and $\overrightarrow{c, y} = f(y)$, the above definition says that the metric of V is carried over to $X(c)$ by the linear isomorphism $f: X(c) \to V$. Consequently, this definition makes $X(c)$ into a metric vector space and makes the linear isomorphisms f and g into isometries (see Exercise 1 below). Hence $X(c)$ and V are isometric and it follows that, if b is another point of X, the metric vector spaces $X(b)$ and $X(c)$ are isometric.

Exercises

In these exercises (X, V, k) is a metric affine space.
*1. Let c be a point in X.
    a. Prove that Definition 374.1 makes the vector space $X(c)$ into a metric vector space, that is, prove that xy satisfies the axioms of an inner product.
    b. Prove that the linear isomorphisms

$$f: X(c) \to V \quad \text{and} \quad g: V \to X(c)$$

are isometries. Observe that, consequently, all tangent spaces of X are isometric. In particular, if $X(c)$ is nonsingular for one point c of X, all tangent spaces are nonsingular. This happens if and only if V is nonsingular and, in that case, X is called a **nonsingular metric affine space**. If V is singular, X is called a **singular metric affine space**.
*2. Let $c \in X$. If x and y are points of X, $x - y$ is a vector of the metric vector space $X(c)$ and hence $(x - y)^2$ is a scalar. Prove that $(x - y)^2$ is the square distance between the points x and y. (Hint: By Exercise 61.6, $x - y = (\overrightarrow{y, x})c$.) Observe that the scalar $(x - y)^2$ does not

depend on the choice of the point c even though the vector $x - y$ varies with c. In Figure 375.1, $x - y$ is drawn as a vector in both of the spaces $X(c)$ and $X(b)$.

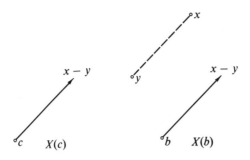

Figure 375.1

3. Let b and c be points of X and put $A = \overrightarrow{c, b}$. Prove that the translation T_A is an isometry of $X(c)$ onto $X(b)$. (Hint: Use Proposition 63.1.)
4. Let $c \in X$. The radical of the metric vector space $X(c)$ is a linear subspace of $X(c)$ and hence is an affine subspace of X passing through c.
   a. Prove that rad $X(c) = S(c, \text{rad } V)$.
   b. Prove that the family of affine subspaces

$$\{\text{rad } X(c) | c \in X\}$$

   is a family of parallel affine subspaces of the same dimension.
5. Let $c \in X$. The light cone of $X(c)$ is defined by

$$\{x | x \in X(c), \quad x^2 = 0\}$$

and has the point c as vertex.
   a. Prove that the isometry $g: V \to X(c)$ (defined by $g(A) = Ac$) maps the light cone of V onto the light cone of $X(c)$.
   b. Let $b \in X$ and put $A = \overrightarrow{c, b}$. If C is the light cone of $X(c)$ and C' is the light cone of $X(b)$, prove that $T_A(C) = C'$.
      Thus, at each point of X, there is a light cone with the point as vertex. One considers these cones as a family of parallel cones since each one is obtained from the other by a translation. In Figure 376.1, the light cones have been drawn at a, b, and c where V is the three-dimensional vector space over $\mathbf{R}$ whose metric is given by the matrix

$$\begin{pmatrix} 1 & 0 & 0 \\ 0 & 1 & 0 \\ 0 & 0 & -1 \end{pmatrix}.$$

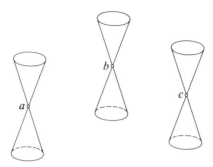

Figure 376.1

*6. Let $c \in X$ and U be a linear subspace of V. We already know that the affine subspace $S = S(c, U)$ is an affine space (X, U, k) (Proposition 14.2). Prove that (S, U, k) is a metric affine space.

7. We denote the square distance between the points x and y of X by $\delta(x, y)$. Prove that
   a. $\delta(x, y) = \delta(y, x)$.
   b. $\delta(x, x) = 0$.
   c. $\delta(x, y) = 0$ if and only if the vector $\overrightarrow{x, y}$ lies on the light cone of V.

*8. Let x, y, b, and c be points of X where $x \neq y$ and $a \neq b$. We study the oriented line segments (x, y) and (a, b) and the nonoriented line segments $\{x, y\}$ and $\{a, b\}$. Observe that the unique vector $\overrightarrow{x, y}$ is associated with the oriented line segment (x, y).
   a. Define the notion of **orthogonal (perpendicular) oriented line segments**. Notation: $(x, y) \perp (b, c)$.
   b. Define the notion of **orthogonal (perpendicular) nonoriented line segments**. Notation: $\{x, y\} \perp \{b, c\}$.
   c. Prove that $(x, y) \perp (b, c)$, equivalently, $\{x, y\} \perp \{b, c\}$ if and only if the line which contains the points x and y is perpendicular to the line which contains the points b and c.

9. As before, three noncollinear points of X are the vertices of a triangle.
   a. Define a right triangle.
   b. Define the hypotenuse of a right triangle.
   c. Prove the Pythagorean theorem. Precisely, if a, b, and c are the vertices of a right triangle with $\{a, b\}$ as hypotenuse, prove that

$$\delta(a, c) + \delta(c, b) = \delta(a, b).$$

Of course, this is the same as

$$\overrightarrow{a, c}^2 + \overrightarrow{c, b}^2 = \overrightarrow{a, c}^2$$

which looks much more familiar. Observe that the Pythagorean theorem holds in all metric affine spaces and is in no way restricted to Euclidean geometry.

10. Prove the parallelogram law. Precisely, if the points x_1, x_2, x_3, x_4 of X are the vertices of a parallelogram (see page 78), prove that

$$\delta(x_1, x_2) + \delta(x_2, x_3) + \delta(x_3, x_4) + \delta(x_4, x_1) = \delta(x_1, x_3) + \delta(x_2, x_4).$$

This says that the sum of square lengths of the four sides of a parallelogram is equal to the sum of the square lengths of the two diagonals. (Hint: In a tangent space at one of the vertices, say $X(x_1)$, apply the parallelogram law of metric vector spaces discussed in Exercise 118.7b.)

11. Define a rhombus and prove that the diagonals of a rhombus are perpendicular.

*12. Let b and c be points of X. The midpoint of b and c is defined as the point m such that

$$\overrightarrow{b, m} = \tfrac{1}{2}(\overrightarrow{b, c}).$$

See Figure 377.1. Observe that the notion of midpoint is an affine notion and not a metric notion.

$$\begin{array}{ccc} c & m & b \end{array}$$

Figure 377.1

a. If $b = c$, prove that $m = b = c$.

b. If $b \neq c$, prove that m lies on the unique line through the points b and c.

c. Prove that $\overrightarrow{b, m} = \overrightarrow{m, c}$.

d. We now bring in the metric of X. Prove that $\delta(m, b) = \delta(m, c)$. Thus m is equidistant from b and c.

We assume for the remainder of this exercise that $b \neq c$ and X is nonsingular. Then $\overrightarrow{b, c} \neq 0$ and $\langle \overrightarrow{b, c} \rangle^*$ is a hyperplane of V. If the vector $\overrightarrow{b, c}$ is isotropic, this hyperplane is singular and contains $\overrightarrow{b, c}$. The hyperplane

$$H = S(m, \langle \overrightarrow{b, c} \rangle^*)$$

of X is called the **perpendicular bisector** of the line segment $\{b, c\}$. In Figure 378.1, $n = 3$.

e. If $\delta(b, c) = 0$, prove that H contains the points b and c. Observe that Figure 378.1 is misleading in this case.

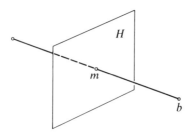

Figure 378.1

f. Prove that H consists of the points which are equidistant from b and c. Precisely, prove that

$$H = \{x \mid x \in X, \quad \delta(x, b) = \delta(x, c)\}.$$

(Hint: If $x \in H$, $x = Am$ where $A \perp \overrightarrow{b, c}$. Apply the theorem of Pythagoras to the right triangles bmx and cmx (see Figure 378.2).)

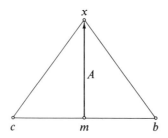

Figure 378.2

Conversely, assume that $\delta(x, b) = \delta(x, c)$ for $x \in X$. Put $x = Am$ and show that $A \perp \overrightarrow{b, c}$.)

*13. The metric affine space X is called **anisotropic** if V is anisotropic.
  a. Prove that X is anisotropic if and only if the tangent space $X(c)$ is anisotropic for all points c of X.
  b. Prove that X is anisotropic as soon as $X(c)$ is anisotropic for one point c of X.
  c. Prove that X is anisotropic if and only if $\delta(x, y) \neq 0$ for all pairs of points $x \neq y$ of X.
14. The metric affine space X is called a **null space** if V is a null space. Formulate and prove questions analogous to those of Exercise 13 above.

Remark on High School Teaching. How can one teach Euclidean geometry axiomatically in high school if the proper axioms for this geometry are those which describe the metric affine space $(X, V, \mathbf{R})$ where V is the Euclidean vector space? Of course, one can't! It is impractical to assume that, in the tenth or eleventh grade, a geometry course can be based on linear algebra. The authors recommend that high school geometry be taught nonaxiomatically and that proofs be written in a sound deductive way, but not be put in the straitjacket of "formal proofs." Instead of spending a great deal of time on "logic" and on dull theorems which fit neatly in the Euclid–Hilbert system, one should concentrate on pretty geometry such as the nine-point circle of a triangle, Morley's theorem for triangles, the parallelogram law, spherical geometry where the sum of the angles of a triangle exceeds 180°, and so on ad infinitum.

Moreover, geometry should be not only pretty, but also serious. Realizing that the foundations of geometry are most efficiently formulated in terms of the algebraic structures inherent in geometry, a high school course should bring these structures to the surface as much as possible. To do this, one chooses a point c in the Euclidean plane and discusses the vector space $X(c)$ in terms of arrows with their initial point at c. One should also discuss the groups of symmetries of triangles and squares, the rotation group, and the Euclidean group of the plane. A little light group theory should go over very well in the tenth and eleventh grades. Why shouldn't it? The notion of a group is so much simpler than many other concepts which are taught in the high schools. Which is easier to grasp, the concept of a group or that of the logarithm of the arccotangent?

Some people defend the teaching of axiomatic geometry in the high schools by claiming that it teaches students how to write proofs. This defense crumbles in the cruel light of experience. There is no number small enough to describe the percentage of college freshmen who can write the kind of fluid deductive proof which is the mainstay of both undergraduate and professional mathematics.

67. RIGID MOTIONS

In the remainder of this chapter, (X, V, k) stands for a metric affine space.
For the moment, we allow X to be singular; we recall that, by definition, X is singular if and only if V is singular. How to define an isometry from X to X should be clear. It should be a one-to-one function from X onto itself which preserves both the affine structure and the metric structure. To say that a function preserves the affine structure means that it is an affine transformation (Definition 71.1); a function preserves the metric structure if it preserves the square distance between points. As before, we denote the

square distance between the points x and y by $\delta(x, y)$ and usually refer to isometries of X as rigid motions.

Definition 380.1. A **rigid motion (isometry)** u of X is an affine transformation of X onto itself with the property that

$$\delta(u(x), u(y)) = \delta(x, y)$$

for all points x and y of X.

Let c be a point of X. For each linear automorphism λ of V, the affine transformation $L(c, \lambda)$ of X was defined by $L(c, \lambda)(x) = (\lambda(\overrightarrow{c, x}))c$ (page 75). If $A \in V$, $T_A L(c, \lambda)$ is an affine transformation and we proved that all affine transformations of X are of this form where the point c is kept fixed. We shall see in Proposition 381.1 that $T_A L(c, \lambda)$ is a rigid motion of X if and only if λ is an isometry of V. The if part of this result is worked out in Exercises 1, 2, and 3 below.

Exercises

*1. If $A \in V$, prove that the translation T_A is a rigid motion of X. (Hint: We know from Chapter 1 that T_A is an affine transformation and $\overrightarrow{T_A(x), T_A(y)} = \overrightarrow{x, y}$.)

*2. Let $\lambda \in O(V)$ and $c \in X$; we study the function $L(c, \lambda)$. Prove that $L(c, \lambda)$ is a rigid motion of X which leaves the point c fixed. Observe that an isometry of V gives rise to many rigid motions of X, namely, one for each point of X.

*3. Rigid motions are multiplied as affine transformations. In other words, if u and w are rigid motions, the product uw is defined by first w and then u.

   a. Prove that uw is a rigid motion of X.

   b. Prove that the rigid motions of X form a subgroup of the group Af of affine transformations of X. We denote this group by Mo.

   c. Prove that the group Tr of translations of X is a subgroup of the group of rigid motions. Hence we have the tower of groups

$$Tr \subset Mo \subset Af.$$

*4. This exercise will be used in the proof of Proposition 381.1. Let $\lambda: V \to V$ be a linear automorphism of V with the property that

$(\lambda B)^2 = B^2$ for all $B \in V$. Prove that λ is an isometry of V. (Hint: If $A, B \in V$, $AB = \frac{1}{2}((A + B)^2 - A^2 - B^2)$.)

In Exercises 1, 2, and 3 above, we proved the if part of the following proposition.

Proposition 381.1. Let $c \in X$, $A \in V$, and λ a linear automorphism of V. Then $T_A L(c, \lambda)$ is a rigid motion of X if and only if λ is an isometry of V.

Proof. There only remains to prove that, if $T_A L(c, \lambda)$ is a rigid motion, $\lambda \in O(V)$. For this, we put $u = T_A L(c, \lambda)$ and choose a vector $B \in V$. We must show that $(\lambda B)^2 = B^2$; then $\lambda \in O(V)$ by Exercise 4 above.

Since u is a rigid motion,

$$\delta(u(c), u(Bc)) = \delta(c, Bc) = B^2.$$

Let us show that

$$\delta(u(c), u(Bc)) = (\lambda B)^2.$$

Now

$$\overrightarrow{u(c), u(Bc)} = \overrightarrow{T_A L(c, \lambda)(c), T_A L(c, \lambda)(Bc)}$$

$$= \overrightarrow{L(c, \lambda)(c), L(c, \lambda)(Bc)} = \lambda \overrightarrow{(c, Bc)} = \lambda B.$$

Consequently, $(\lambda B)^2 = B^2$. Done.

The following theorem describes the rigid motions of X. It requires no proof since it is an immediate corollary of the above proposition and our results on affine transformations.

Theorem 381.1. Let $c \in X$. If u is a rigid motion of X, there exists a unique vector A in V and a unique isometry λ of V such that $u = T_A L(c, \lambda)$. For every $A \in V$ and $\lambda \in O(V)$, $T_A L(c, \lambda)$ is a rigid motion of X.

Our next theorem considers the rigid motions which leave a point of X fixed. It states that these motions form a group and describes the group fully. It follows immediately from the theorem that the structure of the group of motions which leave a point fixed does not depend on the choice of that point. The proof of the theorem is straightforward and is contained in Exercises 5–7 following the statement of the theorem.

Theorem 382.1.   The rigid motions of X which leave a point c fixed form the group $\{L(c, \lambda) \mid \lambda \in O(V)\}$. This group is the orthogonal group of the tangent space $X(c)$ and is isomorphic to the orthogonal group $O(V)$.

Exercises

*5. Prove that the rigid motions of X which leave a point fixed form a subgroup of the group of rigid motions of X.

*6. Let $c \in X$ and u a rigid motion of X.
   a. Prove that the vector A of V, such that $u = T_A L(c, \lambda)$ for $\lambda \in O(V)$, is equal to $\overrightarrow{c, u(c)}$.
   b. If u leaves c fixed, prove that $A = \mathbf{0}$ and $u = L(c, \lambda)$ for $\lambda \in O(V)$.
   c. Prove that $\{L(c, \lambda) \mid \lambda \in O(V)\}$ is the group of rigid motions which leave c fixed.

*7. Let $c \in X$ and consider the metric vector space $X(c)$.
   a. If $\lambda \in O(V)$, the function $L(c, \lambda)$ maps the vector space $X(c)$ onto itself, the reason being that the vectors of $X(c)$ are the points of X. Prove that $L(c, \lambda)$ is an isometry of $X(c)$.
   b. Prove that the mapping $\lambda \to L(c, \lambda)$ is an isomorphism from the orthogonal group $O(V)$ onto the orthogonal group of $X(c)$. We conclude that each isometry of $X(c)$ is equal to $L(c, \lambda)$ for a unique $\lambda \in O(V)$.
   c. Prove that the group of rigid motions of X which leave the point c fixed is identical with the orthogonal group of $X(c)$.

If $(X, V, \mathbf{R})$ is Euclidean n-space, the group of rigid motions is the classical n-**dimensional Euclidean group** and the rigid motions are the classical **Euclidean transformations**. The two-dimensional Euclidean group is also called the **Euclidean plane group** or, simply, the **Euclidean group**. However, we warn those readers who are familiar with the classical Lorentz group that this group is not the group of rigid motions of the Lorentz plane.

In Euclidean geometry, it is customary to define a Euclidean transformation as a one-to-one function of X onto itself which preserves distance. In other words, one never mentions that a Euclidean transformation has to be an affine transformation. Indeed, one gets the affine part free in this case. In fact, whenever X is nonsingular, a distance preserving function is automatically a rigid motion and hence an affine transformation (see Theorem 384.1).

In the two lemmas which prepare the way for the proof of this result, observe that the function u is *not* assumed to be linear.

Lemma 383.1. Let V and W be metric vector spaces over k. Assume $u: V \to W$ satisfies the following two conditions.

1. $(u(A) - u(B))^2 = (A - B)^2$ for all $A, B \in V$.
2. $u(0) = 0$, that is, u sends the origin of V onto the origin of W.

Then u preserves inner products, equivalently,

$$u(A)u(B) = AB$$

for all $A, B \in V$.

Proof. If we put $B = 0$ in Condition 1 and use the fact that $u(0) = 0$, we conclude that $u(A)^2 = A^2$ for all $A \in V$. Expansion of both sides of the equation in Condition 1 gives

$$u(A)^2 - 2u(A)u(B) + u(B)^2 = A^2 - 2AB + B^2.$$

Since $u(A)^2 = A^2$, $u(B)^2 = B^2$, and char $k \neq 2$, it follows that $u(A)u(B) = AB$. Done.

Lemma 383.2. Let V and W be metric vector spaces of the same dimension n. Assume, further, that V is nonsingular and $u: V \to W$ satisfies

$$u(A)u(B) = AB \qquad \text{for all} \quad A, B \in V.$$

Then W is nonsingular and u is a linear isomorphism from V onto W and hence an isometry.

Proof. Let $A_1, \ldots, A_n$ be a rectangular coordinate system of V. Since V is nonsingular, $A_i^2 \neq 0$ for $i = 1, \ldots, n$. The vectors $u(A_1), \ldots, u(A_n)$ of W are mutually orthogonal and $u(A_i)^2 = A_i^2 \neq 0$ for $i = 1, \ldots, n$. This implies the vectors $u(A_1), \ldots, u(A_n)$ are n linearly independent vectors and hence form a rectangular coordinate system of W (Proposition 154.1). It follows immediately that W is nonsingular.

We now show that u is a linear transformation. Let $A, B \in V$ and let us show that

$$u(A + B) = u(A) + u(B).$$

To do this, we choose a third vector $C \in V$ and use the given condition on u twice to get

$$u(A + B)u(C) = (A + B)C = AC + BC = u(A)u(C) + u(B)u(C),$$

equivalently,

$$(u(A + B) - u(A) - u(B))u(C) = 0.$$

In particular, this last equality holds for the coordinate vectors $u(A_1), \ldots, u(A_n)$ and hence the vector $u(A + B) - u(A) - u(B)$ is orthogonal to all vectors of W. Since W is nonsingular, we conclude that

$$u(A + B) = u(A) + u(B).$$

We leave it to the reader to prove in a similar way that $u(tA) = tu(A)$ for all $A \in V$ and $t \in k$. Thus u is a linear transformation.

Finally, u is a linear transformation which maps a coordinate system of V onto a coordinate system of W and, therefore, must be a linear isomorphism. Done.

The next corollary specializes to the case when $V = W$ and is an immediate consequence of Lemmas 383.1 and 383.2.

Corollary 384.1.   Let V be a nonsingular metric vector space and $u: V \to V$ which satisfies

1. $(u(A) - u(B))^2 = (A - B)^2$ for all $A, B \in V$.
2. $u(0) = 0$.

Then u is an isometry of V.

We recall that X is nonsingular if V is nonsingular and a function $u: X \to X$ is said to preserve distance if $\delta(u(x), u(y)) = \delta(x, y)$ for all points x and y of X.

Theorem 384.1.   Let X be nonsingular. Assume that $u: X \to X$ and satisfies $\delta(u(x), u(y)) = \delta(x, y)$ for all x and y in X. Then u is a rigid motion of X and hence is an affine transformation.

Proof. Choose a point c in X. The function u can be regarded as a function of the nonsingular metric vector space $X(c)$ into itself.

Case 1. u leaves the point c fixed. Then $u: X(c) \to X(c)$ satisfies the two conditions of Corollary 384.1, namely, Condition 2 is the same as $u(c) = c$ and the assumption

$$\delta(u(x), u(y)) = \delta(x, y)$$

is the same as

$$(u(x) - u(y))^2 = (x - y)^2$$

which is precisely Condition 1 (Exercise 374.2). It follows from Theorem 382.1 that u is an isometry of $X(c)$ and hence a rigid motion of X.

Case 2. u does not leave the point c fixed. Denote the vector $\overrightarrow{c, u(c)}$ by A.

The translation T_{-A} maps the point $u(c)$ onto c and hence $T_{-A}u$ is a function of X to itself which leave the point c fixed. Since T_{-A} and u both satisfy the hypothesis of our theorem, it is trivial to show that $T_{-A}u$ also satisfies this condition. We conclude from Case 1 that $w = T_{-A}u$ is a rigid motion of X. Since w and T_A are both rigid motions, their product $u = T_A w$ is also a rigid motion. Done.

It is interesting to note that u is not assumed to be either one-to-one or onto in Theorem 384.1, but these properties are a consequence of the fact that u preserves distance and X is nonsingular. The condition that X is nonsingular cannot be removed from the hypothesis of the theorem. For instance, if X is a null space, every function from X to X satisfies the condition of Theorem 384.1. It is now very easy to find functions which do not map each line on a line and hence are not affine. Any function which is not one-to-one will do.

Exercises

If $c \in X$, the isometry $g: V \to X(c)$, defined by $g(A) = Ac$ for $A \in V$, enables one to carry all notions of metric vector spaces from V to $X(c)$. It is clearest to do this first abstractly for arbitrary metric vector spaces (Exercise 8 below) and then apply it to the case in which we are interested (Exercise 9 below).

*8. Let V' be a second metric vector space over k and assume that $u: V \to V'$ is an isometry. If $\sigma \in O(V)$, define $\sigma' = u\sigma u^{-1}$.
   a. Prove that σ' is an isometry of V'.
   b. If σ is a rotation (reflection) of V, prove that σ' is a rotation (reflection) of V'.
   c. Let H be a nonsingular hyperplane of V and τ the symmetry of V with respect to H. Prove that $u(H)$ is a nonsingular hyperplane of V' and τ' is the symmetry of V' with respect to $u(H)$.
   d. More generally, let W be a nonsingular subspace of V of dimension p and assume that $\sigma = -1_W \oplus 1_{W*}$ is the corresponding involution of type p. Prove that $u(W)$ is a nonsingular subspace of V' of dimension p and σ' is the corresponding involution $-1_{u(W)} \oplus 1_{u(W)*}$ of V' of type p.

*9. Let $c \in X$. We study the isometry $g: V \to X(c)$. We recall that, if $\lambda \in O(V)$, $L(c, \lambda)$ is the isometry $g\lambda g^{-1}$ of X (page 75). The following questions follow almost immediately by letting $V' = X(c)$ and $u = g$ in the previous exercise.
   a. If λ is a rotation (reflection) of V, prove that $L(c, \lambda)$ is a rotation (reflection) of $X(c)$.

b. Let H be a nonsingular hyperplane of V and τ the symmetry of V with respect to H. Prove that the affine subspace $S(c, H)$ of X is a nonsingular hyperplane of $X(c)$ and $L(c, \tau)$ is the symmetry of $X(c)$ with respect to $S(c, H)$.

c. More generally, let W be a nonsingular subspace of dimension p of V and assume that $\lambda = -1_W \oplus 1_{W*}$ is the corresponding involution of type p. We denote the affine subspace $S(c, W)$ by S. Prove that S is a nonsingular linear subspace of $X(c)$ of dimension p and $L(c, \lambda)$ is the corresponding involution $-1_S \oplus 1_{S*}$ of $X(c)$ of type p.

If $c \in X$, a rotation of the metric vector space $X(c)$ is also called a **rotation of X**. Consequently, if λ is a rotation of V, $L(c, \lambda)$ is a rotation of X. In the same way, one defines **reflections, symmetries, involutions of type p, 180° rotations, etc., of X** as isometries of the same kind of the metric vector space $X(c)$. Observe that all rigid motions of this kind leave the point c fixed. The point c can be chosen at random (see Exercise 10 below) and, if one wants to indicate the reference point, expressions such as "rotations with the point c as center" are used.

Exercises

*10. This exercise shows that the notions of rotation of X, reflection of X, etc., do not depend on the choice of the reference point c. Assume that u is a rigid motion of X which leaves both points c and b of X fixed. Then there are isometries ρ and σ in $O(V)$ such that $u = L(c, \sigma) = L(b, \rho)$. Would it be possible that ρ is a rotation and σ a reflection of V? In that case, u would be a rotation of X relative to b and a reflection relative to c. Luckily, this situation does not occur. It is actually a consequence of Section 18 that $\rho = \sigma$. Nevertheless, a direct proof that $\rho = \sigma$ will be outlined in this exercise. The proof has nothing to do with isometries, but holds when ρ and σ are arbitrary linear transformations.

Choose a vector $B \in V$; we must show that $\rho B = \sigma B$, knowing that $L(c, \sigma)(Bc) = L(b, \rho)(Bc)$. We denote the vector $\overrightarrow{b, c}$ by A and have

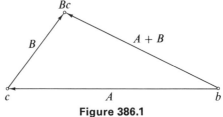

Figure 386.1

Figure 386.1. Prove that
a. $L(c, \sigma)(Bc) = (A + \sigma B)b$.
b. $L(b, \rho)(Bc) = (\rho A + \rho B)b$.
c. $\rho A = A$. (Hint: Use the fact that $u(c) = c$.)
d. $\rho B = \sigma B$.

In the next four exercises, the concept of rigid motion is developed for different metric affine spaces. In the case of different spaces, we say isometry instead of rigid motion.

11. Let (X', V', k) be a second metric affine space.
    a. Define an isometry from X onto X'.
    b. If u is an isometry from X onto X', prove that u has an inverse u^{-1} and that u^{-1} is an isometry from X' onto X.
*12. Two metric affine spaces are called **isometric** if there exists an isometry from one onto the other. Prove that the metric affine spaces (X, V, k) and (X', V', k) are isometric if and only if the metric vector spaces V and V' are isometric. (Hint: Let $\sigma: V \rightarrow V'$ be an isometry. Choose points $c \in X$ and $c' \in X'$ and study the mapping $u: X \rightarrow X'$ defined by $u(x) = \sigma(\overrightarrow{c, x})c'$. Similarly, let $u: X \rightarrow X'$ be an isometry. Choose a point $c \in X$ and study the mapping $\sigma: V \rightarrow V'$ defined by $\sigma A = \overrightarrow{u(c), u(Ac)}$.)
13. We recall that affine subspaces of X are metric affine spaces in their own right (Exercise 376.6).
    a. Prove that parallel affine subspaces of X of the same dimension are isometric.
    b. If W is a linear subspace of V and b, c are points of X, describe an isometry from $S(b, W)$ to $S(c, W)$ which maps b onto c.
14. The metric affine space (X', V', k) has a tangent space $X'(c')$ at each point $c' \in X'$.
    a. Let $c \in X$ and $c' \in X'$. Prove that a function u from X to X' is an isometry which maps c onto c' if and only if u is an isometry from the metric vector space $X(c)$ onto the metric vector space $X'(c')$.
    b. Use Part a to extend the theory of rigid motions of X as much as possible to isometries from X onto X'.

68. INTERRELATION AMONG THE GROUPS *Mo, Tr,* AND *O(V)*

For each point c in X, there is an obvious mapping from the group Mo of rigid motions of X to the orthogonal group $O(V)$. Namely, each rigid motion u of X can be written $T_A L(c, \lambda)$ for a unique vector A and a unique isometry

λ of V. The mapping we have in mind maps u onto λ. This mapping is denoted by ϕ_1 and does not depend on the point c although it may, at first, seem so. Already in Section 18 of Chapter 2, λ was defined in terms of u alone without the choice of a reference point c. There, u was an arbitrary semiaffine transformation. Consequently, the mapping

$$\phi_1 : Mo \to O(V)$$

does not depend on the choice of the point c. We shall prove (Proposition 389.1) that ϕ_1 is a homomorphism onto $O(V)$ and the kernel of ϕ_1 is the group Tr of translations of X. The following exercises prepare for the proof of this proposition.

Exercises

*1. Let $c \in X$, $A \in V$, and $\lambda \in O(V)$. We know that each point $x \in X$ can be written uniquely as Bc where $B \in V$ and $L(c, \lambda)(Bc) = (\lambda B)c$. (The results proved in Parts a, b, c hold just as well when λ is an arbitrary linear transformation of V.)

   a. Prove that $T_A L(c, \lambda)(Bc) = (A + \lambda B)c$.

   b. We now interchange the rigid motions T_A and $L(c, \lambda)$. Prove that $L(c, \lambda)T_A(Bc) = (\lambda A + \lambda B)c$.

   c. Prove that $L(c, \lambda)T_A = T_{\lambda A}L(c, \lambda)$.

   d. If $n \geq 1$, prove that the group Mo is never commutative. Observe that this is also true when X is the exceptional plane (equivalently, when V is the exceptional plane), even though $O(V)$ is then commutative. This is another example of the fact that noncommutativity is the natural thing for groups in geometry.

*2. Let $c \in X$. This exercise combines the results of Exercise 1 above with the fact that the mapping $\lambda \to L(c, \lambda)$ is an isomorphism from $O(V)$ onto the group of rigid motions of X which leaves the point c fixed.

   a. If $\lambda, \sigma \in O(V)$, prove that

   i. $L(c, \sigma)L(c, \lambda) = L(c, \sigma\lambda)$.

   ii. $L(c, \lambda)^{-1} = L(c, \lambda^{-1})$.

   iii. $L(c, 1_V) = 1_X$ where 1_X is the identity mapping of X.

   b. If $A, B \in V$, prove that

   $$T_A L(c, \sigma)T_B L(c, \lambda) = T_{A + \sigma B} L(c, \sigma\lambda).$$

   We now return to the mapping $\phi_1 : Mo \to O(V)$ defined by

   $$\phi_1(T_A L(c, \lambda)) = \lambda$$

   and recall that ϕ_1 does not depend on the choice of the point c.

Proposition 389.1.   Let $c \in X$. The mapping

$$\phi_1: Mo \to O(V)$$

is an epimorphism which has the group *Tr* of translations as kernel. Consequently, *Tr* is a commutative, normal subgroup of *Mo* and $Mo/Tr \simeq O(V)$.

Proof. If $u, w \in Mo$, we must prove that $\phi_1(uw) = \phi_1(u)\phi_1(w)$. There are unique vectors $A, B \in V$ and isometries $\sigma, \lambda \in O(V)$ such that

$$u = T_A L(c, \sigma) \quad \text{and} \quad w = T_B L(c, \lambda).$$

By definition, $\phi_1(u) = \sigma$ and $\phi_1(w) = \lambda$. Furthermore,

$$uw = T_A L(c, \sigma) T_B L(c, \lambda) = T_{A+\sigma B} L(c, \sigma\lambda).$$

It follows that

$$\phi_1(uw) = \sigma\lambda = \phi_1(u)\phi_1(w).$$

If $\lambda \in O(V)$, $L(c, \lambda)$ is a rigid motion such that $\phi_1(L(c, \lambda)) = \lambda$. Thus far, we have shown that ϕ_1 is an epimorphism.

Next, we show that the kernel of ϕ_1 is *Tr*. The unit element of $O(V)$ is 1_V and $\phi_1(T_A L(c, \lambda)) = 1_V$ means that $\lambda = 1_V$, equivalently, that $L(c, \lambda) = 1_X$. Thus

$$\phi_1(T_A L(c, \lambda)) = 1_V \Leftrightarrow T_A L(c, \lambda) = T_A;$$

we conclude that *Tr* is the kernel of ϕ_1. It now follows from general principles of group theory that *Tr* is a normal subgroup of *Mo* and $Mo/Tr \simeq O(V)$. Finally, we know from Proposition 10.2 that the group *Tr* is commutative. Done.

The homomorphism $\phi_1: Mo \to O(V)$ is clearly the restriction of the natural homomorphism $\phi: Sa \to Sem$ which was investigated in Proposition 90.1. The following diagram extends Figure 92.1. Every double arrow indicates an epimorphism which is the restriction of every epimorphism "above" it.

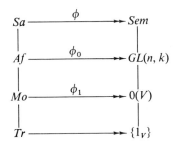

Figure 389.1

Proposition 389.1 explains why, for applications of metric geometry to group theory, it is sufficient to restrict oneself to metric vector spaces and ignore metric affine spaces. The new group which metric affine geometry produces is the group of rigid motions Mo. However, the isomorphism $Mo/Tr \simeq O(V)$ shows that the structure of Mo is so close to the structure of the orthogonal group $O(V)$ that no real algebraic information is obtained by the transition from $O(V)$ to Mo. We elaborate on this point in the following remark which is meant for readers who are familiar with the notion of semidirect products of groups.

Remark on the Isomorphism $Mo/Tr \simeq O(V)$. This isomorphism is quite special in the sense that the group Mo contains subgroups which are isomorphic to $O(V)$. Namely, for each point $c \in X$, the orthogonal group $O(X(c))$ of the tangent space $X(c)$ is such a group. Since $O(X(c))$ consists of those rigid motions which leave the point c fixed,

$$Tr \cap O(X(c)) = \{1_X\}.$$

Consequently, we have the diagram in Figure 390.1. Since Tr is a normal

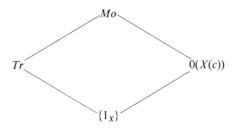

Figure 390.1

subgroup of Mo and each $u \in Mo$ is the product $u = T_A L(c, \lambda)$ of an element $T_A \in Tr$ and $L(c, \lambda) \in O(X(c))$, the group Mo is the semidirect product of the groups Tr and $O(X(c))$ (see Hall, [13], pp. 88–90). Consequently, the structure of Mo is completely determined by the structure of the groups Tr and $O(X(c))$, together with the "action" of $O(X(c))$ on Tr. If $\lambda \in O(V)$ and $A \in V$, it follows from Exercises 388.1 and 388.2 that

$$L(c, \lambda)T_A L(c, \lambda)^{-1} = T\lambda_A L(c, \lambda)L(c, \lambda^{-1}) = T\lambda_A.$$

Therefore, $L(c, \lambda)$ acts on T_A by sending it to $T\lambda_A$. Replacing $O(X(c))$ by its isomorphic copy $O(V)$, we conclude that:

The group Mo is the semidirect product of the groups Tr and $O(V)$, relative to the action $T_A \to T\lambda_A$ of $O(V)$ on Tr.

Exercises

3. Let X be the exceptional plane.
   a. Prove that X has 36 rigid motions. (Hint: The isomorphism $Mo/Tr \simeq O(V)$ enables one to compute the order of the group Mo since the orders of Tr and $O(V)$ are known.)
   b. Rotations, reflections, symmetries, etc., of X were defined on page 386. Prove that X has ten rotations, namely, the identity mapping 1_X and nine $180°$ rotations, one with each of the nine points of X as center.
   c. Prove that X has twelve lines which are divided into four families of three parallel lines each. Also prove that the six lines of two of these families are nonsingular lines while the six lines of the remaining two families are null lines.
   d. Prove that X has six symmetries and no other reflections.
   e. Prove that X has nine translations.
   f. Prove that the total number of rotations, reflections, and translations of X is 24. The other twelve rigid motions are products $T_A L(c, \lambda)$ of a translation T_A with a rotation or reflection $L(c, \lambda)$.
4. We extend the results of the previous exercise by assuming X is an Artinian plane over k where k is an arbitrary finite field with q elements (of course, char $k \neq 2$). Prove that
   a. $O(V)$ consists of $2(q - 1)$ isometries.
   b. X has $2(q - 1)q^2$ rigid motions.
   c. X has $(q - 2)q^2 + 1$ rotations. (Warning: For each of the q^2 points of X, 1_X counts as a rotation with that point as center.)
   d. X has $q(q + 1)$ lines which are divided into $q + 1$ families of q parallel lines each. Prove that $2q$ lines of two of these families are null lines while $q^2 - q$ lines of the remaining families are all nonsingular.
   e. X has $q^2 - q$ symmetries and no other reflections.
   f. X has q^2 translations.
   g. The total number of rotations, reflections, and translations is $q^3 - q$. The other $q^3 - 2q^2 + q$ rigid motions are products $T_A L(c, \lambda)$ of a translation T_A with a rotation or reflection $L(c, \lambda)$.
5. Assume k has q elements and X is the anisotropic plane. We say "the" because, except for isometries, there is only one anisotropic, two-dimensional metric vector space over a finite field; see Artin, [2], p. 143. Therefore, up to isometry, there is only one anisotropic affine plane (Exercise 387.12). Prove that
   a. $O(V)$ has $2(q + 1)$ elements.
   b. X has $2(q + 1)q^2$ rigid motions.

c. X has $q^3 + 1$ rotations.

d. All $q^2 + q$ lines of X are nonsingular.

e. X has $q^2 + q$ symmetries and no other reflections.

f. X has q^2 translations.

g. The total number of rotations, reflections, and translations is $q^3 + 2q^2 + q$. The other $q^3 - q$ rigid motions are products $T_A L(c, \lambda)$ of translations T_A with rotations or reflections $L(c, \lambda)$.

*6. Assume that X is nonsingular. Let S be a hyperplane of X and u a rigid motion of X which leaves S pointwise fixed.

a. If S is nonsingular, prove that either $u = 1_X$ or u is the symmetry with respect to S. (Hint: Select a point $c \in S$ and consider u as an isometry of $X(c)$.)

b. If S is singular, prove that $u = 1_X$.

c. Assume that S is nonsingular and u is the symmetry of X with respect to S. If $x \in X$, prove that $\overrightarrow{x, u(x)} \in H^*$ where H is the direction space of S. If $x \notin S$, prove that S is the perpendicular bisector of the line segment $\{x, u(x)\}$. (See Exercise 377.12d for the definition of perpendicular bisectors.)

*7. Assume that X is a nonsingular plane. Let u be a rigid motion of X which leaves the point c fixed and has no other fixed points. Prove that u is a rotation of X with c as center. (Hint: Consider u as an isometry of $X(c)$.)

In Exercises 8–13, we study rigid motions in terms of coordinate systems. We begin by deriving the equations of an arbitrary affine transformation. This was done on page 95 over an arbitrary division ring as scalar domain. Now that our scalar domain is a field, these equations can be derived much more easily as is shown in the exercise below.

8. Let $A_1, \ldots, A_n$ be a coordinate system of V and c be a point of X. Then $c, A_1, \ldots, A_n$ is a coordinate system of X. The notation $x = (x_1, \ldots, x_n)$ means that the point x has coordinates $(x_1, \ldots, x_n)$ in the coordinate system $c, A_1, \ldots, A_n$, equivalently, the vector $\overrightarrow{c, x}$ has coordinates $(x_1, \ldots, x_n)$ in the coordinate system $A_1, \ldots, A_n$ of V.

Assume that λ is an arbitrary linear automorphism of V whose matrix relative to $A_1, \ldots, A_n$ is (a_{ij}), $i, j = 1, \ldots, n$. We recall that this means

$$(\lambda A_1, \ldots, \lambda A_n) = (A_1, \ldots, A_n)(a_{ij})$$

and the matrix (a_{ij}) is nonsingular.

a. If the vector A has coordinates $(a_1, \ldots, a_n)$ and the vector λA has coordinates $(b_1, \ldots, b_n)$, prove that

$$\begin{pmatrix} b_1 \\ \vdots \\ b_n \end{pmatrix} = (a_{ij}) \begin{pmatrix} a_1 \\ \vdots \\ a_n \end{pmatrix}.$$

b. If $x = (x_1, \ldots, x_n)$ and $L(c, \lambda)(x) = (y_1, \ldots, y_n)$, prove that

$$\begin{pmatrix} y_1 \\ \vdots \\ y_n \end{pmatrix} = (a_{ij}) \begin{pmatrix} x_1 \\ \vdots \\ x_n \end{pmatrix}.$$

c. Assume, furthermore, that A is a vector of V with coordinates $(a_1, \ldots, a_n)$. If $x = (x_1, \ldots, x_n)$ and $T_A L(c, \lambda)(x) = (y_1, \ldots, y_n)$, prove that

$$\begin{pmatrix} y_1 \\ \vdots \\ y_n \end{pmatrix} = (a_{ij}) \begin{pmatrix} x_1 \\ \vdots \\ x_n \end{pmatrix} + \begin{pmatrix} a_1 \\ \vdots \\ a_n \end{pmatrix}. \tag{I}$$

d. Equation I, which may be expressed as

$$y_i = \sum_{j=1}^{n} a_{ij} x_j + a_i \qquad \text{for} \qquad i = 1, \ldots, n$$

shows that, relative to a coordinate system of X, every affine transformation of X is given by a system of n linear equations in n unknowns with a nonsingular matrix (a_{ij}). Prove that, conversely, every system of n linear equations in n unknowns with nonsingular matrix represents an affine transformation of X. Observe that this result has nothing to do with metrics.

*9. Let $c, A_1, \ldots, A_n$ and $T_A L(c, \lambda)$ be as given in the previous exercise where the affine transformation $u = T_A L(c, \lambda)$ is given by Equation I. The matrix $G = (g_{ij})$ where $g_{ij} = A_i A_j$ for $i, j = 1, \ldots, n$ represents the metric of V relative to the coordinate system $A_1, \ldots, A_n$. Prove that u is a rigid motion if and only if

$$G = (a_{ij})^{\mathrm{T}} G (a_{ij})$$

where $(a_{ij})^{\mathrm{T}}$ denotes the transpose of (a_{ij}). As always, the matrix criterion

$$G = (a_{ij})^{\mathrm{T}} G (a_{ij})$$

is not very practical. In a concrete situation, usually the easiest way to determine that u is a rigid motion is to prove by direct geometric means that λ is an isometry of V. The following exercises show this.

*10. Let $n = 2$ and assume that A_1, A_2 is a rectangular coordinate system of V. Consider the coordinate system c, A_1, A_2 of X where $c \in X$. Such a coordinate system is called a rectangular coordinate system of X with c as origin.

a. Prove that the equations

$$y_1 = x_2 - 4, \qquad y_2 = x_1 + 4$$

determine an affine transformation u of X. (Hint: Put these equations in the form I and show that (a_{ij}) is nonsingular.)

b. Find the linear automorphism τ of V and the vector $A \in V$ such that $u = T_A L(c, \tau)$. Precisely, say how τ acts on the coordinate vectors A_1 and A_2 and give the coordinates of A.

c. Prove that u is a rigid motion of X if and only if $A_1{}^2 = A_2{}^2$. (Who would ever use the matrix criterion of Exercise 9 above for this?)

Assume for the remainder of this exercise that $A_1{}^2 = A_2{}^2 \neq 0$ and, therefore, X is nonsingular.

d. Prove that τ is the symmetry of V with respect to the nonsingular line $\langle A_1 + A_2 \rangle$. Also prove that $A \perp (A_1 + A_2)$.

e. Prove that $L(c, \tau)$ is the symmetry of X with respect to the non-singular line $m = S(c, \langle A_1 + A_2 \rangle)$.

f. Prove that u leaves the line l with equation

$$x_2 - x_1 - 4 = 0$$

pointwise fixed and has no other fixed points. Also prove that l and m are nonsingular parallel lines and both are perpendicular to the line through the points c and $Ac = (-4, 4)$ (see Figure 394.1).

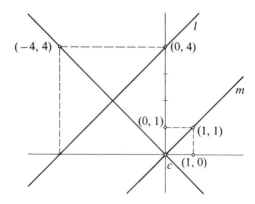

Figure 394.1

g. Prove that u is the symmetry of X with respect to the nonsingular line l. (Hint: This is a special case of Exercise 392.6a.) Observe that the line m plays no special role for the symmetry u; it is the fixed line l that matters.

h. It follows from Part g above and Exercise 392.6c that, if $x \in X$ and $x \notin l$, the line l is the perpendicular bisector of the line segment $\{x, u(x)\}$. Prove this directly using the equations of u given in Part a. (Hint: If $x = (x_1, x_2)$, compute the midpoint of the line segment $\{x, u(x)\}$.)

11. As in Exercise 10 above, assume that $n = 2$ and c, A_1, A_2 is a rectangular coordinate system of the plane X.

a. Prove that the equations

$$y_1 = x_2 - 2, \qquad y_2 = -x_1 + 3$$

determine an affine transformation u of X.

b. Find a linear automorphism ρ of V and the vector A in V such that $u = T_A L(c, \rho)$. Say how ρ acts on the coordinate vectors A_1 and A_2 and find the coordinates of A.

c. Prove that u is a rigid motion of X if and only if $A_1{}^2 = A_2{}^2$. Assume for the remainder of this exercise that $A_1{}^2 = A_2{}^2 \neq 0$ and, therefore, X is nonsingular.

d. Prove that ρ is a rotation of V different from 1_V. If $k = \mathbf{R}$ and the coordinate system is positively oriented, ρ would be a clockwise rotation over how many degrees?

e. Prove that $L(c, \rho)$ is a rotation of X with center c and different from 1_X.

f. Show that $b = (\frac{1}{2}, \frac{5}{2})$ is the only fixed point of u.

g. Prove that u is a rotation of X with b as center. (Hint: This is a special case of Exercise 392.7.) If $k = \mathbf{R}$ and the coordinate system A_1, A_2 is positively oriented, u would be the clockwise rotation of X with b as center over how many degrees?

h. If $x, y \in X$, prove that $\overrightarrow{x, y} \perp \overrightarrow{u(x), u(y)}$.

*12. In this exercise, we invert the procedure of Exercise 10 above. Let $n = 2$ and assume that V has a rectangular coordinate system A_1, A_2 where $A_1{}^2 = A_2{}^2 \neq 0$. Choose a point c in X and consider the rectangular coordinate system c, A_1, A_2 of X.

a. Prove that the line l with the equation

$$x_2 - x_1 - 3 = 0$$

is a nonsingular line of X. (Hint: A line is nonsingular if and only if its direction space is nonsingular.)

We denote the symmetry of X with respect to l by u and put

$$u = T_A L(c, \tau).$$

b. Prove that τ is the symmetry of V with respect to the nonsingular line $\langle A_1 + A_2 \rangle$.

c. Prove that $L(c, \tau)$ is the symmetry of X with respect to the nonsingular line m whose equation is $x_2 - x_1 = 0$. Also show that l is parallel to m.

d. Find the coordinates of the vector A and prove that $A \perp (A_1 + A_2)$. Also prove that the lines l and m are perpendicular to the line through the points c and Ac.

e. Find the equations of u.

*13. In Exercises 10 and 12 above, τ is a symmetry of V with respect to a nonsingular hyperplane H, $T_A L(c, \tau)$ is a symmetry of X and A belongs to H^*. In the present exercise, we choose a vector $A \notin H^*$ and show that $T_A L(c, \tau)$ is not a symmetry of X. We shall see in Exercise 15 below that $T_A L(c, \tau)$ is a symmetry of X if and only if $A \in H^*$.

Let $n = 2$ and assume that V has a rectangular coordinate system A_1, A_2 where $A_1{}^2 = A_2{}^2 \neq 0$. We denote the symmetry of V with respect to the nonsingular line $H = \langle A_1 + A_2 \rangle$ by τ and choose $A = 4A_1 - 3A_2$.

a. Prove that $A \notin H^*$.

b. Let $c \in X$ and consider the rectangular coordinate system c, A_1, A_2 of X. Prove that the equations of the rigid motion

$$u = T_A L(c, \tau)$$

are

$$y_1 = x_2 + 4, \qquad y_2 = x_1 - 3.$$

c. Prove that u leaves no point of X fixed.

d. Since a symmetry of X leaves a line pointwise fixed, u cannot be a symmetry. A rotation of X has precisely one fixed point or all points are fixed, therefore, u is not a rotation. Of course, translations different from 1_X have no fixed points. Prove, however, that u is not a translation. (Hint: If $T_A L(c, \tau) = T_B$, then $L(c, \tau) = T_{B-A}$; show this is impossible by comparing the fixed point sets of $L(c, \tau)$ and T_{B-A}.) In Figure 397.1, m depicts the line $S(c, H)$ and hence $L(c, \tau)$ is the symmetry of X with respect to m. Furthermore, l is an arbitrary line, $l' = L(c, \tau)(l)$, and $l'' = T_A(l') = u(l)$. Observe that, in general, $l \nparallel u(l)$ and hence u is not a translation. For which lines l is $l \parallel u(l)$?

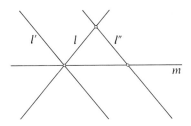

Figure 397.1

e. Prove that the equation of the line $m = S(c, H)$ is

$$x_2 - x_1 = 0.$$

f. From Figure 397.1, we expect $u(l)$ to be parallel to l if and only if $l \parallel m$. Prove that this is correct.

g. A rigid motion of a nonsingular plane which has no fixed points and is not a translation is called a **glide reflection**. Although glide reflections leave no points fixed, they do leave a unique line invariant. Find the equation of the line m' which is left invariant by u. (Hint: It follows from Part f that $m' \parallel m$ and hence the equation of m' is

$$x_2 - x_1 + r = 0$$

for some $r \in k$. Determine r from the equations of u given in Part b and then check that m' is invariant.)

h. Draw a figure analogous to Figure 394.1, displaying the lines m, m', and the vector A.

i. The line m plays no particular role for the glide reflection u. It is the invariant line m' which is important. Choose a point b on m' and put $u = T_B L(b, \tau)$. Prove that B is a nonzero vector which belongs to H. In other words, as long as the point b lies on the invariant line of the glide reflection, the vector B "translates the plane in a direction parallel to m'"; precisely, the line $S(x_1 \langle B \rangle)$ is parallel to m' for all $x \in X$.

14. Prove that a rigid motion of a nonsingular plane is either a symmetry, a rotation, a translation, or a glide reflection. (Hint: Consider the different possibilities for the fixed point set.)

*15. Let X be a nonsingular space of arbitrary dimension and $c \in X$. Assume that H is a nonsingular hyperplane and A a vector of V. We denote the symmetry of V with respect to H by τ and the rigid motion $T_A L(c, \tau)$ of X by u.

a. If u has a fixed point, prove that $A \in H^*$. (Hint: If x is a fixed point of u, then

$$T_A L(c, \tau)(x) = x \Leftrightarrow L(c, \tau)(x) = (-A)x.$$

Now apply Exercise 392.6c.)

b. If $A \in H^*$, prove that the fixed point set of u is the nonsingular hyperplane $S(b, H)$ where $b = (\frac{1}{2}A)c$. (Hint: Each point $y \in S(b, H)$ is of the form $y = (\frac{1}{2}A)x$ where $x \in S(c, H)$. Furthermore,

$$A \in H^* \Leftrightarrow \tau A = -A.$$

If $y \in S(b, H)$, then

$$u(y) = T_A L(c, \tau) T_{\frac{1}{2}A}(x).$$

By Exercise 388.1c,

$$T_A L(c, \tau) T_{\frac{1}{2}A} = T_{A + \tau(\frac{1}{2}A)} L(c, \tau).$$

Now show that $u(y) = y$ and hence $S(b, H)$ is contained in the fixed point set of u. The diagram below makes it geometrically evident that, if $A \in H^*$ and $x \in S(c, H)$, the point $y = (\frac{1}{2}A)x$ is left fixed by u. Figure 398.1 represents the situation in the special case that $n = 2$. Observe that

$$y = (\tfrac{1}{2}A)x, \qquad y' = L(c, \tau)(y), \qquad T_A(y') = u(y) = y.$$

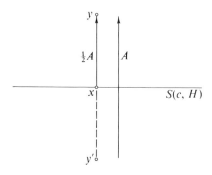

Figure 398.1

To show that the fixed set of u is contained in $S(b, H)$ assume $y \in X$ is such that $u(y) = y$. Then $y = (tA)x$ for some $x \in S(c, H)$. Use the equation

$$T_A L(c, \tau) T_{(tA)}(x) = y = (tA)x$$

to show that $t = \frac{1}{2}$ and, therefore, $y \in S(b, H)$.)

c. Prove that u is a symmetry of X if and only if $A \in H^*$. Also prove that, in this case, $u = L(b, \tau)$ where $b = (\frac{1}{2}A)c$. (Hint: Use Exercise 392.6a.) Figure 399.1 for $n = 2$ makes it geometrically evident that $u = L(b, \tau)$ where

$$y' = L(c, \tau)(y), \qquad y'' = T_A(y') = u(y) = L(b, \tau)(y).$$

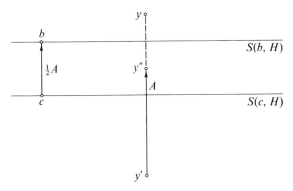

Figure 399.1

16. Let X, c, H, A, τ, and u be as given in the previous exercise. We again put $b = (\frac{1}{2}A)c$.

   The previous exercise tells us that, if $A \in H^*$, u is the symmetry of X with respect to the nonsingular hyperplane $S(b, H)$. If $A \notin H^*$, u is called a **glide reflection** of X.

   a. If $n = 2$, prove that this notion of glide reflection coincides with the one given in Exercise 397.13g.

   We assume for the remainder of this exercise that $A \notin H^*$ and investigate the glide reflection u.

   b. According to Exercise 15a above, u leaves no points of X fixed. Prove that u leaves the nonsingular hyperplane $S(b, H)$ invariant and leaves no other hyperplane invariant. (Hint: $A = B_1 + B_2$ where $B_1 \in H$ and $B_2 \in H^*$. Then each $y \in S(b, H)$ is of the form $y = (\frac{1}{2}B_2)x$ where $x \in S(c, H)$. Now proceed according to the hint given in Part b of Exercise 15 above.) Observe that the condition $A \notin H^*$ was not used in proving the invariance of $S(b, H)$ under u, but only to show that u has no fixed points. The invariance of $S(b, H)$ is evident from Figure 400.1 where $n = 2$ and

   $$y' = L(c, \tau)(y), \qquad y'' = T_A(y') = u(y) \in S(b, H).$$

   c. Let $y \in S(b, H)$ and $B \in V$ such that $u = T_B L(y, \tau)$. Prove that $B \neq 0$ and $B \in H$. Consequently, whenever the point y lies on the

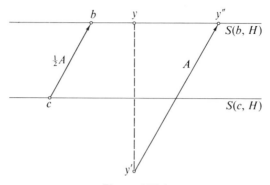

Figure 400.1

invariant hyperplane of the glide reflection, the vector B "translates the space X in the direction parallel to the invariant hyperplane." Precisely, the line $S(x, \langle B \rangle)$ is parallel to $S(b, H)$ for all $x \in X$.

d. If $x \in X$, prove that the midpoint of the line segment $\{x, u(x)\}$ lies on the invariant hyperplane $S(b, H)$. The figure below shows the situation when $n = 2$ and B is the vector such that $u = T_B L(b, \tau)$. In Figure 400.2,

$$x' = L(b, \tau)(x) \qquad \text{and} \qquad T_B(x') = u(x).$$

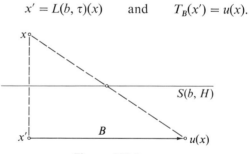

Figure 400.2

69. THE CARTAN–DIEUDONNÉ THEOREM FOR AFFINE SPACES

Each theorem for metric vector spaces has an analogue for metric affine spaces. Usually, all one has to do to find the analogue is to investigate the behavior of the translations of X. In the present section, we prove the Cartan–Dieudonné theorem for affine spaces and urge the reader to do the same for the other major theorems of metric vector spaces.

We assume for the remainder of the chapter that the metric affine space (X, V, k) is nonsingular.

Thus, from now on, V is a nonsingular metric vector space. It is convenient to refer to symmetries of X with respect to parallel nonsingular hyperplanes as **parallel symmetries.** We shall prove in Proposition 401.1 that the product of two parallel symmetries is always a translation T_A where either $A = 0$ or A is not isotropic. The following exercise is used in the proof of that proposition.

Exercise

*1. Let τ be a symmetry of V with respect to a nonsingular hyperplane H and let c, x be points of X. Since $V = H \oplus H^*$,

$$\overrightarrow{c, x} = A + B$$

where $A \in H$ and $B \in H^*$. Prove that

$$\overrightarrow{x, L(c, \tau)(x)} = -2B.$$

(Hint: Use the definition of $L(c, \tau)$ and Exercise 8.10 which says that $\overrightarrow{Cx, Dy} = -C + \overrightarrow{x, y} + D$ where $C, D \in V$ and $x, y \in X$.) Observe that this result agrees with Exercise 392.6c where it was shown that $\overrightarrow{x, L(c, \tau)(x)} \in H^*$. The following figure where $n = 2$ makes this result geometrically evident.

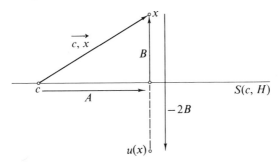

Figure 401.1

Proposition 401.1.

Proposition 401.1.   Let H be a nonsingular hyperplane of V and τ a symmetry of V with respect to H. Assume $L(b, \tau)$ and $L(c, \tau)$ are (parallel) symmetries with respect to the respective hyperplanes $S(b, H)$ and $S(c, H)$ where $b, c \in X$. Finally, put

$$\overrightarrow{b, c} = A + B$$

where $A \in H$ and $B \in H^*$. Then

$$L(c, \tau)L(b, \tau) = T_{2B}.$$

Furthermore, if $S(b, H) = S(c, H)$, then $B = \mathbf{0}$; otherwise, $B^2 \neq 0$.

Proof. Figure 402.1 for $n = 2$ makes the proposition geometrically evident. Here

$$x' = L(b, \tau)(x) \quad \text{and} \quad x'' = L(c, \tau)(x') = L(c, \tau)L(b, \tau)(x) = T_{2B}(x).$$

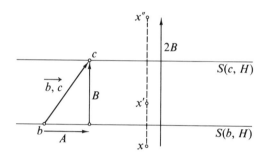

Figure 402.1

Put

$$L(c, \tau) = T_D L(b, \tau)$$

where $D \in V$. Then

$$L(c, \tau)L(b, \tau) = T_D L(b, \tau^2) = T_D 1_X = T_D.$$

Furthermore, the definition of D gives that

$$D = \overrightarrow{b, L(c, \tau)(b)}$$

and it follows from Exercise 1 above that $D = 2B$. Finally, $S(b, H) = S(c, H)$ if and only if $c \in S(c, H)$ which is the same as $\overrightarrow{b, c} \in H$; this is equivalent to $B = \mathbf{0}$. If $B \neq \mathbf{0}$, B is a nonzero vector of the nonsingular line H^* and hence $B^2 \neq 0$. Done.

It is an immediate consequence of the above proposition that not every translation of X is the product of two parallel symmetries. Namely, if T_A is the product of two parallel symmetries, A cannot be a nonzero isotropic vector. We now show that, if A is not a nonzero isotropic vector, the translation T_A is the product of two parallel symmetries. Furthermore, the hyperplane of one of these symmetries may be chosen to pass through any preassigned point.

Proposition 403.1.   Let $A \in V$ where either $A = 0$ or $A^2 \neq 0$ and let $c \in X$. If $A = 0$,

$$T_A = 1_X = L(c, \tau)L(c, \tau)$$

for all symmetries τ of V. If $A \neq 0$, $\langle A \rangle^*$ is a nonsingular hyperplane H of V and we denote the symmetry with respect to H by τ. Then

$$T_A = L(c, \tau)L(b, \tau) = L(a, \tau)L(c, \tau)$$

where $b = (-\frac{1}{2}A)c$ and $a = (\frac{1}{2}A)c$.

Proof. Figure 403.1 pictures the various points and hyperplanes in the case that $n = 2$ and $A^2 \neq 0$.

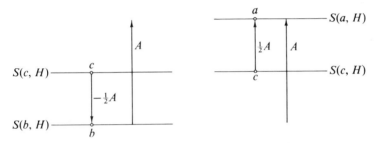

Figure 403.1

The proposition is trivial if $A = 0$; thus we assume that $A^2 \neq 0$ and H, τ, b, a are as stated above. Since

$$\vec{b, c} = \vec{c, a} = \tfrac{1}{2}A \in H^*,$$

it follows from Proposition 401.1 that

$$L(c, \tau)L(b, \tau) = L(a, \tau)L(c, \tau) = T_A.$$

Done.

When X is anisotropic, every translation is the product of two parallel symmetries. In Exercise 2 below, we learn that the product of two symmetries which are not parallel is never a translation.

Exercises

*2. Let S and S' be nonsingular hyperplanes of X which are not parallel. Let u be the symmetry of X with respect to S and u' the symmetry with respect to S'. Prove that uu' is not a translation. (Hint: $S \cap S' \neq \emptyset$ and all points in this intersection are left fixed by uu'. Hence, if uu' is a translation, it must be 1_X. Show this is not possible since $S \not\parallel S'$.)

3. We assume that $n = 1$ and study the rigid motions of the nonsingular line X.

a. Prove that a rigid motion of X is either a translation or a symmetry. (Hint: The only isometries of V are $\pm 1_V$; hence a rigid motion of V either has the form $T_A L(c, 1_V)$ or $T_A L(c, -1_V)$ where $A \in V$ and $c \in X$. Now apply Exercise 399.15c to show that $T_A L(c, -1_V)$ is a symmetry.)

b. According to Part a, $T_A L(c, -1_V)$ is a symmetry. Find a point $b \in X$ and a vector B such that

$$T_A L(c, -1_V) = L(b, -1_V) \qquad \text{and} \qquad b = Bc.$$

c. Choose a coordinate system c, A_1 of X. Prove that a rigid motion of X is given by one of the equations

$$x' = x + t \qquad \text{or} \qquad x' = -x + t$$

where $t \in k$.

d. Which of the equations in Part c represents a translation and which one a symmetry?

e. The hyperplanes of X are its points and they are all nonsingular and parallel. Consequently, all symmetries of X are parallel; furthermore, X is anisotropic. Relative to the coordinate system c, A_1 of Part a, assume that the translation

$$T: x' \to x + t$$

is the product of the two symmetries

$$u: x' \to -x + r \qquad \text{and} \qquad w: x' \to -x + s$$

where $r, s, t \in k$. How are r, s, t related if $T = uw$? if $T = wu$?

Now that we have seen how translations in nonisotropic directions of X can be factored into products of two symmetries, we can investigate the Cartan–Dieudonné theorem for n-dimensional nonsingular affine space X. Is every rigid motion of X the product of symmetries? Since every isometry of V is the product of at most n symmetries, it is reasonable to conjecture that the answer is affirmative. However, it may require more than n symmetries to factor a given rigid motion because of the presence of translations. By far, the simplest theorem is obtained when X is anisotropic because every translation is the product of two parallel symmetries.

Theorem 405.1 (the Cartan–Dieudonné theorem for anisotropic affine spaces). Let X be anisotropic. Then a rigid motion u of X is the product of $r \le n + 1$ symmetries. The hyperplanes of $r - 1$ of these symmetries can be chosen to pass through a preassigned point c.

Proof. There exists $A \in V$ and $\sigma \in O(V)$ such that $u = T_A L(c, \sigma)$. The rigid motion $L(c, \sigma)$ is an isometry of the metric vector space $X(c)$ and hence, by the Cartan–Dieudonné theorem for metric vector spaces,

$$L(c, \sigma) = L(c, \tau_1) \cdots L(c, \tau_h)$$

where $\tau_1, \ldots, \tau_h$ are symmetries of V and $h \le n$. Since V is anisotropic, $T_A = L(b, \tau)L(c, \tau)$ for some point $b \in X$ and symmetry τ of V (Proposition 403.1). Consequently,

$$u = L(b, \tau)L(c, \tau)L(c, \tau_1) \cdots L(c, \tau_h).$$

Since h could be n, we are not quite done. The rigid motion

$$L(c, \tau)L(c, \tau_1) \cdots L(c, \tau_h)$$

is an isometry of $X(c)$ and hence can be written as

$$L(c, \tau_1') \cdots L(c, \tau_q')$$

where $\tau_1', \ldots, \tau_q'$ are symmetries of V and $q \le n$. Therefore,

$$u = L(b, \tau)L(c, \tau_1') \cdots L(c, \tau_q').$$

Done.

If the rigid motion u has a fixed point, it is the product of at most n symmetries and the hyperplanes of these symmetries can be chosen so as to pass through any of the fixed points of u. This is true even when X is not anisotropic.

Theorem 405.2. Let u be a rigid motion of X which has the point c as fixed point. Then u is the product of $r \le n$ symmetries whose hyperplanes can be chosen to pass through c.

Proof. Since u is an isometry of the metric vector space $X(c)$, this is an immediate consequence of the Cartan–Dieudonné theorem for metric vector spaces. Done.

If X is not anisotropic and the rigid motion u has no fixed points, u is still the product of symmetries, but it may take $n + 2$ symmetries to factor u (see Theorem 407.1). The following exercises are used in the proof of this theorem.

Exercises

4. Let H_1 and H_2 be nonsingular hyperplanes of V with the property that the nonsingular lines $H_1{}^*$ and $H_2{}^*$ are orthogonal. Denote the symmetry of V with respect to H_i by τ_i for $i = 1, 2$. If $b, d \in X$, prove that

$$L(b, \tau_1)L(d, \tau_2) = L(d, \tau_2)L(b, \tau_1),$$

that is, these symmetries commute. (Hint: First show that $S(b, H_1)$ and $S(d, H_2)$ have a point c in common; therefore, $L(b, \tau_1) = L(c, \tau_1)$ and $L(d, \tau_2) = L(c, \tau_2)$. Finally, apply Exercise 259.7.) Figure 406.1 depicts the situation for $n = 2$ where

$$y = L(b, \tau_1)L(d, \tau_2)(x) = L(d, \tau_2)L(b, \tau_1)(x).$$

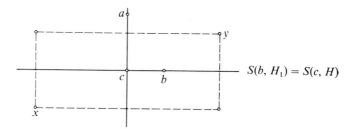

Figure 406.1

5. Prove that the product of an odd number of symmetries of X is never a translation. (Hint: A product of r symmetries of X can be written as

$$T_A L(c, \tau_1) \cdots L(c, \tau_r)$$

where $c \in X$, $A \in V$, and $\tau_1, \ldots, \tau_r$ are symmetries of V. If this product is a translation T_B, then

$$L(c, \tau_1) \cdots L(c, \tau_r) = T_{B-A}$$

which, if r is odd, gives rise to a contradiction in $X(c)$.)
*6. Let A be a nonzero isotropic vector of V and let us study the factorization of the translation T_A into symmetries.
    a. Prove that V contains two orthogonal, nonisotropic vectors B and C such that $A = B + C$. We put $H_1 = \langle B \rangle^*$, $H_2 = \langle C \rangle^*$, and denote the symmetry of V with respect to the nonsingular hyperplane H_i by τ_i for $i = 1, 2$.
    b. Choose a point $c \in X$. Prove that there exist points b and d in X such that

$$T_A = L(b, \tau_1)L(c, \tau_1)L(d, \tau_2)L(c, \tau_2).$$

(Hint: $T_A = T_B T_C$.) Observe that

$$S(b, H_1) \parallel S(c, H_1) \qquad \text{and} \qquad S(d, H_2) \parallel S(c, H_2).$$

The hyperplanes $S(c, H_1)$ and $S(c, H_2)$ pass through the preassigned point c and $H_1^* \perp H_2^*$. The situation is pictured in Figure 407.1 when $n = 2$; then $y = T_A(x)$.

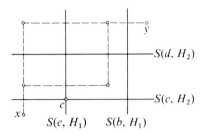

Figure 407.1

c. Prove that

$$T_A = L(b, \tau_1)L(d, \tau_2)L(c, \tau_1)L(c, \tau_2).$$

The commutativity of $L(c, \tau_1)$ and $L(d, \tau_2)$ is indicated in Figure 407.1 above by means of the dotted rectangle.

d. Prove that T_A is the product of four and not less then four symmetries.

Now that we know how translations in isotropic directions of X can be factored into the product of symmetries, we can prove the most general form of the Cartan–Dieudonné theorem for affine spaces. We call a rigid motion u of X a **null motion** if the square distance $\delta(x, u(x)) = 0$ for all points $x \in X$. Equivalently, a rigid motion u of X is a null motion if and only if the vector $\overrightarrow{x, u(x)}$ is isotropic for all $x \in X$. Examples of null motions are 1_X and translations T_A where A is an isotropic vector. The three cases of Theorem 407.1 are exhaustive. Case 1 repeats Theorem 405.2. Since 1_X is the only null motion if X is anisotropic, Case 2 contains Theorem 405.1.

Theorem 407.1   (the Cartan–Dieudonné theorem for affine spaces).   Let u be a rigid motion of X.

Case 1. u has a fixed point. Then u is the product of $r \leq n$ symmetries. If c is a fixed point of u, the hyperplanes of these r symmetries can be chosen to pass through c.

Case 2. u is not a null motion. Then u is the product of $r \le n + 1$ symmetries. If c is a point of X such that the vector $\overrightarrow{c, u(c)}$ is not isotropic, the hyperplanes of $r - 1$ of these symmetries can be chosen to pass through c.

Case 3. u is a null motion without fixed points. Then u is the product of $r \le n + 2$ symmetries. The hyperplanes of $r - 2$ of these symmetries can be chosen to pass through any preassigned point.

Proof. Case 1 is identical to Theorem 405.2. For Case 2, choose a point c in X such that the vector $\overrightarrow{c, u(c)}$ is not isotropic. Put $u = T_A L(c, \sigma)$ where $A = \overrightarrow{c, u(c)}$ and $\sigma \in O(V)$. By Proposition 403.1,

$$T_A = L(b, \tau)L(c, \tau)$$

for some point $b \in X$ and symmetry τ of V. The remainder of the proof of Case 2 is identical to the proof of Theorem 405.1.

We now prove Case 3. Select a point $c \in X$ and put $u = T_A L(c, \sigma)$ where $A = \overrightarrow{c, u(c)}$ and $\sigma \in O(V)$. Then $A \ne \mathbf{0}$ since u has no fixed points, and $A^2 = 0$ since u is a null motion. We factor T_A according to Exercise 406.6 and factor the isometry $L(c, \sigma)$ of $X(c)$ by the Cartan–Dieudonné theorem for metric vector spaces. This gives

$$u = L(b, \tau_1)L(d, \tau_2)L(c, \tau_1)L(c, \tau_2)L(c, \tau_1') \cdots L(c, \tau_h')$$

where $b, d \in X$ and $\tau_1, \tau_2, \tau_1', \ldots, \tau_h'$ are symmetries of V. The rigid motion

$$L(c, \tau_1)L(c, \tau_2)L(c, \tau_1') \cdots L(c, \tau_h')$$

is an isometry of $X(c)$ and hence can be written as

$$L(c, \tau_1^*) \cdots L(c, \tau_q^*)$$

where $\tau_1^*, \ldots, \tau_q^*$ are symmetries of V and $q \le n$. Consequently,

$$u = L(b, \tau_1)L(d, \tau_2)L(c, \tau_1^*) \cdots L(c, \tau_q^*).$$

Done.

In general, one likes to factor a rigid motion in as few symmetries as possible. Sometimes, the maximum number of $n + 2$ is required (see Exercise 8 below). Since this only happens when the rigid motion is a null motion, it is convenient to know a priori that certain rigid motions are not null motions (see Exercise 12 below).

Exercises

7. a. Prove that a translation T_A is a null motion if and only if A is isotropic. This includes the case when $A = \mathbf{0}$ and hence $T_A = 1_X$.
  b. If X is anisotropic, prove that 1_X is the only null motion of X.

*8. Assume that X is an Artinian plane. Let M, N be an Artinian coor-
dinate system of V and $c \in X$. The coordinate system c, M, N is called
an **Artinian coordinate system** of X. Determine all pairs of scalars
(a_1, a_2) such that the translation T_A, where $A = a_1 M + a_2 N$, is the
product of four and not less than four symmetries. (Warning: The
pair $(0, 0)$ does not have this property.)

9. Prove that symmetries are never null motions. This fact also follows
from the criteria of Exercise 11 below, but it is best to prove this
simple fact directly.

10. In order to find criteria for determining when a given rigid motion u
has a fixed point or is a null motion, we determine the structure of the
set

$$\{\overrightarrow{x, u(x)} \mid x \in X\}.$$

Let $u = T_A L(c, \sigma)$ where $A \in V$, $c \in X$, and $\sigma \in O(V)$.
  a. For each $x \in X$, prove that

$$\overrightarrow{x, L(c, \sigma)(x)} = (\sigma - 1_V)(\overrightarrow{c, x}).$$

(Hint: Use the definition of $L(c, \sigma)$ and Exercise 7.4 which states
that $\overrightarrow{x, By} = \overrightarrow{x, y} + B.$)
  b. Prove that

$$\{\overrightarrow{x, L(c, \sigma)x} \mid x \in X\} = \operatorname{im}(\sigma - 1_V).$$

  c. Prove that

$$\{\overrightarrow{x, u(x)} \mid x \in X\} = A + \operatorname{im}(\sigma - 1_V).$$

We recall that $\operatorname{im}(\sigma - 1_V)$ is a linear subspace and $A + \operatorname{im}(\sigma - 1_V)$
is an affine subspace of V which is not a linear subspace unless
$A \in \operatorname{im}(\sigma - 1_V)$.
  d. Prove that u has a fixed point if and only if $A \in \operatorname{im}(\sigma - 1_V)$.

*11. Let $u = T_A L(c, \sigma)$ be as in the previous exercise. The most convenient
criterion for determining when u is a null motion is given in Part b of
this exercise. The linear subspace

$$\{rA + B \mid r \in k, \quad B \in \operatorname{im}(\sigma - 1_V)\}$$

of V, spanned by the vector A and $\text{im}(\sigma - 1_V)$, is denoted by $\langle A, \text{im}(\sigma - 1_V) \rangle$.

a. Prove that u is a null motion of X if and only if $\langle A, \text{im}(\sigma - 1_V) \rangle$ is a null space. (Hint: u is a null motion if and only if every vector in the affine space $A + \text{im}(\sigma - 1_V)$ is isotropic. First, conclude that $A^2 = 0$. Then, if $B \in \text{im}(\sigma - 1_V)$, conclude from $(A + B)^2 = 0$ and $(A - B)^2 = 0$ that $B^2 = 0$ and $AB = 0$.)

b. Prove that u is a null motion of X if and only if the following three conditions are satisfied.

   i. A is isotropic.
   ii. $\text{im}(\sigma - 1_V)$ is a null space.
   iii. $A \in \text{im}(\sigma - 1_V)^*$.

c. Prove that u is a null motion without fixed points if and only if, in addition to Conditions i–iii, $A \in \text{im}(\sigma - 1_V)$.

12. Use the criteria of Exercise 11 above and prove that symmetries and glide reflections are never null motions. (Hint: Symmetries and glide reflections can be written as $T_A L(c, \tau)$ where $A \in V$, $c \in X$, and τ is a symmetry of V with respect to a nonsingular hyperplane H. Show that $\text{im}(\tau - 1_V) = H^*$.)

13. Let X be an Artinian space of dimension $4r$. Assume that ρ is an isometry of V with the property that $\rho B - B$ is a nonzero isotropic vector for every nonisotropic vector B of V. See Lemma 249.1 for isometries of this kind; we recall in particular, that ρ is a rotation and the fixed space of ρ is a null space of dimension $2r$. Choose a point c in X.

a. Prove that $L(c, \rho)$ is a rotation of X which is a null motion. This null motion has fixed points, for example, the point c.

b. If $A \in V$, prove that $T_A L(c, \rho)$ is a null motion if and only if $\rho A = A$.

c. If $u = T_A L(c, \rho)$ is a null motion, prove that u has a fixed point and hence is of the form $L(b, \rho)$ for some $b \in X$.

14. Let X be an anisotropic plane. Every rigid motion u of X is the product of 0, 1, 2, or 3 symmetries.

a. Prove that u is a translation if and only if u is the product of two parallel symmetries.

b. Prove that u is a rotation different from 1_X if and only if u is the product of two symmetries which are now parallel.

c. Prove that u is a glide reflection if and only if u is the product of three symmetries with respect to lines which have no point in common.

Two subsets Y and Y' of X are called **congruent** if there exists a rigid motion u such that $u(Y) = Y'$. This includes the standard notion of congruent triangles in the Euclidean plane. Needless to say, all criteria for congruent triangles which are usually discussed in high school can easily be obtained from the present material. These criteria are not very interesting since they are too obvious from a geometric point of view. The reason so much attention was paid to them in old-fashioned high school curricula is that these criteria fall conveniently within the axioms of Euclid and these curricula were mostly concerned with showing how Euclid's axioms work. These criteria are discussed in the exercises below only to stress again that geometry, based on linear algebra with $k = \mathbf{R}$ and a positive definite inner product, is the same as geometry based on the axioms of Euclid and Hilbert. Since the criteria in question work for all nonsingular planes, the scalar domain k is never mentioned.

Exercises

In these exercises, $n = 2$. As always, a triangle consists of three non-collinear points called the vertices of the triangle.

15. Prove the law of cosines. Precisely, if a, b, c are the vertices of a triangle, prove that

$$\delta(a, b) = \delta(c, a) + \delta(c, b) - 2(\overrightarrow{c, a})(\overrightarrow{c, b}).$$

The law holds just as well when the three points a, b, c are collinear since all that is used is

$$\overrightarrow{a, b} = -\overrightarrow{c, a} + \overrightarrow{c, b}.$$

If X is the Euclidean plane, $(\overrightarrow{c, a})(\overrightarrow{c, b})$ may be replaced by

$$(\text{length } \overrightarrow{c, a})(\text{length } \overrightarrow{c, b}) \cos \theta$$

where θ is the angle between the vectors $\overrightarrow{c, a}$ and $\overrightarrow{c, b}$ (Exercise 294.12). Of course, the length of the vector $\overrightarrow{c, a}$ is, by definition, the length of the side ca and, similarly, for $\overrightarrow{c, b}$. Finally, θ is then the angle between the sides ca and cb.

16. Let abc and $a'b'c'$ be triangles of X. Assume that

$$\delta(c, a) = \delta(c', a'), \qquad \delta(c, b) = \delta(c', b'), \qquad \delta(a, b) = \delta(a', b').$$

Prove that the two triangles are congruent. This is the case " *SSS*."
(Hint: Conclude from the law of cosines that

$$(\overrightarrow{c, a})(\overrightarrow{c, b}) = (\overrightarrow{c', a'})(\overrightarrow{c', b'}).$$

It then follows that the two coordinate systems

$$\overrightarrow{c, a}, \quad \overrightarrow{c, b} \quad \text{and} \quad \overrightarrow{c', a'}, \quad \overrightarrow{c', b'}$$

of V are related by the isometry σ of V defined by

$$\sigma(\overrightarrow{c, a}) = \overrightarrow{c', a'} \quad \text{and} \quad \sigma(\overrightarrow{c, b}) = \overrightarrow{c', b'}.$$

Put $A = \overrightarrow{c, c'}$ and show that $T_A L(c, \sigma)$ is the desired rigid motion.)

17. Let abc and $a'b'c'$ be triangles of X. Assume that

$$\delta(c, a) = \delta(c', a'), \quad \delta(c, b) = \delta(c', b'), \quad (\overrightarrow{c, a})(\overrightarrow{c, b}) = (\overrightarrow{c', a'})(\overrightarrow{c', b'}).$$

Prove that the two triangles are congruent. This is the case " *SAS*."

18. Readers who are interested in high school geometry should also investigate the remaining cases of congruent triangles. Concentration should be on the Euclidean plane when

$$(\overrightarrow{c, a})(\overrightarrow{c, b}) = (\text{length } \overrightarrow{c, a})(\text{length } \overrightarrow{c, b})\cos \theta.$$

70. SIMILARITIES OF AFFINE SPACES

In high school, two triangles in the Euclidean plane are called similar if they have congruent angles. However, it is better to handle the concept of similarity the same way as that of congruence. Namely, one should define first what is meant by a similarity transformation of the plane and then define two triangles to be similar if they correspond under such a transformation. It then becomes a theorem that two triangles are similar if and only if they have congruent angles. As to be expected, the concept of similarity transformation is very important while that of similar triangles is only of secondary importance.

We begin by defining the notion of similarity transformation for affine spaces of arbitrary dimension n. In view of investigations of similarities of metric vector spaces in Section 65, the reader should find the present section to be of a rather trivial nature. In fact, we urge the reader to write out the whole story on similarities for himself before reading this section. Go by the advice which Carl Friedrich Gauss (1777–1855) gave to his students in Göttingen, Germany, in the first lecture of his course in astronomy in 1808:

> ... certainly it is a hundred times more valuable if one solves a difficulty through independent exertion, than if one always needs correction from someone else, just as successful results of one's own meditation are always a hundred times more valuable than all borrowed wisdom. (G. Waldo Dunnington, *Carl Friedrich Gauss*, p. 53, [10]).

We usually say "similarity" instead of "similarity transformation" and recall that (X, V, k) stands for a nonsingular metric affine space of dimension n.

Definition 413.1. A **similarity** s of X is an affine transformation of X for which there exists a nonzero scalar q such that

$$\delta(s(x), s(y)) = q\delta(x, y)$$

for all points $x, y \in X$. The scalar q is called the **square ratio** of the similarity.

It is clear that the rigid motions are precisely the similarities with square ratio one. We leave it to the reader to show that in Euclidean geometry, the square ratio of a similarity is always a positive real number and that the similarity multiplies distances by the square root of that ratio. Similarities are multiplied as affine transformations and we ask the reader to show that the similarities form a group which is a subgroup of the group Af of affine transformations and contains the group Mo or rigid motions as a subgroup. Denoting the group of similarities by Sim, we, therefore, have the following tower of groups

$$Tr \subset Mo \subset Sim \subset Af.$$

If $c \in X$, an affine transformation s can be written uniquely as

$$s = T_A L(c, \lambda)$$

where $A \in V$ and λ is a linear automorphism of V. In particular, every similarity is of the form $T_A L(c, \lambda)$. In order to show that s is a similarity of X if and only if λ is a similarity of V, we first do the following exercise.

Exercise

*1. This exercise is the analogue of Exercise 380.4. Let λ be a linear automorphism of V with the property that $(\lambda B)^2 = qB^2$ for all $B \in V$ where

q is a nonzero scalar. Prove that λ is a similarity of V with square ratio q.

Proposition 414.1.   Let $c \in X$, $A \in V$, and λ be a linear automorphism of V. Then $T_A L(c, \lambda)$ is a similarity of X with square ratio q if and only if λ is a similarity of V with square ratio q.

Proof. Assume that λ is a similarity of V with square ratio q; we must prove that $T_A L(c, \lambda)$ is a similarity of X with square ratio q. Choose points x and y in X and denote $L(c, \lambda)$ by L. Then

$$\overrightarrow{T_A L(x), T_A L(y)} = \overrightarrow{L(x), L(y)} = \lambda \overrightarrow{(x, y)}$$

(Exercise 77.7) and hence

$$\delta(T_A L(x), T_A L(y)) = (\lambda \overrightarrow{(x, y)})^2 = q \overrightarrow{(x, y)}^2 = q\delta(x, y).$$

Conversely, assume $T_A L(c, \lambda)$ is a similarity of X with square ratio q. Put $s = T_A L(c, \lambda)$ and choose a vector B in V. By Exercise 1 above, it is sufficient to prove that $(\lambda B)^2 = qB^2$. Since

$$\delta(s(c), s(Bc)) = q\delta(c, Bc) = qB^2,$$

we are done if we show that

$$\delta(s(c), s(Bc)) = (\lambda B)^2.$$

This last equality was shown in the proof of Proposition 381.1. Done.

As a first application of the above proposition, we observe that dilations are similarities because a dilation is either a translation or a magnification (see Classification of Dilations, page 41). Translations are rigid motions and hence similarities with square ratio 1. The magnification $M(c, r)$ of X where $c \in X$ and $r \in k^*$ is, by definition, equal to $L(c, \lambda)$ where λ is the magnification $M(0, r)$ of V. Consequently, $M(c, r)$ is a similarity of X with square ratio r^2.

Exercise

2. It is clear from the previous paragraph that the group Di of dilations of X is a subgroup of the group Sim of similarities.
   a. The group Sim contains both groups Mo and Di. Prove that the group $Mo \cap Di$ consists of translations and $180°$ rotations.
   b. If $n = 1$, prove that $Di = Sim$.

Consider the product $M(c, r)u$ of the magnification $M(c, r)$ and the rigid motion u. This product is again a similarity of X. It happens in many affine geometries that every similarity is the product of a magnification and a rigid motion. In order to settle precisely when this happens, we denote the set of scalars which are square ratios of similarities of X again by $\mathscr{S}$. This is justified by Proposition 414.1 which tells us that $\mathscr{S}$ is also the set of square ratios of similarities of V. In Exercise 360.7c, it was proved that $\mathscr{S}$ is a subgroup of $k*$. We shall see in Theorem 416.1 that every similarity of X is the product of a magnification and a rigid motion if and only if $\mathscr{S} = k*^2$. The proof of the next theorem makes use of Exercise 3 below.

Exercises

These exercises are the analogues of Exercises 1, 3, 5, and 6 on pages 359–360.

*3. If s_1 and s_2 are similarities of X with square ratios r_1 and r_2, respectively, prove that the similarity $s_1 s_2$ has square ratio $r_1 r_2$. (Hint: It is best to prove this directly without the use of Proposition 414.1.)

4. If s is a similarity of X with square ratio q, prove that the similarity s^{-1} has square ratio q^{-1}.

5. Prove that the mapping

$$s \rightarrow \text{square ratio of } s$$

is an epimorphism from Sim to $\mathscr{S}$.

6. Prove that Mo is a normal subgroup of Sim which contains the commutator subgroup of Sim.

The following theorem is the analogue of Theorem 365.1.

Theorem 415.1. A similarity of X is the product of a magnification and a rigid motion if and only if the square ratio of that similarity is a square in $k*$.

Proof. The product $M(c, r)u$ of a magnification $M(c, r)$ and a rigid motion u of X has r^2 as square ratio. Conversely, assume that s is a similarity of X whose square ratio is a square r^2. Choose a point $c \in X$ and put

$$s = T_A L(c, \lambda)$$

where $A \in V$ and λ is a similarity of V with square ratio r^2 (Proposition 414.1). By Theorem 365.1,

$$\lambda = M(0, r)\sigma$$

where $M(0, r)$ is a magnification and $\sigma \in O(V)$. It follows that

$$s = T_A M(c, r) L(c, \sigma)$$

and hence s is the product of the dilation $D = T_A M(c, r)$ and the rigid motion $L(c, \sigma)$. If D is a translation, s is, in fact, a rigid motion which is the product of itself with the magnification $1_X = M(c, 1)$; if D is the magnification $M(b, r)$ for $b \in X$, then

$$s = M(b, r)L(c, \sigma).$$

Done.

The following theorem is the analogue of Theorem 366.1 and is proved precisely the same way.

Theorem 416.1.   Every similarity of X is the product of a magnification and a rigid motion if and only if $\mathscr{S} = k*^2$.

For example, let $k = \mathbf{R}$. Then $\mathscr{S} = \mathbf{R}*^2$, unless X is an Artinian space (Exercise 365.23c). Thus, in all Euclidean spaces, negative Euclidean spaces, Minkowski space, all odd dimensional spaces, and many others, every similarity is the product of a magnification and a rigid motion. However, in an Artinian space over $\mathbf{R}$, a similarity is the product of a magnification and a rigid motion only if the square ratio of that similarity is positive.

Exercises

*7. Let $\mathscr{S} = k*^2$. Assume that s is a similarity of X and $c \in X$. Prove that s is the product of a dilation and either a rotation or a reflection of X having c as fixed point.

*8. Let $c \in X$. Each similarity s of X can be written as

$$s = T_A L(c, \lambda)$$

for a unique $A \in V$ and a unique similarity λ of V.

a. Prove that the mapping $s \to \lambda$ is an epimorphism from the group Sim to the group $S(V)$ of the similarities of V. We denote this epimorphism by ψ.

b. Prove that the group *Tr* of translations of *X* is a normal subgroup of *Sim*. (Hint: What is the kernel of the epimorphism ψ ?)

The epimorphism

$$\psi: Sim \to S(V)$$

of Exercise 8 above is nothing but the restriction of the epimorphism

$$\phi: Sa \to Sem$$

studied in Section 20 and hence does not depend on the choice of the point *c*. The main groups which control affine geometry and their homomorphisms onto the corresponding groups of vector space geometry are depicted in the following diagram. Each group is a subgroup of every group above it. An arrow with two heads denotes an epimorphism and each epimorphism is the restriction of every epimorphism above it. The group *Tr* is a normal subgroup of every group above it and is the kernel of each of the epimorphisms ϕ, ϕ_0, ψ, and ϕ_1. The group *Mo* is a normal subgroup of *Sim* and contains the commutator subgroup of *Sim*. The group $O(V)$ is a normal subgroup of $S(V)$ and contains the commutator subgroup of $S(V)$. Figure 417.1 extends Figure 389.1.

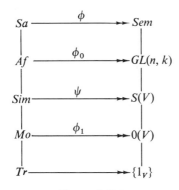

Figure 417.1

We return to the notion of line segment $\{x, y\}$ where *x* and *y* are, by definition, distinct points of *X*, and to the notion of orthogonal line segments $\{x, y\} \perp \{b, c\}$ studied in Exercise 376.8. Of course, an affine transformation of *X* is said to **preserve orthogonality** if $\{x, y\} \perp \{b, c\}$ implies that $\{s(x), s(y)\} \perp \{s(b), s(c)\}$ for all $x \neq y$ and $b \neq c$ in *X*. The following exercise is used in the proof that an affine transformation of *X* is a similarity if and only if it preserves orthogonality.

Exercise

*9. Let s be an affine transformation of X and $c \in X$. Put $s = T_A L(c, \gamma)$ where $A \in V$ and γ is a linear automorphism of V. Prove that s preserves orthogonality if and only if γ preserves orthogonality. (Hint: Use the fact that $\overrightarrow{s(x), s(y)} = \gamma(\overrightarrow{x, y})$ for all $x, y \in X$.)

We are ready for the analogue of Theorem 362.1.

Theorem 418.1. An affine transformation of X is a similarity if and only if it preserves orthogonality.

Proof. Let s be an affine transformation of X and $c \in X$. Put $s = T_A L(c, \gamma)$ where $A \in V$ and γ is a linear automorphism of V. Then, by Exercise 9 above,

s preserves orthogonality $\Leftrightarrow \gamma$ preserves orthogonality;

by Theorem 362.1,

γ preserves orthogonality $\Leftrightarrow \gamma$ is a similarity of V;

finally, by Proposition 414.1,

γ is a similarity of $V \Leftrightarrow s$ is a similarity of X.

Done.

Assume that k is ordered and $\mathscr{S} = k^{*2}$. Readers who are not well versed in ordered fields should again assume that $k = \mathbf{R}$ and X is not Artinian space. The epimorphism

$$\psi : Sim \to S(V)$$

now maps every similarity of X on either a direct similarity or an opposite similarity of V. A similarity s of X is called a **direct (opposite) similarity** if $\psi(s)$ is a direct (opposite) similarity of V.

Exercises

10. Assume that k is an ordered field and $\mathscr{S} = k^{*2}$. Choose $c \in X$. Prove that an affine transformation s of X is a direct (opposite) similarity if and only if

$$s = M(c, r)T_A u$$

where $r > 0$, $A \in V$, and u is a rotation (reflection) of X with c as fixed point.

11. Let k be ordered.
   a. Define the notion of orientation for the affine space X. (Hint: If V is oriented and $c, A_1, \ldots, A_n$ is a coordinate system of X, the coordinate system $A_1, \ldots, A_n$ of V is either positively or negatively oriented.)
   b. Assume that X is oriented and $\mathscr{S} = k^{*2}$. Prove that direct similarities preserve the orientation of X while opposite similarities reverse the orientation.

Theorem 418.1 may be interpreted as saying that similarities preserve $90°$ angles. Actually, they preserve all angles. In order to see this clearly, we devote the remainder of this section to a few remarks on rays and angles in affine space.

We assume for the remainder of this section that the field k is ordered.

If c is a point in X and A is a nonzero vector of V, the **half line** or **ray with vertex** c **and direction vector** A is the set of points

$$\{(tA)c \,|\, t \in k, \quad t \geq 0\}.$$

The fact that a ray determines its vertex uniquely but has many direction vectors is brought out in the following exercises.

Exercise

12. Let $c, c' \in X$ and A, A' be nonzero vectors of V. To say that the ray with vertex c and direction vector A is the same as the ray with vertex c' and direction vector A' can only mean that

$$\{(tA)c \,|\, t \in k, \quad t \geq 0\} = \{(tA')c' \,|\, t \in k, \quad t \geq 0\}.$$

Prove that these rays are the same if and only if $c = c'$ and $A' = qA$ for some positive scalar q. Thus, as a special case, the vectors A and B are direction vectors for the same ray if and only if $B = qA$ for $q > 0$.

The oriented line segment (c, a) of X determines the ray with vertex c and direction vector $\overrightarrow{c, a}$. This ray is denoted by $c \nabla a$ (see Figure 419.1).

Figure 419.1

Exercise

13. If (c, a) and (c', a') are oriented line segments, prove that $c \nabla a = c' \nabla a'$ if and only if $c = c'$ and $a \in c' \nabla a'$ or, equivalently, $a' \in c \nabla a$. Observe that it may very well happen that $a \neq a'$ (see Figure 420.1).

Figure 420.1

An **angle** of X is a pair of rays (not an ordered pair of rays) with the same vertex, called the **vertex** of the angle. The rays are called the **sides** of the angle. A pair of oriented segments (c, a) and (c, b) with the same initial point c determines the angle with vertex c and sides $c \nabla a$ and $c \nabla b$. We denote this angle by $\angle acb$ or $\angle bca$; whenever we write one of these symbols, it is understood that $c \neq a$ and $c \neq b$. An angle is pictured in Figure 420.2.

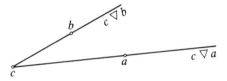

Figure 420.2

Since angles are subsets of X, two angles are congruent if there exists a rigid motion of X which maps one angle onto the other (see page 410 for the definition of congruent sets). It is not a matter of definition, but a matter of proof that this rigid motion maps vertex onto vertex and side onto side. In fact, we learn in Exercise 16 below that all affine transformations have this property.

Exercise

14. Let u be an affine transformation of X, not necessarily a rigid motion.
    a. If l is a ray with vertex c, prove that $u(l)$ is a ray with vertex $u(c)$.
    b. If θ is an angle with vertex c and sides l_1 and l_2, prove that $u(\theta)$ is the angle with vertex $u(c)$ and sides $u(l_1)$ and $u(l_2)$.
    c. Prove that
    $$u(\angle acb) = \angle u(a)u(c)u(b).$$

We assume for the remainder of this section that
1. *$n = 2$.*
2. *All lines of V through $\mathbf{0}$ are isometric.*
3. *V represents the squares of k.*

Recall that we previously made the convention that the field k is ordered. These assumptions allow us to employ the results of Section 53 on plane trigonometry and speak about the angle between two nonzero vectors A and B of V. We learned that this cosine is equal to

$$\frac{AB}{\sqrt{A^2}\,\sqrt{B^2}}.$$

If (c, a) and (c, b) are oriented line segments of X, the cosine of $\angle acb$ $(= \angle bca)$ is defined as the cosine between the vectors $\overrightarrow{c, a}$ and $\overrightarrow{c, b}$. Consequently,

$$\cos(\angle acb) = \frac{(c, a)(c, b)}{\sqrt{\overrightarrow{c, a}^2}\,\sqrt{\overrightarrow{c, b}^2}}.$$

Of course, it must be shown that $\cos(\angle acb)$ depends only on the rays $c\,\nabla\,a$ and $c\,\nabla\,b$ and not on the points a and b (see Exercise 17 below). The symbol $\simeq$ is used to denote congruence between angles.

Exercises

15. Prove that all lines of X are isometric.

*16. Since we do not distinguish between $\angle acb$ and $\angle bca$, the result of this exercise depends heavily on the fact that all lines of X are isometric. Prove that

$$\angle acb \simeq \angle a'c'b'$$

if and only if there exists a rigid motion u of X such that

$$u(c) = c', \qquad u(a) \in c'\,\nabla\,a', \qquad u(b) \in c'\,\nabla\,b'.$$

Observe that it is not necessary (in fact, usually not possible) that $u(a) = a'$ or $u(b) = b'$ as Figure 421.1 shows. Here

$$\delta(c, a) = \delta(c', u(a)) \qquad \text{and} \qquad \delta(c, b) = \delta(c', u(b)).$$

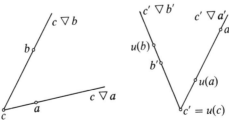

Figure 421.1

*17. Prove that $\cos(\angle acb)$ depends only on the rays $c \nabla a$ and $c \nabla b$ and not on the points a and b chosen on these rays. Hence, in Figure 422.1, $\cos(\angle acb) = \cos(\angle a'cb')$.

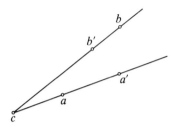

Figure 422.1

*18. This exercise prepares for Exercise 19 below. Let a, b, c, d be points of X and u a rigid motion of X. Prove that

$$\overrightarrow{(a, b)}\overrightarrow{(c, d)} = \overrightarrow{(u(a), u(b))}\overrightarrow{(u(c), u(d))}.$$

This result says that rigid motions preserve inner products between directed line segments. (Hint: Choose a point $p \in X$ and put $u = T_A L(p, \sigma)$ where $\sigma \in O(V)$. Then use the fact, from Exercise 77.7, that $\overrightarrow{u(a), u(b)} = \sigma\overrightarrow{(a, b)}$.)

*19. Prove that

$$\angle acb \simeq \angle a'c'b'$$

if and only if

$$\cos(\angle acb) = \cos(\angle a'c'b').$$

(Hint for the only if part: Assume $\angle acb \simeq \angle a'c'b'$ and apply Exercises 16 and 17 above. Hint for the if part: Assume $\cos(\angle acb) = \cos(\angle a'c'b')$. Since all lines are isometric, there are points $a'' \in c' \nabla a'$ and $b'' \in c' \nabla b'$ such that

$$\overrightarrow{c, a}^2 = \overrightarrow{c', a''}^2 \qquad \text{and} \qquad \overrightarrow{c, b}^2 = \overrightarrow{c', b''}^2 .$$

Conclude that

$$\overrightarrow{(c, a)}\overrightarrow{(c, b)} = \overrightarrow{(c', a'')}\overrightarrow{(c', b'')}.$$

There is now an obvious isometry u from $X(c)$ to $X(c')$. (See Figure 423.1.)

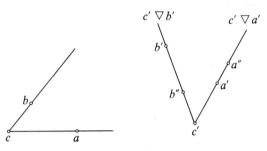

Figure 423.1

We shall say that an affine transformation s of X **preserves angles** if it transforms every angle onto a congruent angle. Explicitly, s preserves angles if and only if

$$\angle acb \simeq \angle s(a)s(c)s(b)$$

for every $\angle acb$. The important related result is that *an affine transformation of X preserves angles if and only if it is a similarity* (see Exercise 22 below).

Exercises

20. Prove than an affine transformation s preserves angles if and only if it preserves the cosine of angles, that is, if and only if

$$\cos(\angle acb) = \cos(\angle s(a)s(c)s(b))$$

    for every $\angle acb$.

21. This exercise is the analogue of Exercise 18 above with rigid motion replaced by similarity. Let a, b, c, d be points of X and s a similarity of X with square ratio q. Prove that

$$(\overrightarrow{s(a), s(b)})(\overrightarrow{s(c), s(d)}) = q(\overrightarrow{a, b})(\overrightarrow{c, d}).$$

*22. Prove that an affine transformation of X preserves angles if and only if it is a similarity.

In Euclidean geometry, two angles are congruent if and only if their measurements in degrees, or in any other unit of angular measurement, are the same. This fact has given rise to the abuse of language of calling congruent angles "equal" or "the same." For instance, one says that the triangles acb and $a'c'b'$ of Figure 424.1 have equal angles even though these angles are only congruent. No angle of triangle acb has even the same vertex as one of the angles of triangle $a'c'b'$.

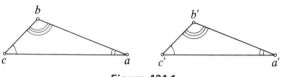

Figure 424.1

This abuse of language has given rise to much confusion in the teaching of angles, simply because it has not always been realized that, when congruent angles are called equal, an abuse of language is involved. Once this abuse is clearly understood, its usage is very convenient and we shall make use of it from now on for angles of triangles in our affine plane X.

Two subsets Y and Y' of X are called **similar** if there exists a similarity transformation s of X such that $s(Y) = Y'$. (Of course, this definition holds for all affine spaces and not only for the very special plane we are dealing with at present.) For example, a triangle consists of three noncollinear points and hence is a subset of X; two triangles are called similar if they are similar as subsets of X. Thus two triangles acb and $a'c'b'$ are similar if there exists a similarity transformation s of S such that, after properly ordering the vertices

$$s(a) = a', \qquad s(c) = c', \qquad s(b) = b'.$$

Since affine transformations preserve angles, it is obvious that similar triangles have equal angles. We ask the reader to prove the converse in the next exercise.

Exercises

*23. Prove that two triangles of X are similar if and only if they have equal angles. (Hint: Assume that triangles acb and $a'c'b'$ have equal angles. Since $\angle acb \simeq \angle a'c'b'$, there exists a rigid motion u such that $u(c) = c'$, $u(a) \in c' \triangledown a'$, and $u(b) \in c' \triangledown b'$. See Figure 424.2. Then

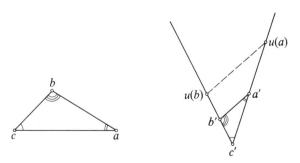

Figure 424.2

$\overrightarrow{c', a'} = r(\overrightarrow{c', u(a)})$ where $r > 0$ and it follows that the similarity $s = M(c', r)u$ has the property that $s(c) = c'$, $s(a) = a'$, and $s(b) = b'' \in c' \bigtriangledown b'$. Now use the fact that $\angle bac \simeq \angle b'a'c'$ and $\angle bac \simeq \angle b''a'c'$ and, therefore, b'' also lies on $a' \bigtriangledown b'$.)

*24. Two triangles acb and $a'c'b'$ are said to have **proportional sides** if there exists a scalar q such that

$$\delta(a', c') = q\delta(a, c), \qquad \delta(c', b') = q\delta(c, b), \qquad \delta(a', c') = q\delta(a, c).$$

It is immediate that $q > 0$. Prove that two triangles are similar if and only if their sides are proportional. (Hint: Trivially, similar triangles have proportional sides. For the converse, show that

$$(\overrightarrow{c', a'})(\overrightarrow{c', b'}) = q(\overrightarrow{c, a})(\overrightarrow{c, b}).)$$

*25. We recall that a figure of X is simply a subset of X and two figures are called homothetic if there exists a dilation which maps one figure onto the other (Exercise 58.12). Since dilations are similarities, homothetic figures are similar. The converse is false and we investigate by how much.

a. Prove that $\mathscr{S} = k^{*2}$ where, as before, $\mathscr{S}$ denotes the set of scalars which are square ratios of similarities. (Hint: V represents only the squares of k.)

b. Let Y and Y' be subsets of X and $c \in X$. Prove that Y and Y' are similar figures if and only if there exists a rotation or a reflection u of X which leaves c fixed and has the property that the figures $u(Y)$ and Y' are homothetic. In short, similar figures are homothetic figures modulo rotations and reflections. (Hint: The if part is practically trivial. For the only if part, use Part a along with Exercise 416.7.)

The pictures behind the last exercise are very suggestive. Figure 425.1 pictures two similar triangles acb and $a'c'b'$. In this case, the rigid motion u of

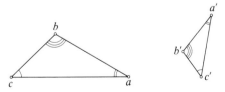

Figure 425.1

Exercise 25 above is a rotation. The two triangles $u(a)u(c)u(b)$ and $a'c'b'$ have parallel sides and hence are homothetic (Proposition 98.1). The dilations

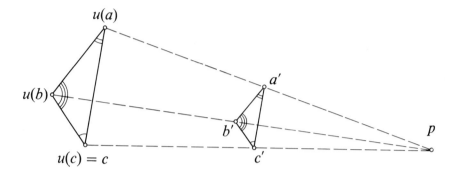

Figure 426.1

which map one triangle onto the other turn out to be magnifications with center p and ratios 2 or $\frac{1}{2}$.

The conclusion that the two triangles of Figure 426.1 are homothetic could also have been made from the theorem of Desargues. This shows how very naturally this theorem could be made a part of high school teaching of similar triangles if only this subject were taught from the viewpoint of similarity transformations.

epilogue

Can one give the reason the Euclid–Hilbert system for Euclidean geometry is so unmanageable for technical geometry while the axiom system, based on linear algebra, is so eminently suited to handle concrete questions in geometry? The answer lies in the way these two systems treat real numbers.

In the system based on linear algebra, the real numbers are never defined. Whenever the field of real numbers is used, it is assumed that this field has been previously discussed and is known. In particular, one does not attempt to define real numbers in terms of the geometry under discussion.

The Euclid–Hilbert system, on the contrary, considers it part of its task to define real numbers in terms of the geometry it is trying to build. Although this task is carried out successfully, the extra burden which it puts on the axioms has proven to be incompatible with a smoothly functioning system for technical geometry.

There are many ways in which real numbers can be defined and we list some of them with a slight indication of their main interest.

1. *Axiomatic method.* (Ordered field with least upper bound axiom.) Best for analysis and geometry.

2. *Cauchy sequences.* Best for modern algebra, algebraic number theory, and algebraic geometry since the method extends immediately to fields with a valuation or rings with a topology.

3. *Dedekind cuts.* Interesting from historical point of view. Dedekind (1831–1916) developed his theory because he was dissatisfied with his freshman course in calculus at the polytechnical university in Zurich, Switzerland. See R. Dedekind, *Essays on the Theory of Numbers* (translated by W. W. Beman), [8]. He taught for 31 years at the high school of his native Brunswick, Germany.

4. *Geometric method* (Euclid–Hilbert). Interesting from a historical point of view and possessing great charm. Harmful to geometry where it has been used the most. Very little is known about Euclid's life. It is not even certain when he lived, but it was probably around 306–283 B.C. It is certain, however, that he was in the foreground of mathematics of his time. If he had lived today, what would he have done? He would have told us to do away with Hilbert's system and use linear algebra instead. And what a book he would have written!

bibliography

1   ARTIN, E., *Galois Theory*, (Notre Dame Mathematical Lecture No. 2), second edition, Univ. of Notre Dame Press, Notre Dame, Indiana, 1944.

2   ARTIN, E., *Geometric Algebra*, Wiley (Interscience), New York, 1957.

3   ARTZY, R., *Linear Geometry*, Addison-Wesley, Reading, Massachusetts, 1965.

4   BELL, E. T., *Men of Mathematics*, Simon and Schuster, New York, 1937.

5   BERGMANN, P. G., *Introduction to the Theory of Relativity*, Prentice Hall, Englewood Cliffs, New Jersey, 1942.

6   BRUCK, R. H., Recent advances in the foundations of Euclidean geometry, *American Mathematical Monthly*, **62** (7), (No. 4 of the Slaught Memorial Papers).

7   BURNSIDE, W., *Theory of Groups of Finite Order*, Cambridge Univ. Press, London and New York, 1897

8   DEDEKIND, R., *Essays on the Theory of Numbers* (trans. W. W. Beman), Open Court, Chicago, 1909

9   DEVRIES, Hk, *Beknopt Leerboek Der Projectieve Meetkunde*.

10 DUNNINGTON, G. W., *Carl Friedrich Gauss*, Louisiana State Univ. Press, Baton Rouge, Louisiana, 1937.

11 EBEY, S., SITRARAM, K., Frobenius groups and Wedderburn's theorem, *American Mathematical Monthly*, **74** (5), 1969.

12 GODEMENT, R., *Algebra*, Houghton Mifflin, Boston, 1968.

13 HALL, M., Jr., *The Theory of Groups*, Macmillan, New York, 1959.

14 KLEIN, F., *Vorlesungen Uber Nicht-Euclidische Geometrie*, Chelsea, New York.

15 LANG, S., *Algebra*, Addison-Wesley, Reading, Massachusetts, 1965.

16 MACLANE, S., and BIRKHOFF, G., *Algebra*, Macmillan, New York, 1967

17 NEWMAN, J. R., *The World Book of Mathematics*, Simon and Schuster, New York, 1956.

18 O'MEARA, O. T., *Introduction to Quadratic Forms*, Berlin, Springer, 1963.

19 PONCELET, J. V., *Traite des proprietes projectives des figures*.

20 SHERK, P., On the decomposition of orthogonalities in symmetries, *Proceedings of the American Mathematical Society*, **1** (4), 1950.

21 STRUIK, D. J., *A Concise History of Mathematics*, Dover, New York, 1948.

22 VAN DER WAERDEN, B. L., *Modern Algebra*, Vol. I, Frederick Ungar, New York, 1949.

23 WITT, E., Theorie der quadratischen Formen in beliebigen Korpen, *Journal fur du reine und angewandte Mathematik*, **176**, October 1936.

INDEX